# The Johnson Solids and Their Duals:

## a Comprehensive Survey

by
Bruce R. Gilson

2014

Copyright 2014 by Bruce R. Gilson
All rights reserved.

## Other books by the same author:

*Construction of Musical Scales: a Mathematical Approach* (2008)
*Units and Measurement Systems* (2008)
*The Fibonacci Sequence and Beyond* (2009)
*Polyhedra: a New Approach* (2012)

All available from the same publisher, for sale on Amazon.com and other retail outlets.

# Table of Contents

Preface..................................................................................................xii
Chapter 1: Introduction..........................................................................1
Chapter 2: The pyramids: Johnson solids 1 and 2..................................7
Chapter 3: The cupolae and rotunda: Johnson solids 3-6....................17
Chapter 4: Elongated and gyroelongated pyramids: Johnson solids 7-11..................................................................................................36
Chapter 5: Dipyramids (bipyramids): Johnson solids 12 and 13....58
Chapter 6: Elongated and gyroelongated dipyramids (bipyramids): Johnson solids 14-17..........................................................65
Chapter 7: Elongated and gyroelongated cupolae and rotundae: Johnson solids 18-25..........................................................78
Chapter 8: Bicupolae, cupolarotundae, and birotundae: Johnson solids 26-34..............................................................................113
Chapter 9: Elongated bicupolae, cupolarotundae, and birotundae: Johnson solids 35-43..........................................................144
Chapter 10: Gyroelongated bicupolae, cupolarotundae, and birotundae: Johnson solids 44-48..........................................................170
Chapter 11: (Laterally) augmented prisms: Johnson solids 49-57.................................................................................................192
Chapter 12: Augmented dodecahedra: Johnson solids 58-61........229
Chapter 13: Diminished icosahedra: Johnson solids 62-64..........251
Chapter 14: "Augmented" Archimedean solids: Johnson solids 65-71..................................................................................................266
Chapter 15: Interlude: An Archimedean solid: the rhombicosidodecahedron, a basis for numerous Johnson solids..................................................................................................316
Chapter 16: Gyrate, diminished, and gyrate diminished rhombicosidodecahedra: Johnson solids 72-83..........................323
Chapter 17: Miscellany: Johnson solids 84-92..................................366
Chapter 18: The Johnson solids and polyhedron fusion..............408
Index.....................................................................................................416

## List of Tables

Table 1: Angles of regular polygons (in degrees)..................................3
Table 2: Properties of Johnson solids 1 and 2 and their duals.......13
Table 3: Dihedral angles of Johnson solids derived from the pyramids..................................................................................................16
Table 4: Properties of Johnson solids 3, 4, and 5 and their duals. ..................................................................................................27
Table 5: Properties of Johnson solid 6 and its dual..........................33
Table 6: Properties of Johnson solids 7, 8, and 9 and their duals.

..................................................................................................49
Table 7: Properties of Johnson solids 10 and 11 and their duals...57
Table 8: Properties of Johnson solids 12 and 13 and their duals..63
Table 9: Properties of Johnson solids 14, 15, and 16 and their duals..................................................................................................72
Table 10: Properties of Johnson solid 17 and its dual........................77
Table 11: Properties of Johnson solids 18, 19, and 20 and their duals..................................................................................................89
Table 12: Properties of Johnson solid 21 and its dual........................95
Table 13: Properties of Johnson solids 22, 23, and 24 and their duals................................................................................................105
Table 14: Properties of Johnson solid 25 and its dual......................112
Table 15: Properties of Johnson solids 26, 27, 28, and 29 and their duals................................................................................................124
Table 16: Properties of Johnson solids 30, 31, 32, and 33 and their duals................................................................................................136
Table 17: Properties of Johnson solid 34 and its dual......................142
Table 18: Properties of Johnson solids 35 through 39 and their duals................................................................................................156
Table 19: Properties of Johnson solids 40, 41, 42, and 43 and their duals................................................................................................168
Table 20: Properties of Johnson solids 44, 45, and 46 and their duals................................................................................................179
Table 21: Properties of Johnson solids 47 and 48 and their duals. ..................................................................................................190
Table 22: Properties of Johnson solids 49, 50, and 51 and their duals................................................................................................204
Table 23: Properties of Johnson solids 52 and 53 and their duals. ..................................................................................................212
Table 24: Properties of Johnson solids 54, 55, and 56 and their duals................................................................................................223
Table 25: Properties of Johnson solid 57 and its dual......................227
Table 26: Properties of Johnson solids 58, 59, and 60 and their duals................................................................................................243
Table 27: Properties of Johnson solid 61 and its dual......................249
Table 28: Properties of Johnson solids 62, 63, and 64 and their duals................................................................................................265
Table 29: Properties of Johnson solids 65, 66, and 67 and their duals................................................................................................278
Table 30: Properties of Johnson solids 68, 69, 70, and 71 and their duals................................................................................................313
Table 31: Johnson solids obtained by altering icosahedrally symmetric Platonic or Archimedean solids..................................319
Table 32: Gyrate, diminished, and gyrate diminished

rhombicosahedra. .................................................................321
Table 33: Properties of Johnson solids 72 through 82 and their duals........................................................................................355
Table 34: Properties of Johnson solid 83 and its dual..................363
Table 35: Properties of Johnson solids 84, 85, and 86 and their duals........................................................................................380
Table 36: Properties of Johnson solids 87, 88, and 89 and their duals........................................................................................392
Table 37: Properties of Johnson solids 90, 91, and 92 and their duals........................................................................................406

# List of Figures

Figure 1: Johnson solid 1: the square pyramid.................................8
Figure 2: The dual of Johnson solid 1: also a square pyramid, but differently proportioned........................................................................9
Figure 3: Johnson solid 2: the pentagonal pyramid.........................10
Figure 4: The dual of Johnson solid 2: also a pentagonal pyramid, but differently proportioned.................................................11
Figure 5: Johnson solid 3: the triangular cupola.............................18
Figure 6: The dual of Johnson solid 3 (the triangular cupola).....19
Figure 7: Johnson solid 4: the square cupola..................................21
Figure 8: The dual of Johnson solid 4 (the square cupola)............22
Figure 9: Johnson solid 5: the pentagonal cupola...........................23
Figure 10: The dual of Johnson solid 5 (the pentagonal cupola).24
Figure 11: Johnson solid 6: the pentagonal rotunda.......................29
Figure 12: The dual of Johnson solid 6 (the pentagonal rotunda). ........................................................................................................31
Figure 13: Johnson solid 7: the elongated triangular pyramid......39
Figure 14: The dual of Johnson solid 7 (the elongated triangular pyramid): a basally augmented triangular prismoid or monofrustum of an asymmetric triangular bipyramid..................41
Figure 15: Johnson solid 8: the elongated square pyramid...........43
Figure 16: The dual of Johnson solid 8 (the elongated square pyramid): a monofrustum of an asymmetric square bipyramid. ........................................................................................................45
Figure 17: Johnson solid 9: the elongated pentagonal pyramid..46
Figure 18: The dual of Johnson solid 9 (the elongated pentagonal pyramid): a monofrustum of an asymmetric pentagonal bipyramid................................................................................47
Figure 19: Johnson solid 10: the gyroelongated square pyramid. ........................................................................................................50
Figure 20: The dual of Johnson solid 10 (the gyroelongated square pyramid): a monofrustum of a square antibipyramid......52

Figure 21: Johnson solid 11: the gyroelongated pentagonal pyramid............53
Figure 22: The dual of Johnson solid 11 (the gyroelongated pentagonal pyramid): a monofrustum of a pentagonal antibipyramid............54
Figure 23: Johnson solid 12: the triangular bipyramid............59
Figure 24: The dual of Johnson solid 12 (the triangular bipyramid): a triangular prism............60
Figure 25: Johnson solid 13: the pentagonal bipyramid............61
Figure 26: The dual of Johnson solid 13 (the pentagonal bipyramid): a pentagonal prism............62
Figure 27: Johnson solid 14: the elongated triangular bipyramid or basally biaugmented triangular prism............66
Figure 28: The dual of Johnson solid 14 (the elongated triangular bipyramid): a bifrustum of a triangular bipyramid............67
Figure 29: Johnson solid 15: the elongated square bipyramid............68
Figure 30: The dual of Johnson solid 15 (the elongated square bipyramid): a bifrustum of a square bipyramid............69
Figure 31: Johnson solid 16: the elongated pentagonal bipyramid.............70
Figure 32: The dual of Johnson solid 16 (the elongated pentagonal bipyramid): a bifrustum of a pentagonal bipyramid.............71
Figure 33: Johnson solid 17: the gyroelongated square bipyramid.............74
Figure 34: The dual of Johnson solid 17 (the gyroelongated square bipyramid)............75
Figure 35: Johnson solid 18: the elongated triangular cupola............79
Figure 36: The dual of Johnson solid 18 (the elongated triangular cupola)............80
Figure 37: Johnson solid 19: the elongated square cupola............82
Figure 38: The dual of Johnson solid 19 (the elongated square cupola)............84
Figure 39: Johnson solid 20: the elongated pentagonal cupola...85
Figure 40: The dual of Johnson solid 20 (the elongated pentagonal cupola)............86
Figure 41: Johnson solid 21: the elongated pentagonal rotunda..91
Figure 42: The dual of Johnson solid 21 (the elongated pentagonal rotunda)............93
Figure 43: Johnson solid 22: the gyroelongated triangular cupola.............96
Figure 44: The dual of Johnson solid 22 (the gyroelongated triangular cupola)............99
Figure 45: Johnson solid 23: the gyroelongated square cupola. 100

Figure 46: The dual of Johnson solid 23 (the gyroelongated square cupola) .................................................................................. 101
Figure 47: Johnson solid 24: the gyroelongated pentagonal cupola .................................................................................................. 102
Figure 48: The dual of Johnson solid 24 (the gyroelongated pentagonal cupola) ........................................................................... 103
Figure 49: Johnson solid 25: the gyroelongated pentagonal rotunda ............................................................................................... 108
Figure 50: The dual of Johnson solid 25 (the gyroelongated pentagonal rotunda) .......................................................................... 109
Figure 51: Johnson solid 26, called by him the gyrobifastigium, the digonal member of the gyrobicupolae ............................... 114
Figure 52: The dual of Johnson solid 26 (the gyrobifastigium), an unnamed but very simple polyhedron ................................... 115
Figure 53: Johnson solid 27: the triangular orthobicupola ............ 116
Figure 54: The dual of Johnson solid 27 (the triangular orthobicupola): a trapezo-rhombic dodecahedron ......................... 118
Figure 55: Johnson solid 28: the square orthobicupola .................. 119
Figure 56: The dual of Johnson solid 28 (the square orthobicupola) ...................................................................................... 120
Figure 57: Johnson solid 29: the square gyrobicupola .................... 121
Figure 58: The dual of Johnson solid 29 (the square gyrobicupola) ...................................................................................... 122
Figure 59: Johnson solid 30: the pentagonal orthobicupola ......... 126
Figure 60: The dual of Johnson solid 30 (the pentagonal orthobicupola) ...................................................................................... 127
Figure 61: Johnson solid 31: the pentagonal gyrobicupola ........... 128
Figure 62: The dual of Johnson solid 31 (the pentagonal gyrobicupola) ...................................................................................... 129
Figure 63: Johnson solid 32: the pentagonal orthocupolarotunda. .......................................................................................................... 130
Figure 64: The dual of Johnson solid 32 (the pentagonal orthocupolarotunda) .......................................................................... 132
Figure 65: Johnson solid 33: the pentagonal gyrocupolarotunda. .......................................................................................................... 133
Figure 66: The dual of Johnson solid 33 (the pentagonal gyrocupolarotunda) ............................................................................ 134
Figure 67: Johnson solid 34: the pentagonal orthobirotunda ...... 138
Figure 68: The dual of Johnson solid 34 (the pentagonal orthobirotunda) .................................................................................. 139
Figure 69: Johnson solid 35: the elongated triangular orthobicupola .................................................................................... 145
Figure 70: The dual of Johnson solid 35 (the elongated triangular orthobicupola) .................................................................................. 146

Figure 71: Johnson solid 36: the elongated triangular gyrobicupola...............147
Figure 72: The dual of Johnson solid 36 (the elongated triangular gyrobicupola)...............148
Figure 73: Johnson solid 37: the elongated square gyrobicupola, also known as Miller's solid or the pseudorhombicuboctahedron...............149
Figure 74: The dual of Johnson solid 37 (the elongated square gyrobicupola)...............150
Figure 75: Johnson solid 37: the elongated pentagonal orthobicupola...............151
Figure 76: The dual of Johnson solid 38 (the elongated pentagonal orthobicupola)...............152
Figure 77: Johnson solid 39: the elongated pentagonal gyrobicupola...............153
Figure 78: The dual of Johnson solid 39 (the elongated pentagonal gyrobicupola)...............154
Figure 79: Johnson solid 40: the elongated pentagonal orthocupolarotunda...............158
Figure 80: The dual of Johnson solid 40 (the elongated pentagonal orthocupolarotunda)...............159
Figure 81: Johnson solid 41: the elongated pentagonal gyrocupolarotunda...............161
Figure 82: The dual of Johnson solid 41 (the elongated pentagonal gyrocupolarotunda). ...............162
Figure 83: Johnson solid 42: the elongated pentagonal orthobirotunda...............163
Figure 84: The dual of Johnson solid 42 (the elongated pentagonal orthobirotunda)...............164
Figure 85: Johnson solid 43: the elongated pentagonal gyrobirotunda...............165
Figure 86: The dual of Johnson solid 43 (the elongated pentagonal gyrobirotunda)...............166
Figure 87: Johnson solid 44: the gyroelongated triangular bicupola...............170
Figure 88: The dual of Johnson solid 44 (the gyroelongated triangular bicupola)...............171
Figure 89: Johnson solid 45: the gyroelongated square bicupola. ...............174
Figure 90: The dual of Johnson solid 45 (the gyroelongated square bicupola)...............175
Figure 91: Johnson solid 46: the gyroelongated pentagonal bicupola...............176
Figure 92: The dual of Johnson solid 46 (the gyroelongated

pentagonal bicupola)......176
Figure 93: Johnson solid 47: the gyroelongated pentagonal cupolarotunda......181
Figure 94: The dual of Johnson solid 47 (the gyroelongated pentagonal cupolarotunda)......183
Figure 95: Johnson solid 48: the gyroelongated pentagonal birotunda. ......185
Figure 96: The dual of Johnson solid 48 (the gyroelongated pentagonal birotunda)......187
Figure 97: Johnson solid 49: the (laterally) augmented triangular prism......195
Figure 98: The dual of Johnson solid 49 (the laterally augmented triangular prism): a triangular bipyramid with one lateral face truncated......196
Figure 99: Johnson solid 50: the (laterally) biaugmented triangular prism......198
Figure 100: The dual of Johnson solid 50 (the laterally biaugmented triangular prism): a triangular bipyramid with two lateral faces truncated......199
Figure 101: Johnson solid 51: the (laterally) triaugmented triangular prism......201
Figure 102: The dual of Johnson solid 51 (the triaugmented triangular prism): the peritruncated triangular bipyramid......202
Figure 103: Johnson solid 52: the (laterally) augmented pentagonal prism......207
Figure 104: The dual of Johnson solid 52 (the laterally augmented pentagonal prism): a pentagonal bipyramid with a lateral vertex truncated......208
Figure 105: Johnson solid 53: the (laterally) biaugmented pentagonal prism......209
Figure 106: The dual of Johnson solid 53 (the laterally biaugmented pentagonal prism): a pentagonal bipyramid with two nonadjacent lateral vertices truncated......210
Figure 107: Johnson solid 54: the (laterally) augmented hexagonal prism......214
Figure 108: The dual of Johnson solid 54 (the laterally augmented hexagonal prism): a hexagonal bipyramid with a lateral vertex truncated......215
Figure 109: Johnson solid 55: the parabiaugmented hexagonal prism......216
Figure 110: The dual of Johnson solid 55 (the parabiaugmented hexagonal prism): a hexagonal bipyramid with two opposite lateral vertices truncated......217
Figure 111: Johnson solid 56: the metabiaugmented hexagonal

prism. ...................................................................................................218

Figure 112: The dual of Johnson solid 56 (the metabiaugmented hexagonal prism): a hexagonal bipyramid with two lateral vertices, neither adjacent nor opposite, truncated........................219

Figure 113: Johnson solid 57: the triaugmented hexagonal prism. ................................................................................................................225

Figure 114: The dual of Johnson solid 57 (the triaugmented hexagonal prism)................................................................................226

Figure 115: Johnson solid 58: the augmented dodecahedron.......232

Figure 116: The dual of Johnson solid 58: a monotruncated icosahedron....................................................................................................235

Figure 117: Johnson solid 59: the parabiaugmented dodecahedron..............................................................................................236

Figure 118: The dual of Johnson solid 59 (the parabiaugmented dodecahedron): an icosahedron truncated at two opposite vertices............................................................................................................237

Figure 119: Johnson solid 60: the metabiaugmented dodecahedron..............................................................................................238

Figure 120: The dual of Johnson solid 60 (the metabiaugmented dodecahedron)............................................................................................239

Figure 121: Johnson solid 61: the triaugmented dodecahedron.. 245

Figure 122: The dual of Johnson solid 61 (the triaugmented dodecahedron)............................................................................................246

Figure 123: Johnson solid 62: the metabidiminished icosahedron. ................................................................................................................253

Figure 124: The dual of Johnson solid 62 (the metabidiminished icosahedron)................................................................................................254

Figure 125: Johnson solid 63: the tridiminished icosahedron.....256

Figure 126: The dual of Johnson solid 63 (the tridiminished icosahedron)................................................................................................259

Figure 127: Johnson solid 64: the augmented tridiminished icosahedron................................................................................................261

Figure 128: The dual of Johnson solid 64 (the augmented tridiminished icosahedron)....................................................................262

Figure 129: Johnson solid 65: the augmented truncated tetrahedron..................................................................................................267

Figure 130: The dual of Johnson solid 65 (the augmented truncated tetrahedron)............................................................................269

Figure 131: Johnson solid 66: the augmented truncated cube.....271

Figure 132: The dual of Johnson solid 66 (the augmented truncated cube)..........................................................................................272

Figure 133: Johnson solid 67: the biaugmented truncated cube. ................................................................................................................273

Figure 134: The dual of Johnson solid 67 (the biaugmented

truncated cube)..................................................................................274
Figure 135: Johnson solid 68, the augmented truncated dodecahedron.........................................................................281
Figure 136: The dual of Johnson solid 68 (the augmented truncated dodecahedron).................................................283
Figure 137: Johnson solid 69: the parabiaugmented truncated dodecahedron.........................................................288
Figure 138: The dual of Johnson solid 69 (the parabiaugmented truncated dodecahedron).................................290
Figure 139: Johnson solid 70, the metabiaugmented truncated dodecahedron.........................................................292
Figure 140: The dual of Johnson solid 70 (the metabiaugmented truncated dodecahedron).................................293
Figure 141: Johnson solid 71: the triaugmented truncated dodecahedron.........................................................302
Figure 142: The dual of Johnson solid 71 (the triaugmented truncated dodecahedron).................................304
Figure 143: A rhombicosidodecahedron, viewed from a square face.........................................................................316
Figure 144: The same rhombicosidodecahedron, viewed from above the pentagonal face that serves as the roof of the marked cupola.........................................................................317
Figure 145: Two overlapping cupolae on a rhombicosidodecahedron.................................................318
Figure 146: Johnson solid 72: the gyrate rhombicosidodecahedron.................................................326
Figure 147: The dual of Johnson solid 72 (the gyrate rhombicosidodecahedron).................................327
Figure 148: Johnson solid 73: the parabigyrate rhombicosidodecahedron.................................................328
Figure 149: The dual of Johnson solid 73 (the parabigyrate rhombicosidodecahedron).................................329
Figure 150: Johnson solid 74, the metabigyrate rhombicosidodecahedron.................................................330
Figure 151: The dual of Johnson solid 74 (the metabigyrate rhombicosidodecahedron).................................331
Figure 152: Johnson solid 75: the trigyrate rhombicosidodecahedron.................................................332
Figure 153: The dual of Johnson solid 75 (the trigyrate rhombicosidodecahedron).................................333
Figure 154: Johnson solid 76, the diminished rhombicosahedron.........................................................................334
Figure 155: The dual of Johnson solid 76 (the diminished rhombicosahedron).................................................335

Figure 156: Johnson solid 77: the paragyrate diminished rhombicosidodecahedron..................337
Figure 157: The dual of Johnson solid 77 (the paragyrate diminished rhombicosidodecahedron)..................338
Figure 158: Johnson solid 78: the metagyrate diminished rhombicosidodecahedron..................339
Figure 159: The dual of Johnson solid 78 (the metagyrate diminished rhombicosidodecahedron)..................340
Figure 160: Johnson solid 79, the bigyrate diminished rhombicosidodecahedron..................342
Figure 161: The dual of Johnson solid 79 (the bigyrate diminished rhombicosidodecahedron)..................343
Figure 162: Johnson solid 80, the parabidiminished rhombicosidodecahedron, one of the two core polyhedra for describing the twelve Johnson solids of this chapter..................345
Figure 163: The dual of Johnson solid 80 (the parabidiminished rhombicosidodecahedron)..................346
Figure 164: Johnson solid 81, the metabidiminished rhombicosidodecahedron..................347
Figure 165: The dual of Johnson solid 81 (the metabidiminished rhombicosidodecahedron)..................348
Figure 166: Johnson solid 82, the gyrate bidiminished rhombicosahedron..................350
Figure 167: The dual of Johnson solid 82 (the gyrate bidiminished rhombicosahedron)..................351
Figure 168: Johnson solid 83, the tridiminished rhombicosahedron, one of the two core polyhedra for describing the twelve Johnson solids of this chapter..................358
Figure 169: The dual of Johnson solid 83 (the tridiminished icosahedron)..................359
Figure 170: Johnson solid 84: the snub disphenoid, dodecadeltahedron, or Siamese dodecahedron..................366
Figure 171: The dual of Johnson solid 84 (the snub disphenoid): an as-yet unnamed polyhedron with trapezoidal and pentagonal faces..................367
Figure 172: Johnson solid 85, named by him the snub square antiprism..................368
Figure 173: The dual of Johnson solid 85 (the snub square antiprism): the square pentagonized globoid..................370
Figure 174: Johnson solid 86: the sphenocorona..................372
Figure 175: The dual of Johnson solid 86 (the sphenocorona)....376
Figure 176: Johnson solid 87: the augmented sphenocorona......382
Figure 177: The dual of Johnson solid 87 (the augmented sphenocorona)..................383

Figure 178: Johnson solid 88: the sphenomegacorona...................384
Figure 179: The dual of Johnson solid 88 (the sphenomegacorona)......................................................................385
Figure 180: Johnson solid 89: the hebesphenomegacorona.........387
Figure 181: The dual of Johnson solid 89 (the hebesphenomegacorona).................................................................388
Figure 182: Johnson solid 90: the disphenocingulum...................396
Figure 183: The dual of Johnson solid 90 (the disphenocingulum).....................................................................397
Figure 184: Johnson solid 91: the bilunabirotunda.......................398
Figure 185: The dual of Johnson solid 91 (the bilunabirotunda). ..................................................................................................399
Figure 186: Johnson solid 92: the triangular hebesphenorotunda. .................................................................................................. 400
Figure 187: The dual of Johnson solid 92 (the triangular hebesphenorotunda)......................................................................403

## Preface.

The idea for this book came to me as a result of having read a book on the Johnson solids by Gregory Zorzos, published by the same firm that I have used to publish this book. I was rather hoping, when I bought it, to have essentially what this book is (with the exception of this book's treating the duals of the Johnson solids as well), and as a result I was rather disappointed, as Zorzos' book has almost no text, nor does it have any of the information which I have included in the several Tables in this book. Perhaps the absence of text can be explained by the fact that Zorzos' first language is not English, so he felt uncomfortable writing much. But in any case, I found it did not meet my needs, and eventually, after completing my book, *Polyhedra: a New Approach*, I decided to write this book to do what I had wished Zorzos had done in his. (My desire to include the duals, however, is simply out of a desire to provide some less-available information.)

Much of the material printed in this book is available in other sources such as *Wikipedia* or *MathWorld* online, or in other books like Anthony Pugh's *Polyhedra: a Visual Approach* or Peter Pearce and Susan Pearce's *Polyhedra Primer*. It is not, however, all gathered together and organized there as I have done, and some of the material is, to my knowledge, not elsewhere to be found. I hope, therefore, that this book will be considered a useful addition to the literature.

In this book I do not define most of my terms. A few, such as "dual," are so commonly used that their definition is easily obtained anywhere where polyhedra are treated. Some, such as "apicobasal polyhedron," are newly coined terms introduced in my book, *Polyhedra: a New Approach*, so that their definitions are only available there. While I would be happy, of course, if the reader were to buy my earlier book to learn these definitions, much information can be obtained from this book without reference to my earlier one.

Bruce R. Gilson

September 22, 2014

# Chapter 1: Introduction.

The Johnson solids are a family of polyhedra named for Norman W. Johnson, an American (originally Canadian) geometer who first described them. He considered all polyhedra whose faces were exclusively regular polygons, then omitted the five Platonic and thirteen Archimedean solids, as well as two infinite classes of polyhedron with regular-polygonal faces: the prisms and antiprisms. What he was left with was a collection of ninety-two polyhedra remaining. These were described, and names given to those which had not been previously named, in a paper published in 1966. He did not, at that time, prove that these ninety-two, together with the Platonic and Archimedean solids, the prisms, and the antiprisms, were the only convex polyhedra whose faces were exclusively regular polygons; this was accomplished in a paper, published three years later, by the Israeli (originally Russian) geometer Victor Zalgaller.

This book will describe and illustrate all ninety-two of the Johnson solids, as well as the polyhedra that can be generated as duals to those solids. While in some cases the duals are themselves Johnson solids (in some cases the *same* polyhedron as the original Johnson solid), any polyhedron that can be so described will be shown in both categories.

The ninety-two Johnson solids are listed in the standard order, as given by Norman Johnson, and each is immediately followed by its dual. In a number of cases, a set of solids which are closely related are listed consecutively by Johnson (for example, the first two are both pyramids), and in such a case they will be treated together in a single chapter in this book. This accounts for fourteen sets, each having a chapter; a fifteenth chapter is devoted to a residual set of polyhedra that do not lend themselves to such grouping, and which Johnson left for the end in his list. For each of the ninety-two solids, Johnson's own name for the solid is given, but in addition, alternatives are in some cases provided. (Most of these alternatives are mentioned in my earlier book, *Polyhedra: a New Approach*.) In general, Johnson's name is the one mentioned in the literature, but in a few cases the alternative names are also common. (While I prefer the name "digonal gyrobicupola" over Johnson's "gyrobifastigium" as describing its structure better, that particular choice does not seem to have taken root in the literature, however.)

Each of the Johnson solids is illustrated with either three or four views. Always included are a front, (right) side, and top view. If the polyhedron has any of the $D$ symmetries or $S_{2nv}$ (which is usually treated as $D_{nd}$ anyway), these will be the only three given, as the bottom is identical to the top in appearance (possibly rotated). If the polyhedron has $C_{nv}$ symmetry, so the top and bottom are significantly different, a bottom view is given as well. It must be noted that, although these views are presented in an arrangement resembling the standard practice used in orthographic projection, these views are not truly orthographic, but are perspective views as the polyhedron would appear from the appropriate directions. It has been my intent to choose the orientation of these views to emphasize the symmetry of the polyhedra; if there is only one axis of rotational symmetry, it will be considered as running from top to bottom, and if there are more than one, but only one with order greater than three, the same choice is made. (This is in keeping with the standard use of the symbols "$v$" and "$h$" [abbreviations for "vertical" and "horizontal"] in the symmetry group symbols in the common Schönflies notation for the orientation of mirror planes.) Similarly, when there are mirror planes including such an axis, the orientation will be chosen to have one of the mirror planes bisect the front view.

The writeup of each Johnson solid and dual will contain a table giving some of the important characteristics of each. The duals have the same symmetry as the original polyhedron, and therefore there will be one line of the table for a polyhedron and its dual. Since all faces of a Johnson solid are regular, the face angles of the solid are always given by Table 1, which is adapted from a similar table in my earlier book, *Polyhedra: a New Approach*. (Only those lines of the table are included that are useful in describing the Johnson solids, as polygons with other numbers of sides are not found as faces of those polyhedra.) The edges are also all equal, so they are not given, but are to be taken as 1. (If a polyhedron has its edges equal to $e$, and

*Introduction.*

the total area and volume in the table are given as $A$ and $V$, the actual area of these scaled-up polyhedra is $Ae^2$ and the volume is $Ae^3$.) It might be noted that since any triangular face with an edge of length 1 has an area of $\frac{1}{4}\sqrt{3}$, any square face has area 1, any pentagonal face has area $\frac{1}{4}\sqrt{(25+10\sqrt{5})}$, any hexagonal face has area $(3\sqrt{3})/2$, any octagonal face has area $2 + 2\sqrt{2}$, and any decagonal face has area $[5\sqrt{(5 + 2\sqrt{5})}]/2$, the areas of the Johnson solids can be figured out for each by referring to the face count by number of sides. Even so, it is given here. However, the edges of the dual will not all be equal, and the actual edges will be tabulated for each dual. The area and volume of the dual will be calculated assuming the original Johnson solid to have edge length 1.

| Number of sides | Sum of angles | Each angle, if equiangular |
|---|---|---|
| 3 | 180 | 60.000 |
| 4 | 360 | 90.000 |
| 5 | 540 | 108.000 |
| 6 | 720 | 120.000 |
| 8 | 1080 | 135.000 |
| 10 | 1440 | 144.000 |

*Table 1: Angles of regular polygons (in degrees).*

Only those numbers of sides are listed in Table 1 which actually occur in the Johnson solids.

Symmetries are all given according to the standard Schönflies notation, since this notation is both the most common one used to describe polyhedron symmetry and the one I am most familiar with, with one departure from its most common form. Polyhedra with an alternating axis and mirror planes through this axis are described in this reference, as in my earlier book, *Polyhedra: a New Approach*, as of symmetry $S_{2nv}$, as was mentioned earlier. The reason, as stated in the book, is that these symmetry elements are easier to visualize on such a polyhedron. Most books and other references would describe their symmetry as $D_{nd}$ instead, but as I have seen some other places where the notation I favor is used, I do not believe I am breaking new ground with this usage. Readers whose familiarity with the Schönflies notation for symmetry groups is not as great as their familiarity with other systems can find comparisons of the notations in numerous places, especially online in places like *Wikipedia*, by doing a Google search for "point group symmetry" or similar words.

Readers who have not read my earlier book, *Polyhedra: a New Approach*, might not be familiar with the concept of *transitivity class*. While I hardly can claim to have invented the term, it is possible to read many books on polyhedra without ever meeting with it, but because the concept is frequently used in this book, it will here be defined. Basically, if two faces of a polyhedron (or two edges, or two vertices, though in this book the classification will mostly be applied to faces) are so located that one of the symmetry transformations of the polyhedron will move one to the other, they belong to the *same* transitivity class. (It does not matter which is transformed to the other, because symmetry transformations form a *group*, and all members of a group have inverses, so if transformation X moves face A to position B, the inverse transformation to X would move face B to position A.)

It is always true that the number of *faces* of the dual will be equal to the number of *vertices* of the original Johnson solid, and the number of *vertices* of the dual will be equal to the number of *faces* of the original Johnson solid. Since it will be of interest, the number of faces of each (the original Johnson solid and its dual) will be broken down into numbers of triangles, squares, etc. In the case of the original Johnson solid, it is not necessary to state anything but the number of sides, so a figure like [4](5) will be understood to mean four regular pentagons, with the number in square brackets being the number of polygons of that type as shown, and the number in parentheses being the number of sides of the polygon. However, if the polygons of one type belong to more than one transitivity class, a breakdown is given. Thus, for a square cupola ($J_4$), it gives [1+4](4) to indicate that the five squares fall into two transitivity classes, one containing a single square (perpendicular to the fourfold axis) and the other containing four squares. In the description of other features, they will be termed 4r and 4L faces (for "roof" and "lateral"), so the dihedral angles will be shown, for example, as 3-4L and 4r-4L. In the case of the *duals*, more information will be necessary. If a set of faces is by necessity

*Introduction.*

composed of isosceles triangles because of symmetry, the notation "3i" will be found, similarly for isosceles polygons (in the sense of my earlier book, *Polyhedra: a New Approach* — which is to say, polygons containing a mirror line bisecting a side). While "4i" therefore will denote an isosceles trapezoid, "4k" will denote a kite-quadrilateral. Sometimes only designations like "4a," "4b," and so forth will be used because there are irregular quadrilaterals, or because there are more than one type of kite-quadrilateral.

As noted, the number of *faces* of the dual will be equal to the number of *vertices* of the original Johnson solid, and the number of *vertices* of the dual will be equal to the number of *faces* of the original Johnson solid. It would therefore not be necessary to give numbers of vertices of either the Johnson solids themselves or their duals. However, it will be easier for the reader to use the tables if these numbers are included, so they are given, but not broken down in any way. Thus, for the square cupola mentioned earlier, the face count is given as [4](3) + [1+4](4) + [1](8). But for the dual, the vertex count is simply shown as 10.

The actual dual of any polyhedron depends on what point is chosen for the center of the polar reciprocation process (see my earlier book, *Polyhedra: a New Approach*, for details). In all these cases, it is assumed that the centroid of the vertices is used. In this way, a consistent approach is followed, and faces, edges, or vertices that belong to a single transitivity class are dualized to vertices, edges, and faces that still belong to a single transitivity class. With any other center, the number of vertices, edges, and faces, and their *types* (e. g., number of faces meeting at a vertex) will be identical, but their sizes and the exact shape of a face will not necessarily be the same. For example, the dual of the pentagonal pyramid $J_2$ is shown as a thin but high pentagonal pyramid. By selecting a reciprocation center nearer to the apex of the pyramid, a dual can be generated much less thin and high; the correct choice will produce one congruent to the original pyramid.

It will be noted repeatedly that any polyhedron and its dual have the same symmetry. In fact, this is *because* the centroid is taken as the center of polar reciprocation. As long as all axes of rotational symmetry and all mirror planes of a polyhedron pass through the center of polar reciprocation, it will be true that the polyhedron and its dual will have the same symmetry, and by their nature, the axes of rotational symmetry and mirror planes will all pass through the centroid. So it should be taken as implied that, when it is stated that any polyhedron and its dual have the same symmetry, this actually means that any polyhedron and its dual *as constructed in this book* have the same symmetry.

The designation of some vertices as "trivalent," "quadrivalent," etc. is not common in the literature, but is useful, especially because so much of this book deals with duals of polyhedra and the dual of a triangular face is a trivalent vertex (and so forth). While the terms suggest chemistry, they are not unknown in this context. (They are more common in connection with graph theory than in books treating polyhedra in a more geometric manner such as this one.) A trivalent vertex is simply one where three faces (and three edges) come together, and similar definitions apply to vertices of higher valency.

# Chapter 2: The pyramids: Johnson solids 1 and 2.

A *triangular* pyramid with all regular-polygon faces would be a Platonic solid, the regular tetrahedron. It is therefore omitted from the list of Johnson solids. A pyramid with all regular-polygon faces cannot be constructed with the base having six or more sides. Thus only square and pentagonal pyramids are found in Johnson's list, and they are the first two, usually designated $J_1$ and $J_2$. This chapter will deal with these two Johnson solids, and is thus one of the shortest in the book, as subsequent chapters will be devoted to larger sets from among the ninety-two Johnson solids.

The duals of all pyramids (whether, as in the regular tetrahedron and the Johnson solids $J_1$ and $J_2$, the triangles are equilateral or, as the more general case, they are simply isosceles) are pyramids of like form (though turned 180° from the originals). The duals of the Johnson solids $J_1$ and $J_2$ are actually not identical to themselves, but have isosceles triangles as their lateral faces, although it is possible to choose a reciprocation center which would give duals that are identical to the original polyhedra, except for being oppositely oriented. Figure 1 shows the square pyramid, $J_1$, giving the top, front, side, and bottom views. It is easy to see that the faces include four triangles and a square; that the triangles are equilateral is not obvious from any of the views, because in no case is a triangular face viewed head-on, but it is to be understood that every Johnson solid has only regular polygons as faces, so the triangular faces cannot be anything but equilateral. The symmetry is clearly $C_{4v}$, with all four triangular faces equivalent, and the polyhedron has its fourfold axis through the centers of the top and bottom views, running vertically in the front and side views. More details of the geometry are shown below in Table 2, one of a set of tables that will summarize all the important geometric parameters for all the ninety-two Johnson solids and their duals.

The orientation of the views in Figure 1 will be typical of the figures in this book, and has been discussed in the previous chapter; the shading of the faces is done in a way intended to emphasize the symmetry of the polyhedron. In some cases in subsequent chapters of the book, it will be hard, even with the figure provided, to visualize the specific polyhedron being discussed, but I am trying the best I can to make visualization possible.

The order in which the ninety-two Johnson solids will be discussed is the standard order originating in Johnson's paper; fortunately, the division of the ninety-two solids into sets for discussion can be done without deviating from Johnson's ordering.

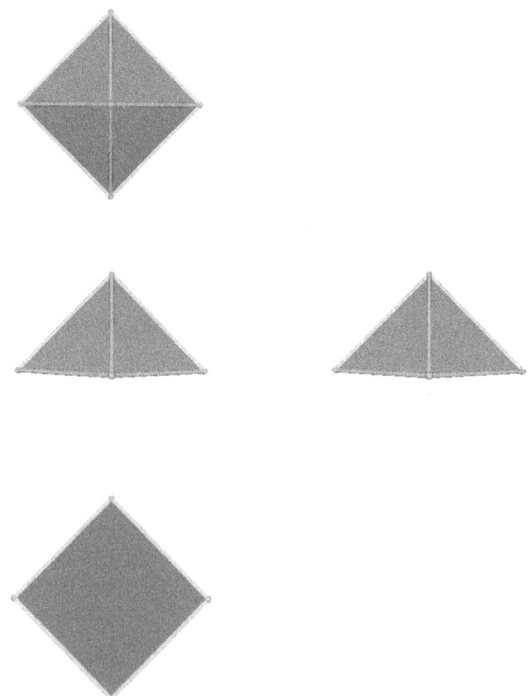

Figure 1: Johnson solid 1: the square pyramid.

Figure 2 shows the dual of Johnson solid 1, with (as stated in Chapter 1) the centroid of the vertices taken as the reciprocation center. Again, top, front, side, and bottom views are shown. It should be noted that any polyhedron and its dual have identical symmetry when the reciprocation center is chosen as it is in this book, so that both Johnson solid 1 and its dual have

$C_{4v}$ symmetry. Because the symmetry is the same for the dual of every Johnson solid as for the original solid, in Table 2 only one line gives both.

Unlike Johnson solid 1 itself, which is a square pyramid all of whose lateral faces are equilateral triangles, the dual has lateral faces which are long, narrow isosceles triangles. The base angles of those triangles are 78.9042° = 78° 54' 15", while the apex angle is 22.1916° = 22° 11' 30"; the edge that forms the base of each lateral face is 0.56568542, while the edges that are the equal sides of the isosceles triangles are 1.46969385, nearly three times as long.

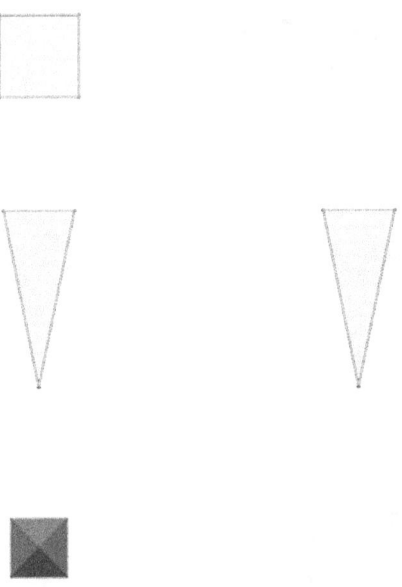

Figure 2: *The dual of Johnson solid 1: also a square pyramid, but differently proportioned.*

Johnson solid 2, the pentagonal pyramid, is illustrated below in Figure 3. As was done in Figure 1 for Johnson solid 1, the top, front, side, and bottom views are shown. Most of what can be said about $J_1$ applies to $J_2$ except that the principal axis of rotation is fivefold rather than fourfold, so the symmetry is $C_{5v}$ rather than $C_{4v}$. In the discussion of the geometry of $J_1$ above, replacing "four" by "five" in all appropriate places (and "square" by "regular pentagon") will give a completely accurate description of $J_2$, and so it will be unnecessary to repeat those sentences here. However, if it is desired to examine the detailed geometric parameters of $J_2$ or to compare them with those of $J_1$, the reader is invited to consult Table 2 on p. 13 below. Similar tables will be found to include every one of the ninety-two Johnson solids discussed in this book, as well as all of their duals.

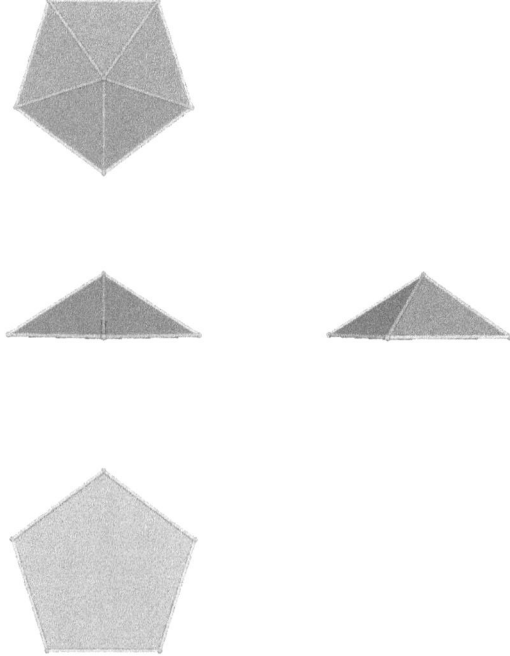

Figure 3: Johnson solid 2: the pentagonal pyramid.

In a similar way to the dual of the square pyramid being another square pyramid, but not proportioned the same way, the dual of the pentagonal pyramid is another pentagonal pyramid, illustrated below in Figure 4. This pyramid is even taller in proportion to its width than the dual of the square pyramid, shown in Figure 2 above, with the very sharp angle, at the apex

*The pyramids: Johnson solids 1 and 2.*

of each triangular face, of only 8.5150° (8° 30′ 54″). The base angles of these isosceles triangles are 85.7425° (85° 44′ 33″), which is so close to a right angle that the ratio of edges is extremely great: the lateral edges are 1.91746853, nearly *seven times as long* as the edges that form the bases of the isosceles triangles, which are 0.28470066 (assuming the edge of the original Johnson solid is taken as 1). Not only the angles within the faces, but also the dihedral angles between those faces and the pentagonal base, become nearly right angles, as can be seen in Table 2 below.

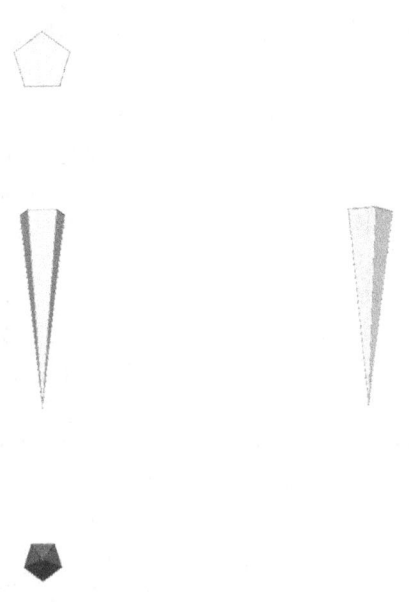

Figure 4: The dual of Johnson solid 2: also a pentagonal pyramid, but differently proportioned.

In comparing Johnson solids 1 and 2, it is more enlightening to consider the regular tetrahedron as well, as all three form a progression of members of the same family of pyramids, with the restriction of equilaterality of the lateral faces serving as a constraint.

At this point, it is appropriate to discuss what might be termed the *3-4-5 trend rule*. There are many sets of regular-polygon-faced solids (a term which is intended to include the Platonic and Archimedean solids, and not merely the Johnson solids) in which three members of the set exist, with threefold, fourfold, and fivefold symmetry. An example would be the regular tetrahedron, as well as $J_1$, and $J_2$, just discussed. In *all* of these, as the order of the principal axis of rotation increases from 3 to 5, the solid becomes flatter, more squat. It should also be noted that the trend among the *duals*, as generated by polar reciprocation, goes the other way; as the order of the principal rotation axis increases from 3 to 5, the solid becomes spikier, more stretched-out. This can be cited as a rule, which I will call the *3-4-5 trend rule*, and which I will mention repeatedly in this book. Thus the dihedral angles between the lateral faces of Johnson solids 1 and 2, which are 109.4712° (109° 28′ 16″) and 138.1897° (138° 11′ 23″), follow a sequence with the 70.5288° (70° 31′ 44″) of the regular tetrahedron, each step increasing by approximately 30° from the preceding one. And the dihedral angles between lateral face and base start with the same tetrahedral figure of 70.5288° (70° 31′ 44″), because *all* edges of a regular tetrahedron are alike, but, rather than increasing steadily, are found to decrease steadily, to 54.7356° (54° 44′ 8″) and 37.3774° (37° 22′ 39″), as the number of sides increases from 3 to 4 and 5. (In the case of the square pyramid, there is no coincidence that the lateral face/lateral face dihedral angle of 109.4712° [109° 28′ 16″] is exactly double the base/lateral face dihedral angle of 54.7356° [54° 44′ 8″]. If two such square pyramids are fused base to base, one obtains a regular octahedron, whose edges are all alike. The base/lateral face dihedral angle of the square pyramid must then be *half of* the octahedral dihedral angle, while the lateral face/lateral face dihedral angle of the square pyramid is identical to the octahedral dihedral angle.) The effects of the 3-4-5 trend rule will be seen repeatedly in this book.

Table 2 below is a summary of the properties of Johnson solids 1 and 2 and their duals.

*The pyramids: Johnson solids 1 and 2.*

| Johnson solid number | 1 | 2 |
|---|---|---|
| Name | Square pyramid | Pentagonal pyramid |
| Symmetry* | $C_{4v}$ | $C_{5v}$ |
| Faces | [4](3) + [1](4) | [5](3) + [1](5) |
| Edges* | 8 | 10 |
| Vertices | 5 | 6 |
| Dihedral angles | 3-3: 109.4712° (109° 28′ 16″), 3-4: 54.7356° (54° 44′ 8″) | 3-3: 138.1897° (138° 11′ 23″), 3-5: 37.3774° (37° 22′ 39″) |
| Area | 2.73205081 = 1+√3 | 3.88554091 = ¼[√(25+10√5) + 5√3] |
| Volume | 0.23570226 = (√2)/6 | 0.30150283 = (5+√5)/24 |
| **Dual** | | |
| Name | Square pyramid‡ | Pentagonal pyramid‡ |
| Faces | [4](3) + [1](4) | [5](3) + [1](5) |
| Vertices | 5 | 6 |
| Face angles† | 3: 22.1916° (28° 11′ 30″), [2]78.9042° (78° 54′ 15″); 4: [4]90° | 3: 8.5150° (8° 30′ 54″), [2]85.7425° (85° 44′ 33″); 5: [5]108° |
| Dihedral angles | 3-3: 92.2042° (92° 12′ 15″), 3-4: 78.6901° (78° 41′ 24″) | 3-3: 108.4376° (108° 26′ 15″), 3-5: 84.1190° (84° 7′ 8″) |
| Area | 1.951686244 | 1.50044768 |
| Volume | 0.15084945 | 0.08841806 |

Table 2: *Properties of Johnson solids 1 and 2 and their duals.*

*The symmetry and number of edges are always the same for each solid and its dual, so are not repeated in the "dual" section.

†Face angles are only given for the duals, as the Johnson solids have regular faces, whose angles can be found in Table 1 on p. 3.

‡Differently proportioned from the Johnson solids; the triangular faces are isosceles, not equilateral.

Some points ought to be made at this time. The two pyramids $J_1$ and $J_2$ (and the regular tetrahedron) can be spotted in the structures of many other polyhedra, and so the dihedral angles appearing in Table 2 (and the tetrahedron dihedral angle of 70° 31′ 44″) can be found all over the place. For example, the bipyramids $J_{12}$ and $J_{13}$ are obtained by fusing two pyramids at their bases. So the triangular bipyramid $J_{12}$ has dihedral angles between two triangles meeting at the apex equal to the tetrahedral dihedral angle (70° 31′ 44″), while between two triangles meeting along the belt around the center the dihedral angle is exactly twice this figure (141° 3′ 27″). Similarly, the pentagonal bipyramid $J_{13}$ has dihedral angles between two triangles meeting at the apex equal to the 3-3 angle shown in Table 2 for the pentagonal pyramid $J_2$ (138° 11′ 23″), while between two triangles meeting along the belt around the center the dihedral angle is exactly twice the 37° 22′ 39″ figure for the $J_2$ 3-5 dihedral angle (74° 45′ 17″).

A consequence of that is that, since two copies of the square pyramid $J_1$ fused together in this manner form a regular octahedron, all of whose dihedral angles are equal, the 3-3 angle in $J_1$ is equal to the octahedral dihedral angle (109° 28′ 16″), while the 3-4 angle must be exactly half this figure (54° 44′ 8″). This is observed in Table 2.

Additionally, when the elongated pyramids $J_{7-9}$ and elongated bipyramids $J_{14-16}$ are examined, the same angles will appear combined with the 90° angles deriving from the prism that is fused to the pyramid(s), as every prism has a 90° dihedral angle between the base and any lateral face.

Whenever three equilateral triangles meet at a vertex (and of course, in any Johnson solid, *all* triangles are equilateral) the dihedral angles between any two of them must be the tetrahedral dihedral angle of 70° 31′ 44″. This is because any three equal edges that form a triangle (even when the triangle is not a face of the polyhedron!) must form an equilateral triangle, and all equilateral triangles are similar. But when four or five triangles meet at a vertex, they need not form equal dihedral angles (for example in $J_{13}$, at the belt, four triangles all meet at each vertex, but the dihedral angles are not the same as the ones in $J_1$). When the vertex in question is the point where a

*The pyramids: Johnson solids 1 and 2.*

fourfold or fivefold axis of rotation meets the surface of the polyhedron, of course, the dihedral angles will agree with the ones in $J_1$ or $J_2$. (An example is $J_{58}$, the augmented dodecahedron, where a pentagonal pyramid is fused to a regular dodecahedron.) And in some cases, even where there is no fourfold or fivefold axis, the polyhedron is still a fragment of an octahedron or icosahedron, so the dihedral angles will match in those cases as well. (An example is $J_{62}$, where the vertices in question are derived from those of an icosahedron, but the fivefold axes of the icosahedron have been lost by the "diminishing" operation.)

Because of the fact that a number of Johnson solids can be decomposed into pyramids and prisms or antiprisms, it is useful to provide Table 3 below. Although this could eliminate the necessity of providing the dihedral angles in the tables corresponding to Table 2 for those Johnson solids which are covered by Table 3, for consistency these dihedral angles will be included. In all those cases, the faces meeting at the apex (or apices, where there are two opposite apices) have a dihedral angle equal to $l_{Py}$, the lateral dihedral angle of the pyramid. For all the solids described as "elongated," each face of the ones which meet at the apex makes a dihedral angle with the lateral faces of the elongation equal to $b_{Py} + 90°$, where $b_{Py}$ is the basal dihedral angle of the pyramid, while adjacent faces of the elongation make dihedral angles of $l_{Pr}$, the lateral dihedral angle of the prism. For all the solids described as "gyroelongated," each face of the ones which meet at the apex makes a dihedral angle with the lateral faces of the elongation equal to $b_{Py} + b_a$, where $b_a$ is the basal dihedral angle of the antiprism, while adjacent faces of the elongation make dihedral angles of $l_a$, the lateral dihedral angle of the antiprism. And in addition, the bipyramids (not elongated or gyroelongated) have central dihedral angles (around the plane halfway between the apices) of $2b_{Py}$, the elongated pyramids have basal dihedral angles of $90°$, and the gyroelongated pyramids have basal dihedral angles of $b_a$. Table 3 displays all these values. (It should be noted that three of the solids that are displayed in Table 3 are non-Johnson solids and their initials are shown in the appropriate cells: the regular tetrahedron, octahedron, and icosahedron. There is no listing for a gyroelongated triangular pyramid or bipyramid because they cannot be constructed so as to have all equilateral triangles: the triangles from the pyramid and antiprism would form a dihedral angle of 180° so they would merge into a rhombus.)

| $n$ | 3 | 4 | 5 |
|---|---|---|---|
| Pyramid | $T$ | $J_1$ | $J_2$ |
| Elongated pyramid | $J_7$ | $J_8$ | $J_9$ |
| Gyroelongated pyramid | (see above) | $J_{10}$ | $J_{11}$ |
| Bipyramid | $J_{12}$ | $O$ | $J_{13}$ |
| Elongated bipyramid | $J_{14}$ | $J_{15}$ | $J_{16}$ |
| Gyroelongated bipyramid | (see above) | $J_{17}$ | $I$ |
| Lateral dihedral angle of pyramid, $l_{Py}$ | 70.5288° (70° 31′ 44″) | 109.4712° (109° 28′ 16″) | 138.1897° (138° 11′ 23″) |
| Basal dihedral angle of pyramid, $b_{Py}$ | 70.5288° (70° 31′ 44″) | 54.7356° (54° 44′ 8″) | 37.3774° (37° 22′ 39″) |
| Lateral dihedral angle of prism, $l_{Pr}$ | 60° | 90° | 108° |
| Lateral dihedral angle of antiprism, $l_a$ | 109.4712° (109° 28′ 16″) | 127.5516° (127° 33′ 6″) | 138.1897° (138° 11′ 23″) |
| Basal dihedral angle of antiprism, $b_a$ | 109.4712° (109° 28′ 16″) | 103.8362° (103° 50′ 10″) | 100.8123° (100° 48′ 44″) |
| $2b_{Py}$ | 141.0576° (141° 3′ 27″) | 109.4712° (109° 28′ 16″) | 74.7547° (74° 45′ 17″) |
| $b_{Py} + 90°$ | 160.5288° (160° 31′ 44″) | 144.7356° (144° 44′ 8″) | 127.3774° (127° 22′ 39″) |
| $b_{Py} + b_a$ | 180° (see above) | 158.5718° (158° 34′ 18″) | 138.1897° (138° 11′ 23″) |

Table 3: Dihedral angles of Johnson solids derived from the pyramids.

# Chapter 3: The cupolae and rotunda: Johnson solids 3-6.

The names "cupola" and "rotunda" were coined by Norman Johnson for two types of what in my earlier book, *Polyhedra: a New Approach*, are called *uni/bigeneral tectal polyhedra*. As explained there, they are *tectal* because there are two planes perpendicular to the axis of rotational symmetry, each including a face of the polyhedron, and between which the entire polyhedron is found. They are *uni/bigeneral* because those faces are one $n$-gon and one $2n$-gon, where $n$ is the order of the axis of rotational symmetry. In the description which follows, it is useful to designate the $2n$-gon as the *base*, and the $n$-gon as the *roof* of the polyhedron; all the other faces are termed *lateral* faces of the polyhedron.

In a cupola, the lateral faces are, in general, isosceles triangles and rectangles. Because Johnson was only interested in polyhedra whose faces are regular polygons, the definition of "cupola" which he adopted required the lateral faces to be *equilateral* triangles and squares. However, many writers, including the author of the article "Cupola (geometry)" in *Wikipedia*, prefer a more general definition in which the lateral faces are, as indicated at the beginning of this paragraph, isosceles triangles (with equilateral as simply a special case) and rectangles (with squares as simply a special case). Because the order of rotational symmetry of a cupola is restricted to no greater than five when the lateral faces are required to be equilateral triangles and squares, only three types of cupola are recognized in Johnson's list: the triangular ($J_3$), square ($J_4$), and pentagonal ($J_5$) cupolae. The more general definition of "cupola," however, permits a greater variety of cupolae, and these are discussed in another book of mine, *Encyclopedia of Polyhedron Families*, which is planned to appear soon. However, in this book we are solely concerned with those cupolae that are Johnson solids.

Each of the sides of the roof polygon is an edge of the polyhedron, shared with one of the square lateral faces. Each of the vertices of the roof is a vertex of the type $n\cdot 4\cdot 3\cdot 4$, where $n$

is the order of the axis of rotational symmetry. Thus the triangular lateral faces share only a vertex, not an edge, with the roof, and are separated by squares on both sides from the edges of the roof. The base polygon, as stated earlier, of order 2*n*, has alternate sides shared with the triangles and squares, and all vertices of type 2*n*•4•3, though alternating between having this arrangement clockwise and counterclockwise. Figure 5 below shows the triangular cupola (*J₃*), the case for *n* = 3.

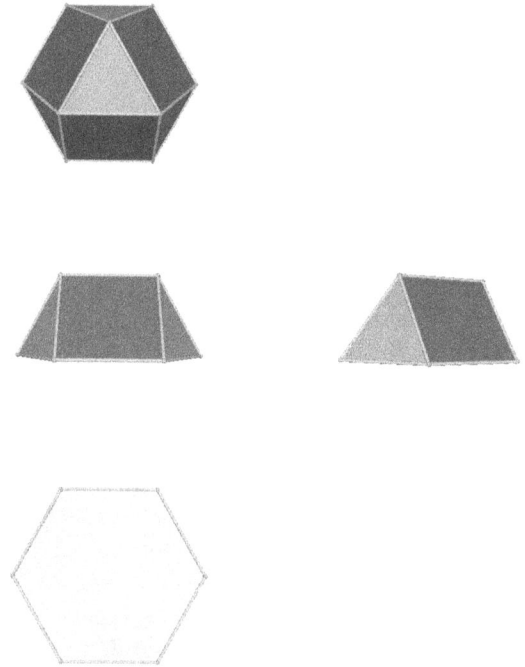

Figure 5: Johnson solid 3: the triangular cupola.

The symmetry of the triangular cupola is $C_{3v}$, and in general, all Johnson cupolae have $C_{nv}$ symmetry, where *n* is the order of the principal axis of rotation.

The duals of cupolae have no common name, but they do have distinctive appearances, and it would be very much appreciated if names were coined for them. They all have a vertex at one end of an *n*-fold principal axis of rotation, with *n* kite-

*The cupolae and rotunda: Johnson solids 3-6.*

quadrilaterals meeting there; at the other end of the principal axis, there are 2*n* (normally scalene) triangles which pair (alternating from one to the opposite orientation). For purposes of this book, it will be useful to reserve the term "apex" for the vertex where the *n* kite-quadrilaterals all come together, and designate the opposite vertex, where the 2*n* triangles come together, as the *antiapex*, a term introduced in my earlier book, *Polyhedra: a New Approach*, in the discussion of diapical polyhedra. The apex of a cupola dual resembles that of an antibipyramid (otherwise known as a trapezohedron, but my reasons for avoiding the latter term are given in *Polyhedra: a New Approach*) in having a number of kite-quadrilaterals all meeting equal to the order of symmetry of the principal axis of rotation. The antiapex resembles the apex of a 2*n*-gonal pyramid, although the triangles are not isosceles, but scalene. They share the same $C_{nv}$ symmetry as the cupolae of which they are duals. Figure 6 below illustrates the dual of the triangular cupola.

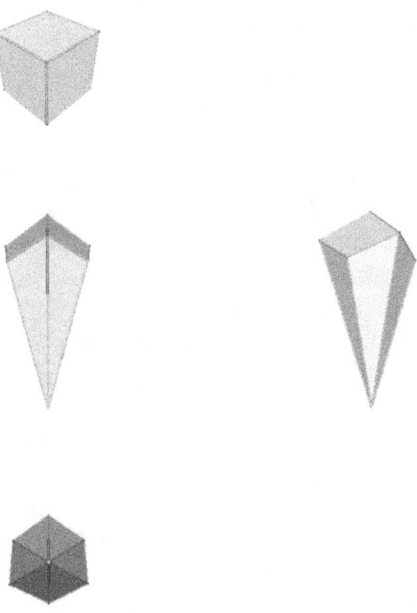

*Figure 6: The dual of Johnson solid 3 (the triangular cupola).*

It might be useful to refer to an apex such as the one of an antibipyramid or cupola dual (where $n$ kite-quadrilaterals all come together, $n$ being the order of the principal axis of rotation) as an *antibipyramid-like apex*, and an antiapex (or apex) such as the one of a cupola dual (where $2n$ alternately-oriented scalene triangles all come together, $n$ being the order of the principal axis of rotation) as a *pseudopyramid-like antiapex* (or apex). It should be noted that because the triangles at a pseudopyramid-like apex/antiapex are all congruent (though alternating in direction), all the $2n$ angles at that vertex are equal in magnitude. To complete the classification of apices and antiapices, we might introduce a third term, to describe an apex (or antiapex) where (where $n$ isosceles triangles all come together, $n$ being the order of the principal axis of rotation), as in a pyramid, and name such an apex/antiapex a *pyramid-like apex* (or *antiapex*). In general, all the Johnson solids and duals which can be described (in the terminology introduced in my earlier book, Polyhedra: a New Approach) as *diapical* or *apicobasal* have apices/antiapices that fall into one of these three categories, and this terminology will be followed in the remainder of this book whenever appropriate.

Much of what has been said of $J_3$ and its dual above applies to all cupolae and cupola duals; the square cupola, $J_4$, is illustrated below in Figure 7 and its dual in Figure 8; the pentagonal cupola is illustrated below in Figure 9 below and its dual in Figure 10. Except for the $C_{4v}$ symmetry of the square cupola and its dual, and the $C_{5v}$ symmetry of the pentagonal cupola and its dual, as opposed to the $C_{3v}$ symmetry of the triangular cupola and its dual, all the cupolae are very similar, and all the cupola duals are very similar, once allowing for the 3-4-5 trend rule described earlier. The general features of the cupolae and their duals were discussed earlier in this chapter, and apply to all three cupolae and their duals, so it will not be necessary to repeat the material separately in the discussion of each cupola or cupola dual. The specific details of the geometry of the cupolae and their duals are given in Table 4 below. It will be helpful to the reader to compare Figures 5, 7, and 9, as well as Figures 6, 8, and 10 to view the similarities and differences among the three cupolae and their duals.

One thing that will be noted, in subsequent chapters as

well as this one, is that many Johnson solids incorporate a cupola (minus its base) as part of their faces, and in every case the dual polyhedron will include an antibipyramid-like apex.

According to *Wikipedia*, the square cupola is sometimes called the *lesser dome*. (What the *greater dome* is, I have not been able to determine.) There is no place, other than that article and reproductions of the *Wikipedia* article, however, where I have seen this term used, and thus it is not used in this book, except only to mention the term here.

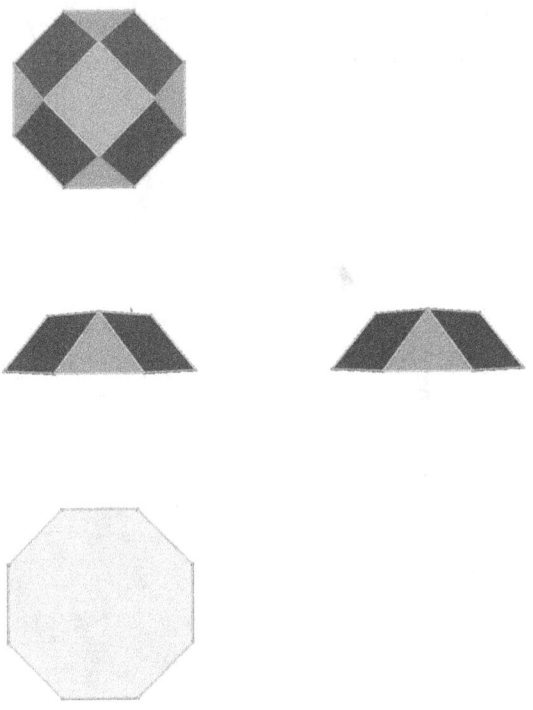

Figure 7: Johnson solid 4: the square cupola.

The 3-4-5 trend rule discussed earlier can be seen in its effects particularly by looking at Table 4 below, specifically at the dihedral angles. The dihedral angle between the roof polygon and the lateral square faces, 125.2644° (125° 15′ 52″) in $J_3$, increases to 135° in $J_4$ and to 148.2825° (148° 16′ 57″) in $J_5$; the dihedral angle between the lateral triangular faces and the base, 70.5288° (70° 31' 44") in $J_3$, decreases to 54.7356° (54° 44' 8") in $J_4$ and to 37.3774° (37° 22' 39") in $J_5$; and the dihedral angle between the lateral square faces and the base, 54.7356° (54° 44' 8") in $J_3$, decreases to 45° in $J_4$ and to 31.7175° (31° 43' 3") in $J_5$. The combination of the increases in the dihedral angle at the roof and the decreases in the dihedral angles at the base clearly shows the transition to squatter and squatter polyhedra.

The dual of the square cupola, shown in Figure 8 below, is much like the dual of the triangular cupola, which was illustrated in Figure 6 earlier in this chapter. The principal axis is fourfold, rather than threefold, and the symmetry is thus $C_{4v}$ rather than $C_{3v}$, but all the properties are similar.

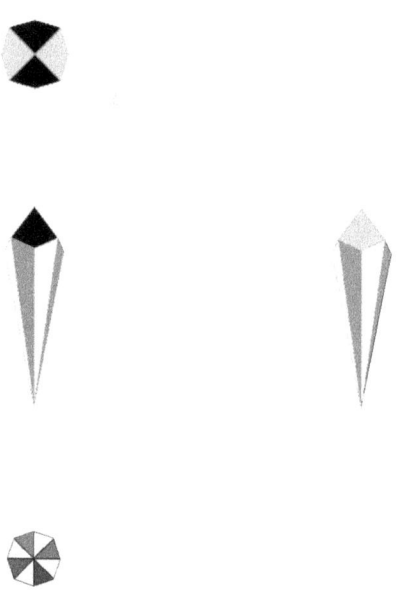

*Figure 8: The dual of Johnson solid 4 (the square cupola).*

Again, one can consult Table 4 to see the effects of the 3-4-5 trend rule, though it is better shown in the face angles (of the kite-quadrilaterals and triangles that comprise the apex and antiapex). The vertex angle of the kite-quadrilaterals at the apex, 99.8304° (99° 49′ 49″) in the dual of $J_3$, decreases to 58.0343° (58° 2′ 3″) in the dual of $J_4$ and to 30.9420° (30° 56′ 31″) in the dual of $J_5$. The vertex angle of the triangles at the antiapex, 17.2430° (17° 14′ 35″) in the dual of $J_3$, decreases to 8.4115° (8° 24′ 42″) in the dual of $J_4$ and to 4.0075° (4° 0′ 27″) in the dual of $J_5$. Both of these decreases are consistent with the thinner but taller ("spikier") appearance of the dual of $J_4$ and especially the dual of $J_5$ as compared with $J_3$.

The pentagonal cupola, Johnson solid 5, is illustrated in Figure 9 below. It has the expected $C_{5v}$ symmetry and continues the trends remarked about in the discussion of $J_4$ previously.

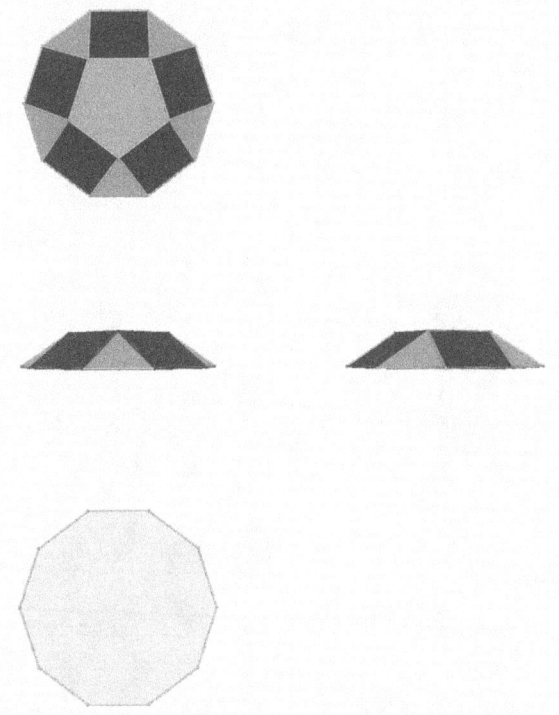

Figure 9: Johnson solid 5: the pentagonal cupola.

The pentagonal cupola is related to an Archimedean solid, the rhombicosidodecahedron, about which much will be said later in this book. (In particular, see Chapters 15 and 16.) Specifically, the configuration of equilateral triangles, squares, and pentagon at the vertices other than the base, 3•4•5•4, is exaxtly what is found at each vertex of the rhombicosidodecahedron, and as a result, the dihedral angles everywhere but at the base are identical to those of the rhombicosidodecahedron. For example, the values of 159.0948° (159° 5′ 41″) for the 3-4 dihedral angle and 148.2825° (148° 16′ 57″) for the 4-5 dihedral angle, shown in Table 4 for $J_5$, are in a sense "inherited" from the rhombicosidodecahedron, and will be seen over and over again in other tables where portions of a particular Johnson solid are similarly arranged.

The dual of the pentagonal cupola is illustrated in Figure 10 below.

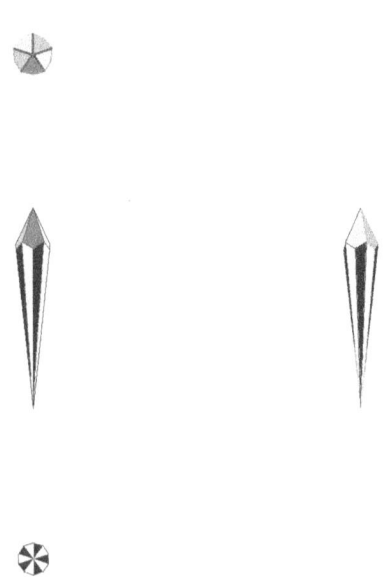

Figure 10: The dual of Johnson solid 5 (the pentagonal cupola).

As was the case for the pyramids (of which one can also consider the regular tetrahedron as an example confirming the

*The cupolae and rotunda: Johnson solids 3-6.*

trends), the 3-4-5 trend rule (introduced on p. 12 in Chapter 2) can be seen to apply to the cupolae. As the order of the principal axis of rotation increases, because of the need to keep the lateral faces constant in size, the polyhedron becomes flatter. Also, as the Johnson solids become flatter, their duals, as generated by polar reciprocation, go the opposite way, becoming spikier. Besides this being a part of the 3-4-5 trend rule, it is an example of another general rule, which applies even when not considering families of different-symmetry-order polyhedra; the flatter a polyhedron, the less flat its dual becomes.

Table 4 below is a summary of the properties of Johnson solids 3, 4, and 5 and their duals.

One might note that although the triangular faces of the duals, which all meet at the $2n$-order vertex corresponding to the $2n$-gonal base of the original cupola, are scalene triangles, and the ten edges meeting at that vertex alternate in length between longer and shorter lengths, the dihedral angles at those edges are all equal. This will be noted as well in Table 5 for the pentagonal rotunda. In general, although the edges meeting at the vertex which is dual to a *regular* $2n$-gonal face of a polyhedron may be unequal, the dihedral angles at those edges will be equal. It also might be noted that both the 3r-4 and 3L-4 dihedral angles of $J_3$ are identical, even though there is no reason to expect this from the symmetry.

| Johnson solid number | 3 | 4 | 5 |
|---|---|---|---|
| Name | Triangular cupola | Square cupola | Pentagonal cupola |
| Symmetry* | $C_{3v}$ | $C_{4v}$ | $C_{5v}$ |
| Faces | [1+3](3)‡ + [3](4) + [1](6) | [4](3) + [1+4](4)‡ + [1](8) | [5](3) + [5](4) + [1](5) + [1](10) |
| Edges* | 15 | 20 | 25 |
| Vertices | 9 | 12 | 15 |
| Dihedral angles | 3r-4: 125.2644° (125° 15′ 52″), 3L-4: 125.2644° (125° 15′ 52″), 3L-6: 70.5288° (70° 31′ 44″), 4-6: 54.7356° (54° 44′ 8″) | 3-4L: 144.7356° (144° 44′ 8″), 3-8: 54.7356° (54° 44′ 8″), 4r-4L: 135°, 4L-8: 45° | 3-4: 159.0948° (159° 5′ 41″), 3-10: 37.3774° (37° 22′ 39″), 4-5: 148.2825° (148° 16′ 57″), 4-10: 31.7175° (31° 43′ 3″) |
| Area | 7.33012701 = 3 + (5√3)/2 | 11.56047793 = 7 + 2√2 + √3 | 16.57974975 = <20 + √[10(80 + 31√5 + √{2175 + 930√5})]>/4 |
| Volume | 1.17851130 = (5√2)/6 | 1.94280904 = 1 + ⅔√2 | 2.32404532 = (5 + 4√5)/6 |
| Dual | | | |
| Name | Unnamed | Unnamed | Unnamed |
| Faces | [6](3)+[3](4k) | [8](3)+[4](4k) | [10](3)+[5](4k) |
| Vertices | 8 | 10 | 12 |
| Face angles† | 3: 17.2430° (17° 14′ 35″), 57.6435° (57° 38′ 37″), 105.1134° (105° 6′ 48″); 4k: [2]72.3457° (72° 20′ 45″), 99.8304° (99° 49′ 49″), 115.4782° (115° 28′ 42″) | 3: 8.4115° (8° 24′ 42″), 60.7789° (60° 46′ 44″), 110.8095° (110° 48′ 34″); 4k: [2]90.9922° (90° 59′ 32″), 58.0343° (58° 2′ 3″), 119.9813° (119° 58′ 53″) | 3: 4.0075° (4° 0′ 27″), 56.8849° (56° 53′ 6″), 119.1076° (119° 6′ 27″); 4k: [2]108.2807° (108° 16′ 50″), 30.9420° (30° 56′ 31″), 112.4967° (112° 29′ 48″) |

*The cupolae and rotunda: Johnson solids 3-6.*

| Johnson solid number | 3 | 4 | 5 |
|---|---|---|---|
| Dihedral angles | 3-3: 122.3088° (122° 18′ 32″), 4-4: 101.8812° (101° 52′ 52″), 3-4: 115.3317° (115° 19′ 54″) | 3-3: 135.7530° (135° 45′ 11″), 4-4: 131.1495° (131° 8′ 58″), 3-4: 107.9202° (107° 55′ 13″) | 3-3: 144.2164° (144° 12′ 59″), 4-4: 146.4298° (146° 25′ 47″), 3-4: 114.1606° (114° 9′ 38″) |
| Area | 3.80239916 | 4.64015575 | 7.57684552 |
| Volume | 0.50362213 | 0.62513426 | 1.08243727 |

*Table 4: Properties of Johnson solids 3, 4, and 5 and their duals.*

*The symmetry and number of edges are always the same for each solid and its dual, so are not repeated in the "dual" section.

†Face angles are only given for the duals, as the Johnson solids have regular faces, whose angles can be found in Table 1 on p. 3.

‡One triangle, in the case of the triangular cupola, and one square, in the case of the square cupola, forms the roof and is designated 3r or 4r in the dihedral angle lists in the table. The remaining ones are lateral faces and are designated 3L or 4L.

Like the cupola, the *rotunda* was named by Johnson; however, it is very odd that he named his $J_6$ (illustrated in Figure 11 below) the "pentagonal rotunda," as no other kind of rotunda exists under Johnson's restriction to all regular-polygonal faces. If this restriction is dropped, however, a whole family of rotundae can be constructed. (These will be discussed in my book, *Encyclopedia of Polyhedron Families*, which is planned to come out shortly.) The edges which constitute the sides of the roof *n*-gon are shared with triangular lateral faces (isosceles in general, equilateral under Johnson's restriction to all regular-polygonal faces). At each vertex of the roof, besides the roof *n*-gon and the two triangles sharing edges with the roof *n*-gon, a fourth face is a pentagon. (This would, of course, be a regular pentagon under Johnson's restriction to all regular-polygonal faces; in general it would merely be what, in my earlier book, *Polyhedra: a New Approach*, I term an isosceles pentagon, though the term is not generally in use.)

The pentagonal rotunda has a fivefold axis through the centers of the roof pentagon and the base decagon, and five mirror planes through that axis, making the symmetry $C_{5v}$. (In general, even if other rotundae are constructed whose triangular and pentagonal faces are merely isosceles, if the base and roof are regular polygons, the resulting rotunda will have $C_{nv}$ symmetry.)

The Johnson (regular-polygon-faced) pentagonal rotunda is related to an Archimedean solid, the icosidodecahedron. If an icosidodecahedron is cut into two halves by a plane passing through its center of symmetry, parallel to any of its pentagonal faces, the resulting halves are two congruent pentagonal rotundae. Because of this, many of the parameters of the pentagonal rotunda show more symmetry than might be expected.

In the same way as the parts of the three cupolae other than the base appear repeatedly as pieces of more complex Johnson solids, so does the pentagonal rotunda appear. As a result, there is often the phenomenon that although triangles, or pentagons, in a particular Johnson solid may belong to different transitivity classes, they have the same dihedral angles, 142.6226° (142° 37′ 21″), as those of the icosidodecahedron, which means the same as each other. Although this will be seen in Table 5 below, this dihedral angle will appear repeatedly in later tables in this book. This icosidodecahedron dihedral angle, 142.6226° (142° 37′ 21″), has the value of

$$\cos^{-1}\{-\sqrt{[(5 + 2\sqrt{5})/15]}\}.$$

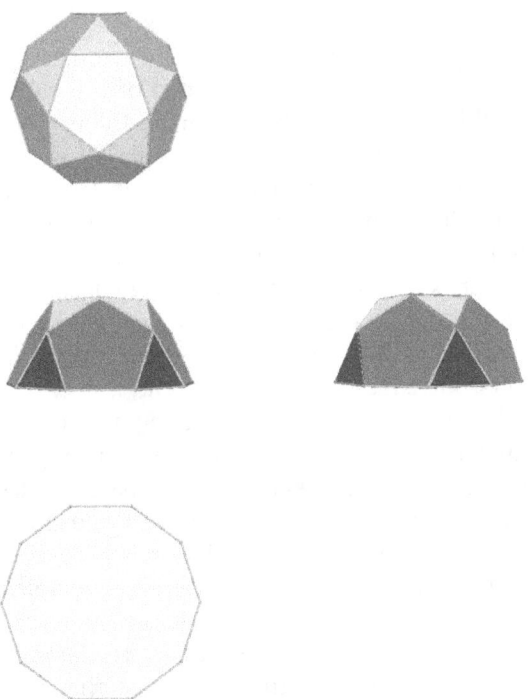

Figure 11: Johnson solid 6: the pentagonal rotunda.

The dual of the pentagonal rotunda, of course, has the same $C_{5v}$ symmetry as the pentagonal rotunda. Like the duals of the cupolae, this solid has an assemblage of (in this case five) kite-quadrilaterals around one apex, and twice the number (thus, in this case, ten) of scalene triangles around the other. (For this purpose, in the following discussion, the vertex where ten triangles meet will be considered the antiapex and the one where five kite-quadrilaterals meet will be considered the apex.) Also, just as in the cupola duals, these scalene triangles are congruent to each other, but alternate in orientation. So the apex is antibipyramid-like, as defined on p. 20 above, and the antiapex is pseudopyramid-like. However, unlike the cupola duals, this polyhedron has an additional set of five kite-quadrilaterals as lateral faces. In this discussion, the kites meeting at the apex will be termed *apical kites*, and the other ones the *lateral kites*. There are five kinds of edges. The five edges that meet at the apex are all equal by symmetry, and form one set of sides of the apical kites, each edge shared by two apical kites. The other edges of the apical kites are also edges of the lateral kites, and these are all equal to each other, again by symmetry, but different from the first set. As was mentioned of the cupolae, it will be seen in subsequent chapters that many Johnson solids incorporate a pentagonal rotunda (minus its base) as part of their faces, and in every case the dual polyhedron will include an antibipyramid-like apex. Additionally, there will be another set of kite-quadrilaterals adjacent to the five which form the antibipyramid-like apex, as the lateral kites are adjacent to the apical kites in the dual of $J_6$. The remaining edges of the lateral kites form a third set of five equal edges, different again from the first two sets but equal by symmetry. The last two sets of five edges are shared by the triangles meeting at the antiapex, and alternate in length between a shorter and a longer. Figure 12 below illustrates this polyhedron. (Compare Figure 12 with Figure 10, and it can be seen that the most conspicuous difference is the ring of lateral kite-quadrilaterals around the middle. The arrangement of ten kite-quadrilaterals, five apical and five lateral, should become familiar as you see some of the other Johnson solid duals in this book, as it is found in a large number of them.) One will generally notice that in every dual of a Johnson solid incorporating a rotunda, both the lateral and apical kites are nearly rhombi, with all edges differing by only a small amount, and the lateral kites have two equal acute angles, with the obtuse angles being only a little bit different; while the apical kites have two equal obtuse angles, with the acute angles being only a little bit different; additionally, the two nearly-equal obtuse angles of the lateral kites are approximately equal to the two equal obtuse angles of the apical kites; while the two nearly-equal acute angles of the apical kites are approximately equal to the two equal acute angles of the lateral kites. This will be repeated in so many places that it should be given a name: the *apical/lateral kite rule*.

Actually, $J_6$ does not illustrate the apical/lateral kite rule as well as most of the others, because the "nearly-equal" angles

*The cupolae and rotunda: Johnson solids 3-6.*

differ by rather more than in other duals of Johnson solids derived from the rotunda, as will be seen as the book proceeds.

Because the dual of the pentagonal rotunda has twenty faces, it could be termed an icosahedron, although the term is generally restricted to the regular icosahedron. But just as one can talk of a regular (or pentagon-) dodecahedron and a rhombic dodecahedron, there is nothing to prevent us from referring to this polyhedron as a *strombo-triangular icosahedron* (or *deltoido-triangular icosahedron*).

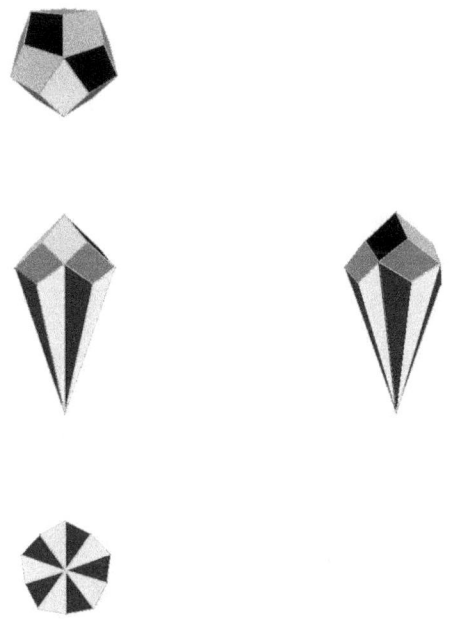

*Figure 12: The dual of Johnson solid 6 (the pentagonal rotunda).*

However, it may actually become more useful to find a name for the assemblage of ten kite-quadrilaterals (five forming an antibipyramid-like apex, plus the five directly adjacent to them) forming the upper half of this polyhedron; this set of ten, with angles and edge-lengths somewhat varying from case to case, is found in the duals of all those Johnson solids which incorporate a rotunda (less its base, which drops out in fusion) among their faces. Essentially, we are talking about all the kite-quadrilateral faces of the dual of the pentagonal rotunda as opposed to the triangular faces which form the pseudopyramid-like antiapex. I will not suggest a name at this point, but encourage others reading this book to devise one if they have a good idea.

Table 5 below summarizes the geometric properties of the pentagonal rotunda and its dual.

*The cupolae and rotunda: Johnson solids 3-6.*

| Johnson solid number | 6 |
|---|---|
| Name | Pentagonal rotunda |
| Symmetry* | $C_{5v}$ |
| Faces | [5+5](3)‡ + [1+5](5)‡ + [1](10) |
| Edges* | 35 |
| Vertices | 20 |
| Dihedral angles | 3-10: 79.1877° (79° 11′ 16″),<br>3-5♭: 142.6226° (142° 37′ 21″),<br>5-10: 63.4349° (63° 26′ 6″) |
| Area | 22.34720027 |
| Volume | 6.91776297 |
| **Dual** | |
| Name | Unnamed (but could be termed a strombo-triangular icosahedron) |
| Faces | [10](3) + [5+5](4k)‡ |
| Vertices | 17 |
| Face angles† | 3: 12.0698° (12° 4′ 11″),<br>52.8201° (52° 49′ 12″),<br>115.1102° (115° 6′ 37″);<br>4: [2]118.5090° (118° 30′ 33″),<br>53.5065° (53° 30′ 23″),<br>69.4754° (69° 28′ 32″);<br>4L: [2]67.2426° (67° 14′ 33″),<br>105.5264° (105° 31′ 35″),<br>119.9884° (119° 59′ 18″) |
| Dihedral angles | 3-3: 146.0203° (146° 1′ 13″),<br>3-4L: 144.2463° (144° 14′ 47″),<br>4-4: 129.9170° (129° 55′ 1″),<br>4-4L: 135.6131° (135° 36′ 47″), |
| Area | 10.97401863 |
| Volume | 2.79098160 |

Table 5: *Properties of Johnson solid 6 and its dual.*

*The symmetry and number of edges are always the same for each solid and its dual, so are not repeated in the "dual" section.

†Face angles are only given for the duals, as the Johnson

solids have regular faces, whose angles can be found in Table 1 on p. 3.

˒Though there are three different types of 3-5 edges, all the 3-5 dihedral angles are equal.

†In the dual polyhedron, the kite-quadrilaterals are of two types: five lateral (designated 4L), and five meeting at the dual vertex to the rotunda base (designated just 4).

*The cupolae and rotunda: Johnson solids 3-6.*

It should be noted that in the pentagonal rotunda, the sum of the 3-10 and 5-10 dihedral angles is equal to the 3-5 dihedral angle. This is not a coincidence. As was stated earlier, two pentagonal rotundae placed base to base, with the triangles of one sharing an edge with the pentagons of the other, would fuse to form an icosidodecahedron, an Archimedean solid, with all identical edges. The symmetry of the icosidodecahedron accounts for this equality, as well as the equality of all three types of 3-5 dihedral angle.

The grouping of the pentagonal rotunda with the three cupolae in a single chapter was done for two reasons. First, it is not desirable, in my belief, to devote a chapter to a single Johnson solid, so the discussion of the pentagonal rotunda was combined with that of the three cupolae (which directly precede it in the ordering Johnson gave to the polyhedra in his list). But the second reason is more significant. In a number of places, it will be seen that there is a sequence of Johnson solids in which three have $C_{nv}$ symmetry, with the three cupolae in order at one end of the principal rotation axis, followed by a fourth which has, like the third, $C_{5v}$ symmetry, with the pentagonal rotunda in the same place that the pentagonal cupola has in the third. These sets of four polyhedra, listed in consecutive positions on the list, occur at numbers 18-21 (Chapter 7) and 22-25 (also in Chapter 7). In addition, there are sets where both ends of a Johnson solid are either cupolae or rotundae, and again these are all grouped together in an order related to this. These are found at numbers 27-34 (Chapter 8), 35-43 (Chapter 9), and 44-48 (Chapter 10). Because of these sets of consecutively listed Johnson solids, it is useful to consider the three cupolae and the pentagonal rotunda as a block of related members of the list.

## Chapter 4:  Elongated and gyroelongated pyramids: Johnson solids 7-11.

Starting with the seventh Johnson solid, we encounter the first of those which can be produced by fusing two or more simpler solids. (The term "fusion" is defined in my earlier book, *Polyhedra: a New Approach*.) Johnson does not use the term "fusion" generally, but instead uses special terminology for the fusion of specific types of polyhedron to a face of a basic polyhedron. Fusing a pyramid (in some cases, something other than a pyramid, but usually a pyramid) to any face of a polyhedron was termed *augmentation* by Johnson; another term found in the literature is *cumulation*, or (specifically when it is a pyramid that is fused) *akisation*. Since the latter two terms usually imply fusing a pyramid to *all* faces of a polyhedron, it is better to use the term "augmentation" when only one or some small number of faces is fused to a pyramid. Johnson introduced two new terms for the fusion of a prism or an antiprism to a polyhedron: an elongated polyhedron involves fusion of a prism to a polyhedron, and a gyroelongated polyhedron involves fusion of an antiprism to a polyhedron. However, Johnson only uses the terms "elongation" and "gyroelongation" when the fusion of a prism or antiprism is to the base of a polyhedron with a rotational symmetry axis (that is to say, preserving the symmetry), while he prefers to use "augmentation" for the cases where a pyramid or other polyhedron is fused to a face which is *not* the base of a polyhedron with a rotational symmetry axis (*i. e.*, either a lateral face of a rotationally symmetric polyhedron, or any face of a polyhedron which fails to have a single principal axis of rotational symmetry). It should be noted that when a polyhedron is one like a bipyramid which itself can be envisioned as produced by fusion of two simpler polyhedra along a face perpendicular to the rotational axis, Johnson assumes that these pieces are separated and the prism or antiprism is sandwiched between those two pieces. In this book, Johnson's restriction of the terms "elongation" and "gyroelongation" to cases when the fusion of a prism or antiprism is to the base of a polyhedron with a rotational symmetry axis will be maintained, but this author prefers to allow the term "augmentation" in a more general sense, as was used in my earlier book,

*Polyhedra: a New Approach*, for the fusion of a pyramid to *any* face of a polyhedron, and use an appropriate adjective to characterize the face being augmented.

Because Johnson wished to restrict the term "augmented" in the manner just mentioned, the seventh through eleventh Johnson solids were considered by him to be based on the pyramids (Chapter 2) and generated by elongating or gyroelongating them, as those terms were just defined. However, it is equally correct to name them by *starting* with a prism or antiprism, and fusing a pyramid to one of its bases (*augmenting* a base, in the terms of the previous paragraph), thus naming these forms "monoaugmented prisms" and "monoaugmented antiprisms." As between the terms "monoaugmented prism" and "elongated pyramid" (or the terms "monoaugmented antiprism" and "gyroelongated pyramid") I can see no reason to prefer one to the other, so despite Johnson's preference, I consider them equally preferred alternatives. In fact, though *Wikipedia* articles do change from time to time, so that the article may not read that way when you have this book in your hand, the article entitled "Johnson solid" did, as of September 2013, when I checked it, give the names "Augmented triangular prism" as an alternative for $J_7$, "Augmented cube" for $J_8$, "Augmented pentagonal prism" for $J_9$, "Augmented square antiprism" for $J_{10}$, and "Augmented pentagonal antiprism" for $J_{11}$, directly underneath Johnson's names in the listing shown. Obviously, there are others than myself to whom these names seem reasonable alternative names. (However, as was just noted in the previous paragraph, as in the cases of such solids as $J_{49}$ and $J_{52}$, Johnson preferred to use the term "augmented prism" for prisms augmented on a *lateral face*, rather than on a base. If it is not clear what was intended, such solids as $J_{49}$ and $J_{52}$ should be described as "laterally monoaugmented prisms" and the elongated pyramids as "basally monoaugmented prisms."

In order to maintain convexity of the resulting polyhedron, when two polyhedra are fused together, it is necessary that the dihedral angles that are combined at each edge where fusion takes place sum to less than 180°. In Table 3 (p. 16), the basal dihedral angles of pyramids and antiprisms with $n=3$ to 5 (the only ones that are relevant here) are given; the basal dihedral angle of a prism (at least, of the kinds of prisms involved in elongation, which are *right prisms*) is always 90°. Since the basal dihedral angle of a pyramid is never larger than 70.5288° (70° 31′ 44″) and that of a prism is 90°, fusion of a prism (at its base) and a pyramid (at its base) is never going to produce a dihedral angle exceeding 180°, so there is no problem. In fact, the appropriate dihedral angles are given in the row of Table 3 with the specification "$b_{Py}$ + 90°." This is not so when combining a pyramid with an antiprism, as can be seen in in the row of Table 3 with the specification "$b_{Py}$ + $b_a$." Therefore only for $n > 3$ can a pyramid and antiprism be fused to produce a gyroelongated pyramid or basally augmented antiprism.

While the pyramids in Chapter 2 include only the square and pentagonal pyramids, because a triangular pyramid would be a (Platonic) regular tetrahedron, the *elongated* pyramids include a triangular one, since fusing it with a prism destroys the identity of all faces, so the resultant solid is a Johnson solid. The elongated triangular pyramid (basally monoaugmented triangular prism) $J_7$ is illustrated in Figure 13 below.

Since $J_7$ is the first on the Johnson solids to be obtained by fusion, it is important to note that in many cases the properties of a polyhedron obtained by fusion can be obtained by considering the properties of the components which have been fused together. All the Johnson solids that can be obtained by fusion have components that are either simpler Johnson solids or among the solids that are excluded from Johnson's list because they can be found among the Platonic or Archimedean solids, regular-polygon-faced prisms, or regular-polygon-faced antiprisms. And, for example, the dihedral angles of the fused polyhedra can be completely determined from the dihedral angles of the components, as was implied in the discussion on p. 15. Specifically, $J_7$ is the product of the fusion of a triangular pyramid (regular tetrahedron) and a triangular prism. As neither of these is a Johnson solid, the dihedral angles are not tabulated here in the regular tables in this book devoted to the properties of the Johnson solids, but the necessary information is found in Table 3: there are three edges of $J_7$ that derive from edges of the regular tetrahedron and which thus have a dihedral angle of 70.5288° (70° 31′ 44″), six edges of $J_7$ that derive from edges of the prism (three from among the six edges that bound the triangular bases of the prism, with dihedral angles of

90°, and three between the lateral square faces of the prism, with dihedral angles of 60°), while the remaining edges of $J_7$ are produced by the fusion of the two polyhedra, so that the dihedral angle is the *sum* of 70.5288° (70° 31′ 44″) from the tetrahedron and 90° from the prism: 160.5288° (160° 31′ 44″). All these numbers are shown in Table 6, but this reasoning shows how they could have been derived if Table 6 had not displayed the data.

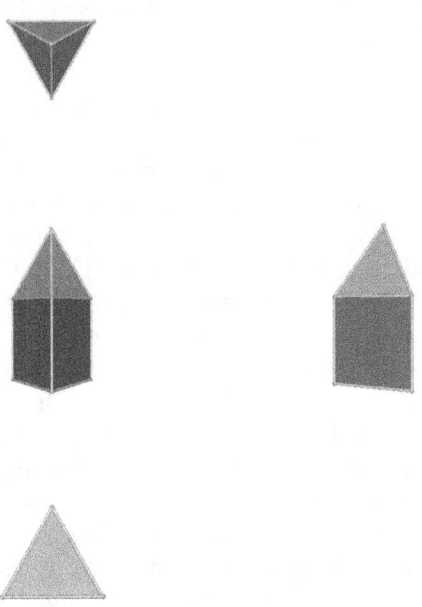

*Figure 13: Johnson solid 7: the elongated triangular pyramid.*

Another name for the fusion of a pyramid to a prism, due to the mathematician John Horton Conway, is "hermaphrodite."

The elongated pyramid (basally monoaugmented prism) has been described as *self-dual*, like the pyramids in Chapter 2. In fact, this is only true in a partial way: comparing Figure 13 with Figure 14, one can see that the dual has the right number of faces, edges, and vertices, and it has quadrilaterals where the original has quadrivalent vertices and triangles where the original has trivalent vertices, but it is in fact a fusion of a pyramid and a *prismoid*, rather than a prism. See Figure 14 below for an example. The polyhedron illustrated in Figure 14 is not an elongated pyramid as defined by Johnson, but might be called a basally augmented triangular prismoid or a monofrustum of an asymmetric triangular bipyramid. (I use the term *asymmetric bipyramid* to denote a fusion of two pyramids, base to base, that are not congruent.)

Actually, in creating the duals of all five of the Johnson solids covered in this chapter, it is much easier to think of them as augmented prisms or antiprisms than to concentrate on the names that Johnson gave them. The dual of a prism is a bipyramid, and the dual of an antiprism is an antibipyramid; the dual of *any* augmented polyhedron is obtained by truncating the dual of the original polyhedron. So it is obvious that the duals of the three monoaugmented prisms ($J_7$ through $J_9$) are obtained by truncating one vertex of a bipyramid (thus, monofrusta of bipyramids; see Figures 14, 16, and 18), and the duals of the two monoaugmented antiprisms ($J_{10}$ and $J_{11}$) are obtained by truncating one vertex of an antibipyramid (thus, monofrusta of antibipyramids; see Figures 20 and 22). (While it is unlikely that you have seen the term "monofrustum of an antibipyramid" anywhere else, it is the proper term, provided that, as I do, you prefer to refer to an antibipyramid by that name rather than a "trapezohedron" or "deltohedron," and my reasons for preferring that term are given in my earlier book, *Polyhedra: a New Approach*, so I will not repeat them here. However, in that book I introduced the term "polytimoid" for this figure, which is a lot less clumsy than "monofrustum of an antibipyramid.")

In general, all the elongated pyramids (Johnson solids 7 through 9) have duals that are basally augmented prismoids (monofrusta of asymmetric bipyramids), and thus the elongated pyramids are not quite self-dual, but close enough to being so that they are often so characterized. It should be noted that, by adjusting the sides of a basally augmented prismoid in an appropriate manner, it is possible to create a polyhedron whose dual is identical to itself (oppositely oriented), and it is on this basis that one might consider an elongated pyramid and a basally augmented prismoid as forms of the same type of polyhedron, so that $J_{7-9}$ are self-dual in the same way as $J_{1-2}$.

The dual of the elongated triangular pyramid has two different types of triangular faces. The one that is in a plane perpendicular to the threefold axis is constrained by the $C_{3v}$ symmetry to be equilateral, and the remaining three (which surround the apex opposite the first triangle) are merely isosceles, in fact being very far from equilateral. The remaining three faces of this polyhedron are very short, squat isosceles trapezoids (See Figure 14, below).

Figure 14: The dual of Johnson solid 7 (the elongated triangular pyramid): a basally augmented triangular prismoid or monofrustum of an asymmetric triangular bipyramid.

The isosceles triangles are obtuse, with an angle of 105.9914° (105° 59′ 29″) at their apex, which is also the apex of the polyhedron. The entire polyhedron is extremely squat; its total height is only ~0.4226, as compared to the parallel sides of the isosceles trapezoid, of which the shorter is ~0.7842 and the longer is ~1.0838.

While (by definition, since all faces are regular polygons) all edges of a Johnson solid are equal, it is easy, looking at the *duals*, to see that the edges between the lateral quadrilaterals are not equal to the edges between the triangular faces. In the dual of $J_7$, they are a lot shorter, with the inter-trapezoid edges being ~0.2357 and the inter-triangle edges being ~0.6786 (assuming the edges on the original $J_7$ are all equal to 1), almost a 1-to-3 ratio. In the case of $J_8$ (see Figure 16 below), the corresponding lengths are ~0.4433 and ~0.9777, a ratio of 1 to ~2.2, and in the case of $J_9$ (see Figure 18 below), the corresponding lengths are ~0.7838 and ~1.4003, a ratio of 1 to ~1.8, so as the order of the principal axis increases, both edge lengths increase, but they become more nearly equal.

In general, the duals of the Johnson solids have faces that are *not* regular polygons. Since the symmetry of the dual is the same as that of the original, where the original Johnson solid possesses a mirror plane, so will the dual, and so triangles will be isosceles, quadrilaterals will be isosceles trapezoids or kite-quadrilaterals, and pentagons will be isosceles (as defined in my earlier book, *Polyhedra: a New Approach*) when the vertices to which they are dual lie on a mirror plane. But only when the Johnson solid has an *apex* (a vertex on an axis of rotational symmetry) is its dual required to have a regular polygon as a face. This was seen previously in the case of $J_1$ and $J_2$, and not again until the polyhedra covered in this chapter. Using the terminology introduced in my earlier book for the classification of polyhedra with axes of rotational symmetry, *diapical* and *apicobasal* polyhedra among the Johnson solids will have duals with a regular polygon as a face; *tectal* polyhedra (and those few without an axis of rotational symmetry) are not required to by symmetry (though they may by coincidence). Coincidence may also cause some duals to have regular faces at points where the original Johnson solid does not have a rotational axis of the maximum order. An example is Johnson solid 57, which has no fourfold axes, but whose dual has three faces that are perfect squares. The $D_{3h}$ symmetry of the polyhedron would permit those faces to be rhombi, but they are in fact perfectly square.

*Elongated and gyroelongated pyramids: Johnson solids 7-11.*

The elongated square pyramid, Johnson solid 8, is essentially the same as the elongated triangular pyramid except that its principal axis of rotation is fourfold, making the symmetry $C_{4v}$. It is illustrated in Figure 15 below.

Since the base is square rather than triangular, there are two types of square face and only one type of triangular, as opposed to the case of the elongated triangular pyramid, which has two types of triangle and only one square. Otherwise, little is new to say about it that was not said of the elongated triangular pyramid.

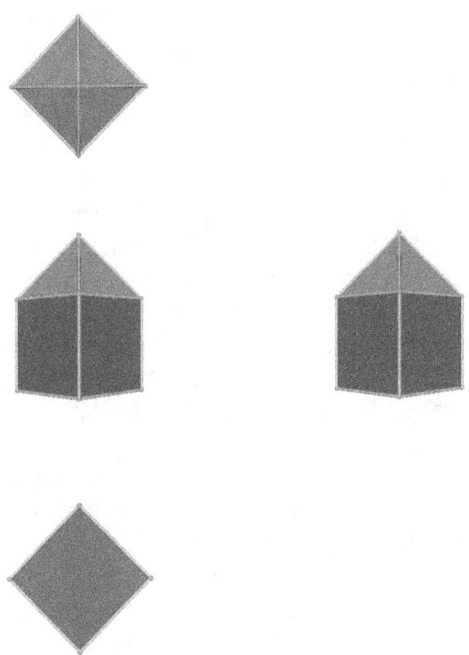

Figure 15: Johnson solid 8: the elongated square pyramid.

One point that can be made specifically about the elongated square pyramid, however, which does not apply to the other elongated pyramids in this chapter, is that it is the first member we meet of what can be termed the *augmented Platonics*. Starting with any of the five Platonic solids, one can augment it by fusing pyramids to anywhere from one face to the number of faces in the entire polyhedron. In most cases, this cannot be done without either losing convexity or flattening the pyramids so that the added faces are not equilateral triangles, so the resulting polyhedra are not Johnson solids. This will be worked out in greater detail in Chapter 12, where those Johnson solids that are normally treated as augmented Platonics rather than as examples of other classes of polyhedron are discussed. (If *all* of the faces of the starting Platonic solid are augmented, a *Catalan solid* is obtained, but in none of those five cases can the triangles be equilateral!) However, there are seven combinations that will yield Johnson solids: From the regular tetrahedron, $J_{12}$; from the cube, $J_8$ and $J_{15}$; and four from the regular dodecahedron: $J_{58-61}$.

In a similar way to what was said of the elongated square pyramid itself, the dual of the elongated square pyramid, illustrated in Figure 16 below, can be compared with the dual of the elongated triangular pyramid. The main difference is that it has a fourfold axis rather than threefold; but that does lead to one major difference. There are two types of quadrilateral face and only one type of triangular, as opposed to the case of the dual of the elongated triangular pyramid, which has two types of triangular face and only one type of quadrilateral. This difference is similar to the properties of the original Johnson solids, where the elongated triangular pyramid has two types of triangular face and only one type of quadrilateral, while the elongated square pyramid has two types of quadrilateral face and only one type of triangular. This is in keeping with the "self-dual" status of the elongated pyramids; although they are not truly their own duals, they are identical to their duals in terms of the numbers of triangles, quadrilaterals, etc., as well as the numbers of trivalent, quadrivalent, etc. vertices. The lengths of edges, angles, and such precise measurements, however, differ between the original and the dual. The angles, for example, can be seen in Table 6. This type of duality, where edges and angles are not the same as obtained by polar reciprocation, is sometimes referrred to as *topological duality*, but the term is not quite right, as topological transformations would allow replacing straight lines by curves. It might be a better choice of a term, for example, to refer to it as *combinatorial duality*, since we are talking about keeping the count of edges, faces (and faces of any number of sides), and vertices (and vertices of any specific valency), but this term is not to my knowledge used anywhere that I have seen.

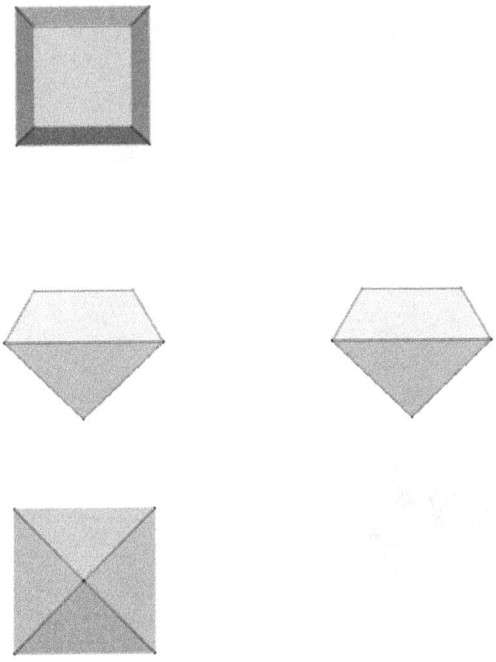

Figure 16: The dual of Johnson solid 8 (the elongated square pyramid): a monofrustum of an asymmetric square bipyramid.

Johnson solid 9, the elongated pentagonal pyramid, is exactly like the previous two except that the principal axis of rotation is fivefold, making the symmetry $C_{5v}$. It is illustrated in Figure 17 below, which can be compared with Figures 13 and 15 above. While the prismatic part of the solid (the elongation) is the same height for all the elongated pyramids (because the need for all faces to be regular means that the prismatic faces are squares whose side lengths are equal to the common length of all the edges in the polyhedron), the pyramidal part matches the sequence of regular tetrahedron and $J_{1-2}$, and in conformity to the 3-4-5 trend rule, $J_9$ is more squat than $J_7$ or $J_8$. (In fact, because these three Johnson solids are obtainable by fusion, the pyramidal parts are *exactly identical* to the sequence of regular tetrahedron and $J_{1-2}$, as can be seen by comparing Table 6 with Table 2 on p. 13 above.)

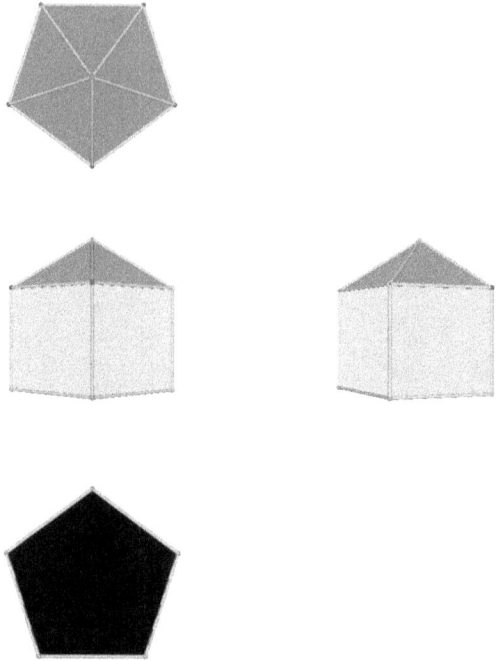

*Figure 17: Johnson solid 9: the elongated pentagonal pyramid.*

The dual of the elongated pentagonal pyramid is illustrated in Figure 18 below. A comparison with Figures 14 and 16 illustrates the portion of the 3-4-5 trend rule that applies to the duals very well. In Figure 14 one sees a very squat polyhedron, with the apex angles of the isosceles triangles very obtuse and the 4-4 edges extremely short. In Figure 16, the isosceles triangles are nearly equilateral (See Table 6 below), and the 4-4 edges are considerably longer, though still much shorter than the other edges in the polyhedron. And in Figure 18, the apex angle of each of the the isosceles triangles is actually the smallest angle of the triangle, and the 4-4 edges are longer than the 4-5 edge of the polyhedron.

It might be noted that in Table 6 it was necessary to distinguish two kinds of triangle in the dual of $J_7$, and two kinds of quadrilateral in the dual of $J_8$, but in $J_9$ there are no transitivity classes needing to be distinguished. This is only because the base of $J_7$ is triangular (so it needs to be distinguished from the triangles around the apex) and that of $J_8$ is square (so it needs to be distinguished from the isosceles trapezoids among the quadrilaterals), but the base of $J_9$, a pentagon, is the only pentagonal face in the polyhedron.

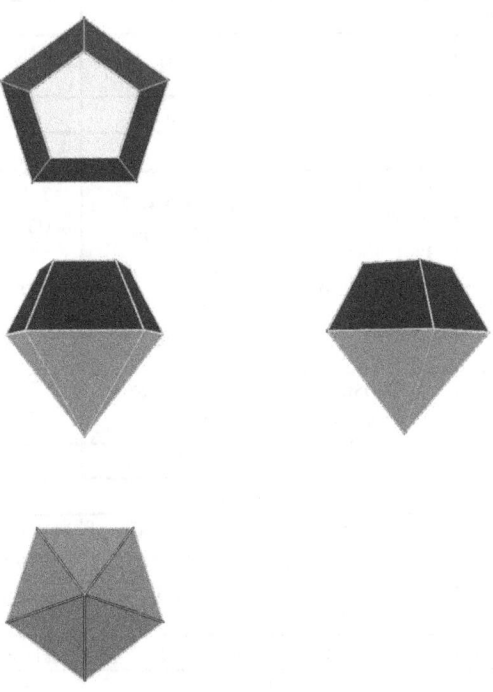

*Figure 18: The dual of Johnson solid 9 (the elongated pentagonal pyramid): a monofrustum of an asymmetric pentagonal bipyramid.*

The properties of the elongated pyramids and their duals are summarized in Table 6 below.

| Johnson solid number | 7 | 8 | 9 |
|---|---|---|---|
| Name | Elongated triangular pyramid | Elongated square pyramid | Elongated pentagonal pyramid |
| Symmetry* | $C_{3v}$ | $C_{4v}$ | $C_{5v}$ |
| Faces | [1+3](3) + [3](4) | [4](3) + [1+4](4) | [5](3) + [5](4) + [1](5) |
| Edges* | 12 | 16 | 20 |
| Vertices | 7 | 9 | 11 |
| Dihedral angles | 3a-3a: 70.5288° (70° 31′ 44″), 3a-4: 160.5288° (160° 31′ 44″), 3b-4: 90°, 4-4: 60°♣ | 3-3: 109.4712° (109° 28′ 16″), 3-4: 144.7356° (144° 44′ 8″), 4-4: 90°△ | 3-3: 138.1897° (138° 11′ 23″), 3-4: 127.3774° (127° 22′ 39″), 4-4: 108°, 4-5: 90° |
| Area | 4.73205081 = 3 + √3 | 6.73205081 = 5 + √3 | 8.88554091 |
| Volume | 0.55086383 | 1.23570226 | 2.02198023 |
| Dual | | | |
| Name | Monofrustum of asymmetric triangular bipyramid | Monofrustum of asymmetric square bipyramid | Monofrustum of asymmetric pentagonal bipyramid |
| Faces | [1+3](3) + [3](4) | [4](3) + [1+4](4) | [5](3) + [5](4) + (5) |
| Vertices | 7 | 9 | 11 |
| Face angles† | 3e: [3]60°; 3: [2]37.0043° (37° 0′ 16″), 105.9914° (105° 59′ 29″); 4: [2]50.5351° (50° 32′ 6″), [2]129.4649° (129° 27′ 54″)* | 3: [2]56.2713° (56° 16′ 17″), 67.4574° (67° 27′ 27″); 4i: [2]114.6813° (114° 40′ 53″), [2]65.3187° (65° 19′ 7″); 4s: [4]90°▫ | 3: [2]67.4319° (67° 25′ 55″), 45.1362° (45° 8′ 10″); 4: [2]72.5980° (72° 35′ 53″), [2]107.4020° (107° 24′ 7″); 5: [4]108° |

*Elongated and gyroelongated pyramids: Johnson solids 7-11.*

| Johnson solid number | 7 | 8 | 9 |
|---|---|---|---|
| Dihedral angles | 3e-3i: 118.3813° (118° 22' 53"), 3i-3i: 112.3491° (112° 20' 57"), 3i-4: 101.6182° (101° 37' 5"), 4-4: 80.7303° (80° 43' 49")* | 3-3: 116.4710° (116° 28' 16"), 3-4i: 110.7564° (110° 45' 23"), 4i-4i: 102.1921° (102° 11' 31"), 4i-4s: 117.3583° (117° 21' 30")▫ | 3-3: 122.3522° (122° 21' 8"), 3-4: 119.5524° (119° 33' 9"), 4-4: 115.9511° (115° 57' 4"), 4-5: 115.5555° (115° 33' 20") |
| Area | 1.44006080 | 3.72873347 | 7.24879864 |
| Volume | 0.10549400 | 0.53213748 | 1.54375177 |

Table 6: *Properties of Johnson solids 7, 8, and 9 and their duals.*

*The symmetry and number of edges are always the same for each solid and its dual, so are not repeated in the "dual" section.

†Face angles are only given for the duals, as the Johnson solids have regular faces, whose angles can be found in Table 1 on p. 3.

♦The designation 3a refers to the three triangles forming the pyramidal apex; 3b refers to the triangle that forms the base.

△Although the squares fall into two different transitivity classes, the 4-4 dihedral angles are all equal regardless of which pair of squares is chosen.

*The designation 3e refers to the equilateral triangle; 3i to the isosceles triangles. Since all quadrilateral faces are identical isosceles trapezoids, they can be designated as 4 rather than as 4i.

▫The designation 4i to the isosceles trapezoids; 4s refers to the square. Since all triangular faces are identical isosceles triangles, they can be designated as 3 rather than as 3i.

There is no gyroelongated triangular pyramid, to begin the set of gyroelongated pyramids, because fusing a triangular pyramid with all equilateral faces (a regular tetrahedron) to a triangular antiprism (a regular octahedron), which would make a gyroelongated triangular pyramid according to the definitions, leads to a solid where two of the triangular faces would in fact fall into the same plane. So the first of the gyroelongated pyramids is the gyroelongated square pyramid, Johnson solid 10, illustrated below in Figure 19. By definition it is a fusion of a square pyramid ($J_1$) to a square antiprism. It has twelve triangular faces, four deriving from the pyramid and eight from the antiprism, and one square face. Since the eight triangular faces consist of four adjacent to the faces derived from the pyramid and four adjacent to the square base, the triangular faces in fact belong to three different transitivity classes, as shown in Table 7 below.

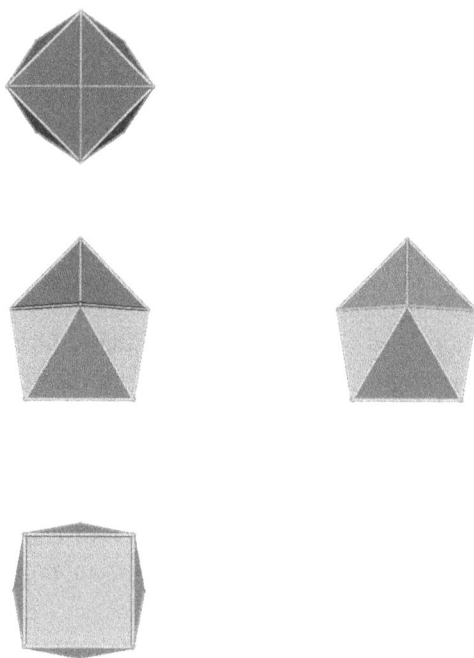

*Figure 19: Johnson solid 10: the gyroelongated square pyramid.*

*Elongated and gyroelongated pyramids: Johnson solids 7–11.*

The symmetry is $C_{4v}$, with a fourfold axis through the center of the square base and the apex (derived from the apex of the pyramid). Four mirror planes all pass through the fourfold axis: two run through the square base diagonally and two bisect pairs of opposite sides of the square.

The dual of the gyroelongated square pyramid, illustrated in Figure 20 below, shares the $C_{4v}$ symmetry of its dual, of course. As the elongated pyramids can be thought of as base-augmented prisms, and with augmentation being the dual operation of truncation and prisms being the duals of bipyramids, the duals of the elongated pyramids were seen to be monofrusta of bipyramids, so the gyroelongated pyramids can be thought of as base-augmented antiprisms, and with augmentation being the dual operation of truncation and antiprisms being the duals of antibipyramids, the duals of the gyroelongated pyramids can be seen to be monofrusta of antibipyramids, with Figure 20 as an example.

The top view shows, face-on, the square face that is produced by truncating the antibipyramid, though also visible are the four faces that started off in the antibipyramid as kite-quadrilaterals but became isosceles pentagons (defining "isosceles" as in my earlier book, *Polyhedra: a New Approach*). The front and side views show the isosceles pentagons (or at least half of them) more clearly, and also the kite-quadrilaterals, remaining from the antibipyramid, which form an antibipyramid-like apex, as defined on p. 20. The bottom view also shows all of the kite-quadrilaterals that form the antibipyramid-like apex.

If the polar reciprocation of $J_{10}$ had been performed with a center coinciding with the centroid of the original antiprism, the pentagons formed by truncation would be identical to the kite-quadrilaterals, except for a vertex being cut off and a side introduced perpendicular to the main axis of the kites. But in fact the centroid of $J_{10}$ does not coincide with the centroid of the antiprism, so the pentagons and kites have different angles. However, the four pentagons have to be isosceles because of the mirror planes implied by the $C_{4v}$ symmetry; they must be congruent because of the fourfold axis. Similarly, the four quadrilaterals at the apex must be kite-quadrilaterals because of the mirror planes, and must be congruent because of the fourfold axis. So the inventory of faces is one square, four identical kite-quadrilaterals, and four identical isosceles pentagons. For more detailed data, consult Table 7 below.

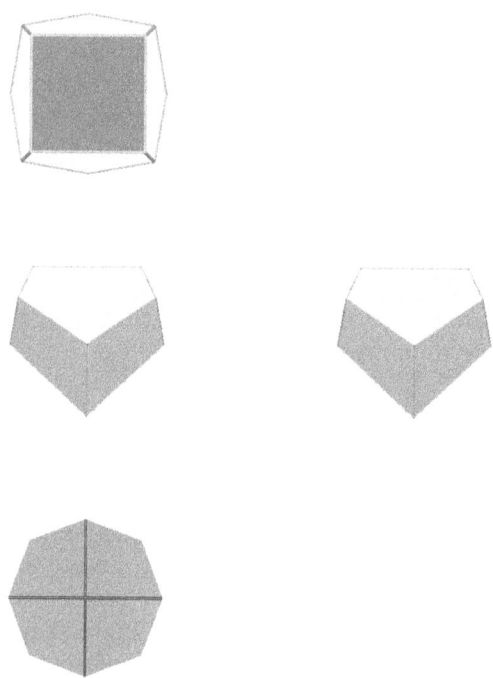

*Figure 20: The dual of Johnson solid 10 (the gyroelongated square pyramid): a monofrustum of a square antibipyramid.*

The gyroelongated pentagonal pyramid, Johnson solid 11, illustrated in Figure 21 below, is much like the gyroelongated square pyramid, except that the symmetry is fivefold rather than fourfold. Everything that occurred in sets of four in $J_{10}$ occurs in sets of five in $J_{11}$. The only difference is that, because five is an odd number rather than an even, the five mirror planes each bisect one side and one vertex angle of the pentagonal base, as opposed to the situation in Johnson solid 10, where half the mirror planes bisect two sides of the base and the other half bisect two vertex angles. Otherwise, the discussion of the properties of $J_{10}$ can be moved verbatim to here, with "four" replaced by "five," "eight" (twice four) by "ten," and

*Elongated and gyroelongated pyramids: Johnson solids 7-11.*

"square" by "regular pentagon," as appropriate.

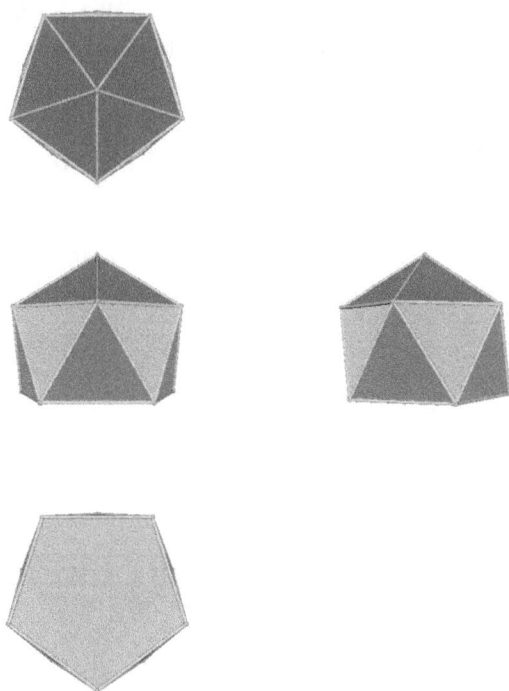

*Figure 21: Johnson solid 11: the gyroelongated pentagonal pyramid.*

Later on, in Chapter 13, there will be a discussion of a set of Johnson solids designated the diminished icosahedra. These are solids obtained by cutting an icosahedron so as to replace a pyramidal set of five triangular faces by a single pentagonal face. In fact, $J_{11}$ can be derived in the same manner. So, in keeping with the nomenclature for Johnson solids 62 and 63 in Chapter 13, $J_{11}$ could also be called the diminished icosahedron; however, this term was not introduced by Johnson and the name "gyroelongated pentagonal pyramid" is regularly used rather than "basally augmented pentagonal antiprism" or "diminished icosahedron" in all literature. It might, however, be noted that because $J_{11}$ is really a diminished icosahedron, all the triangles do in fact form a part of the surface of an icosahedron, and in particular, all the dihedral angles between those triangles are equal to the dihedral angle of an icosahedron (and thus to each other). Thus, although they fall into three different transitivity classes, the triangles are not distinguished in Table 7.

As the description of $J_{11}$ is identical to that of $J_{10}$ except for the differences between fourfold and fivefold symmetry, so id the description of its dual (illustrated in Figure 22), compared to that of the dual of $J_{10}$. The reader is advised to compare Figure 22 with Figure 20 above.

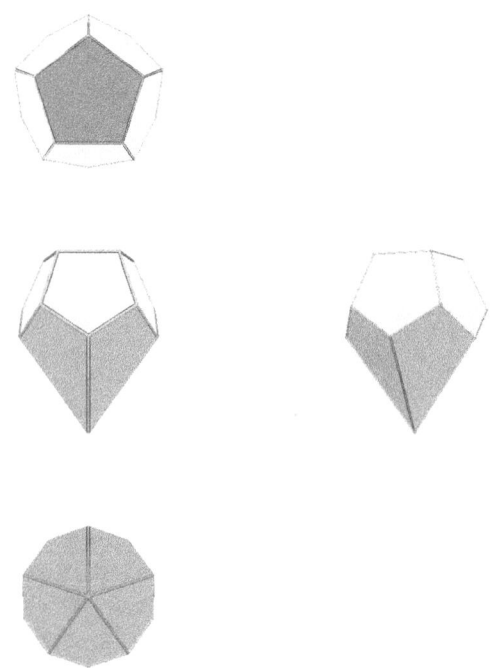

Figure 22: The dual of Johnson solid 11 (the gyroelongated pentagonal pyramid): a monofrustum of a pentagonal antibipyramid.

Because this is the case, no detailed description of the dual of $J_{11}$ is provided, and the reader is advised to look at the discussion about the dual of $J_{10}$, making the appropriate substitutions to account for the fivefold symmetry

Table 7 below is a summary of the properties of the gyro-

*Elongated and gyroelongated pyramids: Johnson solids 7-11.*

elongated pyramids and their duals. As was noted for the pyramids, the 3-4-5 trend rule (introduced on p. 12 in Chapter 2) applies to the elongated and gyroelongated pyramids, a fact which will be apparent on looking at the figures as well as from the numbers in Table 7.

| Johnson solid number | 10 | 11 |
|---|---|---|
| Name | Gyroelongated square pyramid | Gyroelongated pentagonal pyramid |
| Symmetry* | $C_{4v}$ | $C_{5v}$ |
| Faces | [4+4+4](3) + [1](4) | [5+5+5](3) + [1](5) |
| Edges* | 20 | 25 |
| Vertices | 9 | 11 |
| Dihedral angles | 3a-3a: 109.4712° (109° 28′ 16″), 3a-3b: 158.5718° (158° 34′ 18″), 3b-3c: 127.5516° (127° 33′ 6″), 3c-4: 103.8362° (103° 50′ 10″)♦ | 3-3: 138.1897° (138° 11′ 23″), 3-5: 100.8123° (100° 48′ 44″) |
| Area | 6.19615242 = 3 + 3√3 | 8.21566793 |
| Volume | 1.19270224 | 1.88019216 |
| **Dual** | | |
| Name | Monofrustum of square antibipyramid (Square polytimoid%) | Monofrustum of pentagonal antibipyramid (Pentagonal polytimoid%) |
| Faces | [1+4](4) + [4](5) | [5](4) + [1+5](5) |
| Vertices | 13 | 16 |
| Face angles† | 4k: [2]101.4181° (101° 25′ 5″), 62.8481° (62° 50′ 53″), 94.3158° (94° 18′ 57″); 4s: [4]90°; 5: [2]104.6171° (104° 37′ 2″), [2]111.0673° (111° 4′ 2″), 108.6311° (108° 37′ 52″)□ | 4: [2]107.2427° (107° 14′ 34″), 41.0672° (41° 4′ 2″), 104.4474° (104° 26′ 51″); 5i: [2]105.0496° (105° 2′ 58″), [2]109.3540° (109° 21′ 15″), 111.1928° (111° 11′ 34″); 5r: [5]108°△ |

*Elongated and gyroelongated pyramids: Johnson solids 7-11.*

| Johnson solid number | 10 | 11 |
|---|---|---|
| Dihedral angles | 4k-4k: 111.9190° (111° 55' 8"), 4k-5: 106.3332° (106° 20' 0"), 4s-5: 112.6569° (112° 39' 25"), 5-5: 98.5335° (98° 32' 1")[¤] | 4-4: 119.5155° (119° 30' 56"), 4-5i: 116.9505° (116° 57' 2"), 5i-5i: 113.8067° (113° 48' 24"), 5i-5r: 111.7204° (111° 43' 13")[△] |
| Area | 3.02265372 | 5.82107985 |
| Volume | 0.39747330 | 1.10946491 |

Table 7: *Properties of Johnson solids 10 and 11 and their duals.*

\*The symmetry and number of edges are always the same for each solid and its dual, so are not repeated in the "dual" section.

†Face angles are only given for the duals, as the Johnson solids have regular faces, whose angles can be found in Table 1 on p. 3.

♦The designation 3a refers to the triangles forming the pyramidal apex; 3b and 3c refer to the triangles forming the antiprismatic band, where the 3b triangles are the ones sharing an edge with the 3a triangles, and the 3c triangles are the ones sharing an edge with the base.

%Name introduced in my earlier book, *Polyhedra: a New Approach.*

¤The designation 4k refers to the four kite-quadrilaterals at the apex; 4s refers to the single square face which forms the base of the polyhedron; it is not necessary to designate the pentagons as 5i, since *all four* are isosceles, according to the definition in my earlier book, *Polyhedra: a New Approach.*

△The designation 5i refers to the set of five isosceles pentagons in the lateral positions; 5r refers to the regular pentagon which forms the base of the polyhedron; it is not necessary to designate the kite-quadrilaterals as 4k, since *all five* quadrilateral faces are congruent kite-quadrilaterals.

# Chapter 5: Dipyramids (bipyramids): Johnson solids 12 and 13.

In lists of Johnson solids, the twelfth and thirteenth are generally referred to as *dipyramids;* the form *bipyramid* is also found, and I prefer it. The argument for "dipyramid" is that "*pyramid*" is a Greek-origin word, and the prefix "*bi-*" is Latin, while "*di-*" is Greek. The argument for "bipyramid" is that in every other case where two identical forms are fused together, such as "bicupola" or "birotunda," the prefix "bi" is found. Since in any case the reader is likely to encounter both terms, it is necessary to make it clear that both are actually in use. As was the case for the pyramids, it is impossible to have an order of rotational symmetry greater than 5, as six equilateral triangles will lie flat and more simply cannot be put together at an apex. However, while it was the triangular pyramid that was omitted from Johnson's list because it is a Platonic solid (a tetrahedron) when all faces are equilateral triangles, it is the *square bipyramid* that must be omitted from Johnson's list because it is a Platonic solid (an octahedron) when all faces are equilateral triangles. The triangular ($J_{12}$) and pentagonal ($J_{13}$) bipyramids are the two that are found, as has already been stated. An illustration of the triangular bipyramid, Johnson solid 12, is in Figure 23 below.

The triangular bipyramid contains only triangular faces, and thus it belongs to the category of polyhedra known as *deltahedra*; three Platonic solids and a number of other Johnson solids also are included. The symmetry of the triangular bipyramid is $D_{3h}$. The bipyramids are alone among the Johnson solids in having all their faces belong to a single transitivity class, as is true of the Platonic solids, although unlike the Platonic solids, the vertices are of two kinds, trivalent and quadrivalent. Of the threefold axis and three twofold axes required by $D_{3h}$ symmetry, every vertex has one of these through it. (The threefold axis passes through two, thus, although there are four axes of rotation, there are five vertices.)

There are six faces and five vertices, so applying the Euler formula $V + F = E + 2$, one can tell that there are nine edges.

*Dipyramids (bipyramids): Johnson solids 12 and 13.*

They too belong in two classes: the six edges that meet in sets of three at the trivalent vertices, and the remaining three edges that run around the middle in the fusion plane between the two regular tetrahedra.

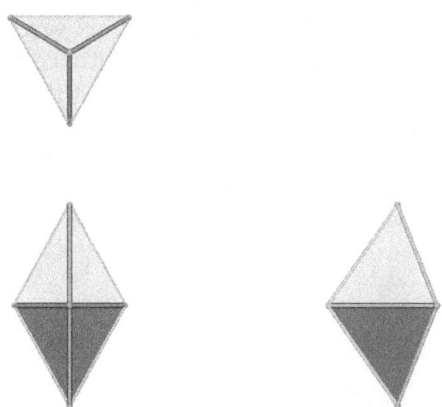

*Figure 23: Johnson solid 12: the triangular bipyramid.*

One does not often think of the triangular bipyramid as an example of an augmented Platonic (see p. 43), but in fact it is; the only point that needs to be made is that which of the two tetrahedra is considered the starting Platonic solid, and which is considered the augmenting pyramid, is arbitrary, since in fact it is two congruent regular tetrahedra that are fused to obtain $J_{12}$.

The duals of the bipyramids are prisms, so that the dual of $J_{12}$ would be a triangular prism, as is seen in Figure 24 below. (Note that the lateral edges are much shorter than the edges that form the base and roof.) Thus, although the prisms were excluded by Norman Johnson from his list, prisms appear among the duals, although these prisms have faces that are far from being regular. (The bases of the triangular prism, of course, are equilateral triangles, forced by the threefold symmetry of the solid, and similarly the bases of the pentagonal prism that is dual to $J_{13}$ are regular pentagons, forced by the fivefold symmetry of that solid; but the lateral faces of the triangular prism are long narrow rectangles, and, as will be seen later, the lateral faces of the pentagonal prism are similarly non-square rectangles, but with their longer sides vertical rather than horizontal.)

Since a polyhedron and its dual have the same symmetry, the dual of $J_{12}$ has the same $D_{3h}$ symmetry as $J_{12}$ itself. And since the numbers of faces and vertices are interchanged by the dualization, the dual of $J_{12}$ has five faces, six vertices, and nine edges. As, in $J_{12}$, there was only one type of face but two types of vertex and two types of edge, in its dual there is only one type of vertex but two types of face and two types of edge, the normal effect of dualization wherein "face" and "vertex" are interchanged, leaving "edge" alone.

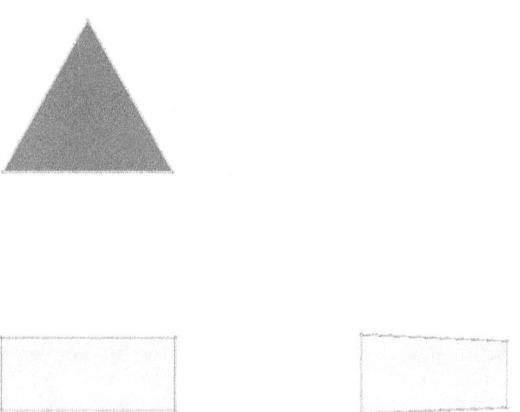

Figure 24: The dual of Johnson solid 12 (the triangular bipyramid): a triangular prism.

As was mentioned previously, the square bipyramid is not a Johnson solid, being a (Platonic) regular octahedron. Johnson solid 13, the pentagonal bipyramid, is illustrated in Figure 25 below, and its dual in Figure 26.

Much of what was said about $J_{12}$ applies to $J_{13}$, with only the fivefold symmetry replacing threefold; $J_{13}$ is a deltahedron, having only equilateral triangles as faces; it has only one class of face, but two of edges and two of vertices; and all that was said about $J_{12}$ can be carried over to this discussion, making only

*Dipyramids (bipyramids): Johnson solids 12 and 13.*

the necessary changes for the fivefold symmetry. One difference is that, since the pyramids fused in $J_{12}$ are regular tetrahedra, the dihedral angles in $J_{12}$ are all either the tetrahedral dihedral angle of 70.5288° (70° 31' 44") or double that value (at the fusion plane); this is not true of $J_{13}$. Instead, the dihedral angle between two faces on the same side of the fusion plane is exactly equal to the dihedral angle between triangular faces in $J_2$, and the dihedral angle between faces on opposite sides of the fusion plane is double the 3-5 dihedral angle in $J_2$.

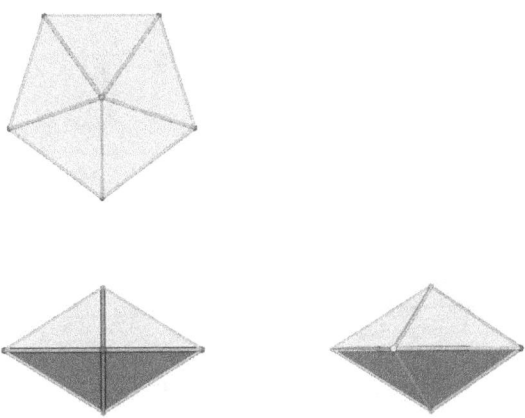

*Figure 25: Johnson solid 13: the pentagonal bipyramid.*

The triangular and pentagonal bipyramids are also encountered in other lists: Since they have only equilateral triangles as faces, they are also members of the set of polyhedra called deltahedra. The deltahedra also include three of the Platonic solids (regular tetrahedron, octahedron, and icosahedron), as well as some other Johnson solids ($J_{17}$, $J_{51}$, and $J_{84}$). Altogether there are eight deltahedra, but as three are Platonic solids (one of those three, the regular octahedron, is also an antiprism, and thus *twice* excluded from the Johnson solid list), only the remaining five are to be found among the Johnson solids.

As was the case for the triangular bipyramid, the pentagonal bipyramid has a dual that is a prism, but while the dual of $J_{12}$ is very short, the dual of $J_{13}$ (illustrated in Figure 26 below) is tall. As was noted for the pyramids, the 3-4-5 trend rule (introduced on p. 12 in Chapter 2) applies to the bipyramids, a fact which will be apparent on looking at the figures. It is useful to recall that the regular octahedron falls between $J_{12}$ and $J_{13}$ in following these trends. For example, the duals of all three polyhedra ($J_{12}$, the regular octahedron, and $J_{13}$) are prisms, but the dual of the regular octahedron is specifically a perfect cube.

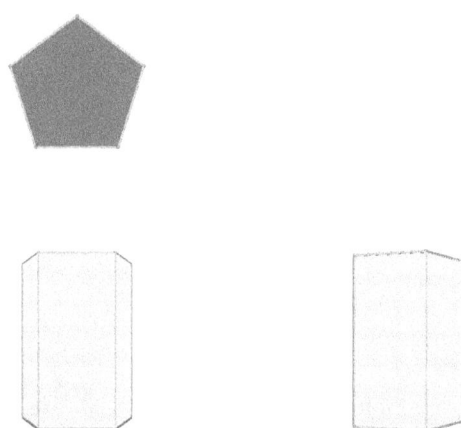

*Figure 26: The dual of Johnson solid 13 (the pentagonal bipyramid): a pentagonal prism.*

Thus, looking at the 3-4-5 trend rule, it can be concluded that the dual of $J_{12}$ would be a very squat triangular prism and the dual of $J_{13}$ would be a very high pentagonal prism. And looking at Figures 24 and 26, this is clearly true.

*Dipyramids (bipyramids): Johnson solids 12 and 13.*

| Johnson solid number | 12 | 13 |
|---|---|---|
| Name | Triangular bipyramid (dipyramid) | Pentagonal bipyramid (dipyramid) |
| Symmetry* | $D_{3h}$ | $D_{5h}$ |
| Faces | [6](3) | [10](3) |
| Edges* | 9 | 15 |
| Vertices | 5 | 7 |
| Dihedral angles | 3a-3a: 70.5288° (70° 31' 44"), 3a-3b: 141.0576° (141° 3' 27")♣ | 3a-3a: 74.7547° (74° 45' 17"), 3a-3b: 138.1897° (138° 11' 23")♣ |
| Area | 2.59807621 = $(3\sqrt{3})/2$ | 4.33012702 = $(5\sqrt{3})/2$ |
| Volume | 0.23570226 | 0.60300566 |
| **Dual** | | |
| Name | Triangular prism | Pentagonal prism |
| Faces | [2](3) + [3](4) | [5](4) + [2](5) |
| Vertices | 6 | 10 |
| Face angles† | 3: [3]60°, 4: [4]90° | 4: [4]90°, 5: [5]108° |
| Dihedral angles | 3-4: 90°, 4-4: 60° | 4-4: 108°, 4-5: 90° |
| Area | 0.52269257 | 1.70130162 |
| Volume | 0.02209709 | 0.15278640 |

Table 8: Properties of Johnson solids 12 and 13 and their duals.

*The symmetry and number of edges are always the same for each solid and its dual, so are not repeated in the "dual" section.

†Face angles are only given for the duals, as the Johnson solids have regular faces, whose angles can be found in Table 1 on p. 3.

♣The designation 3a refers to the triangles around one apex; 3b refers to those around the opposite one. (There is no need to say which is which, because of symmetry; the actual distinction is simply between *pairs* of faces at a common apex and pairs at opposite apices.)

## Chapter 6: Elongated and gyroelongated dipyramids (bipyramids): Johnson solids 14-17.

Johnson's terminology for the polyhedra described in this chapter calls them "elongated and gyroelongated dipyramids." As in the previous chapter, one might use the prefix "bi-" rather than "di-," with the same points in favor of each form as were given for the polyhedra in that chapter. It is equally correct to name them according to another system, as was discussed in my earlier book, *Polyhedra: a New Approach*, and mentioned earlier in Chapter 4. For the general construction can be conceived of as starting with a bipyramid, cutting the two pyramids apart, and fusing a prism (for the elongated forms) or antiprism (for the gyroelongated forms) between them. But one can equally well *start* with the prism or antiprism, and fuse pyramids to both of its bases, analogous to the construction described in Chapter 4. It is therefore completely correct to describe these polyhedra as "biaugmented prisms" and "biaugmented antiprisms." However, as was noted in Chapter 4, such solids as $J_{53}$ are termed "biaugmented prisms" by Johnson, so that terms like "basally" and "laterally" are necessary to make the description clearer on occasion.

Johnson solid 14, which he calls the elongated triangular bipyramid but which can also be termed the basally biaugmented triangular prism, is illustrated below in Figure 27. It has six triangular and three square faces, the six triangular faces being grouped into two sets of three meeting at an apex, and the three square faces forming a belt separating the two sets of triangular faces. It has $D_{3h}$ symmetry, with a threefold axis through the apices where three triangular faces come together, three mirror planes through that threefold axis, and an additional mirror plane, which is perpendicular to the threefold axis and to the three aforementioned mirror planes. The numerical values of such things as dihedral angles can be obtained by a process similar to that described on p. 38, but Table 9 gives them explicitly.

If one wants to inventory the faces, vertices, and edges, first note that there are two triangular pyramids (regular tetrahedra) fused to a triangular prism. Each tetrahedron contributes four faces, four vertices, and six edges (noting the Euler formula $V + F = E + 2$, the edge count does not need to be stated separately, but it might as well be included here). The prism contributes five faces (two bases, three lateral faces), six vertices, and nine edges. So before the fusion, there are 2(4) + 5 = 13 faces, 2(4) + 6 = 14 vertices, and 2(6) + 9 = 21 edges. However, four of those faces are lost in the fusion process: one from each tetrahedron where it is fused to the prism and both of the bases of the prism, leaving nine faces. And the vertex count is decreased by six because three vertices from each tetrahedron are merged with vertices of the prism, so the total number of vertices is reduced to eight. Similarly to the vertices, edges of the tetrahedron are merged with those of the prism, three in each fusion plane for a total of six altogether, reducing the number of edges to fifteen.

*Figure 27: Johnson solid 14: the elongated triangular bipyramid or basally biaugmented triangular prism.*

This will in fact be similar to all the elongated bipyramids in this chapter; each $n$-gonal pyramid provides $n$ lateral faces and a base (a total of $n + 1$ faces), $n + 1$ vertices (one at the apex and $n$ around the base), and $2n$ edges ($n$ meeting at the apex and $n$ surrounding the base). The prism provides two bases and $n$ lateral faces (a total of $n + 2$ faces), $2n$ vertices, and $3n$ edges ($n$ around each base and $n$ joining the bases). The total number

*Elongated and gyroelongated dipyramids (bipyramids): Johnson solids 14-17.*

of faces, $3n + 4$, is reduced by 4 from the fusion, leaving $3n$; the total number of vertices, $4n + 2$, is reduced by $2n$ from the fusion, leaving $2n + 2$; and the total number of edges, $7n$, is reduced by $2n$ from the fusion, leaving $5n$. This computation is included at this point so as to eliminate the need to state it explicitly in the discussions of the other elongated bipyramids.

As was the case for $J_7$ through $J_{11}$, when determining the duals of $J_{14}$ through $J_{17}$, it is more productive to consider them as augmented prisms and antiprisms, and the same reasoning that was used for those solids makes it clear that $J_{14}$ through $J_{16}$, as basally biaugmented prisms, have duals that are bifrusta of bipyramids, while $J_{17}$, a basally biaugmented antiprism, has a dual which is a bifrustum of an antibipyramid.

The dual of Johnson solid 14 is a bifrustum of a triangular bipyramid, with two triangular bases and six lateral isosceles trapezoids, a threefold axis, three mirror planes through that threefold axis, and an additional mirror plane perpendicular to the other three mirror planes and to the threefold axis. The overall symmetry is $D_{3h}$, exactly as the symmetry of $J_{14}$ itself, as any polyhedron and its dual must have identical symmetry. The detailed properties of this polyhedron are given in Table 9.

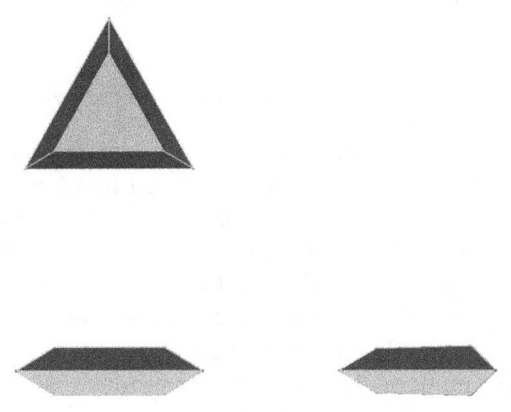

Figure 28: *The dual of Johnson solid 14 (the elongated triangular bipyramid): a bifrustum of a triangular bipyramid.*

It is important to note that the duals of elongated bipyramids are figures that could be generated by fusion, a phenomenon that will be taken up in greater detail in Chapter 9.

The elongated square bipyramid, Johnson solid 15, illustrated in Figure 29 below, is very similar to the elongated triangular bipyramid, allowing for the symmetry being fourfold rather than threefold. The number of triangular faces at each apex is four, rather than three, for a total of eight, rather than six; the number of square faces in the belt is four, rather than three, and the symmetry is $D_{4h}$ rather than $D_{3h}$.

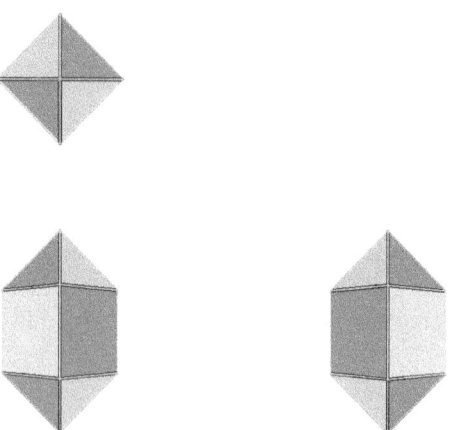

Figure 29: Johnson solid 15: the elongated square bipyramid.

The dual of the elongated square bipyramid, illustrated in Figure 30 below, can also be best described by referring to the dual of the elongated triangular bipyramid, allowing for the change from threefold to fourfold symmetry. It is a bifrustum of a square bipyramid, as the dual of $J_{14}$ is a bifrustum of a triangular bipyramid, with two square bases (replacing the triangular bases of the dual of $J_{14}$) and eight lateral isosceles trapezoids (as the dual of $J_{14}$ had six), a fourfold axis instead of threefold, four mirror planes through that fourfold axis, and an additional mirror plane perpendicular to the other four mirror planes and to the fourfold axis. The overall symmetry is $D_{4h}$, exactly as the symmetry of $J_{15}$ itself, as any polyhedron and its dual must have identical symmetry. This description can be compared with the description of the dual of Johnson solid 14 above, and any specific properties of the dual of $J_{14}$ that do not have analogous ones in this section can be used as a basis for determining corresponding properties of the dual of $J_{15}$. The reader can compare Figure 30 with Figure 28 above for an overview of the differences between these two polyhedra, and also consult Table 9 below for more precise data.

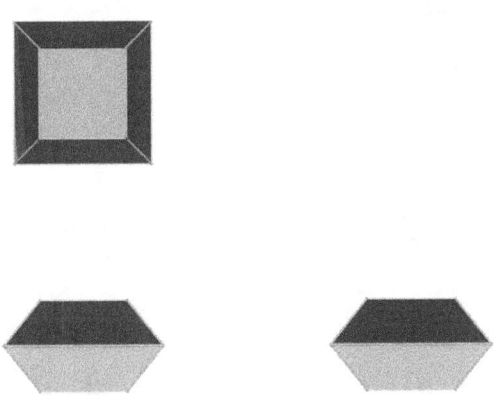

Figure 30: The dual of Johnson solid 15 (the elongated square bipyramid): a bifrustum of a square bipyramid.

The elongated pentagonal bipyramid, Johnson solid 16, illustrated in Figure 31 below, can be described by simply consulting the descriptions above for the elongated triangular bipyramid and the elongated square bipyramid, and making the appropriate changes to accommodate the fivefold symmetry, replacing "three" or "four" by "five," "six" or "eight" by "ten," "triangle" or "square" (insofar as it refers to the base, rather than a lateral face) by "regular pentagon," "threefold" or "fourfold," by "fivefold," etc. The symmetry is $C_{5v}$. Compare Figure 31 with Figures 27 and 29 for the similarities and differences among these three polyhedra.

As is usual, the 3-4-5 trend rule applies, although the central prism itself stays the same height because of its square faces. The overall polyhedron, though, becomes squatter as the order of the principal axis of rotation increases from 3 to 4 to 5, because of the pyramids. In this case, all three cases exist in

Johnson's list, even though the triangular pyramid and square bipyramid were excluded by reason of being Platonic solids; the additional square lateral faces take them out of the isohedral class (all faces alike and in one transitivity class) required for a Platonic solid.

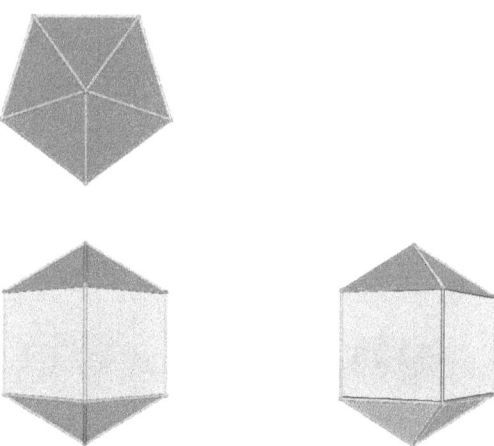

*Figure 31: Johnson solid 16: the elongated pentagonal bipyramid.*

The dual of the elongated pentagonal bipyramid, illustrated in Figure 32 below, is, as expected, a bifrustum of a pentagonal bipyramid. Like the other duals of elongated bipyramids, it has two bases that are regular polygons of the appropriate number of sides, and a lateral belt of two tiers of isosceles trapezoids. The symmetry is $C_{5v}$, like that of its dual.

In the same way that the description of the dual of $J_{15}$ can be derived from the description of the dual of $J_{14}$ (earlier in this chapter) by changing a few numbers, so can the description of $J_{16}$, making the necessary changes to account for the fivefold symmetry.

As was noted for the pyramids, the 3-4-5 trend rule (introduced on p. 12 in Chapter 2) applies to the elongated bipyramids, a fact which will be apparent on looking at Figure 32 and comparing with Figures 28 and 30. In particular, the isosceles trapezoids in the dual of $J_{14}$ are extremely short, while those in the dual of $J_{16}$ are much taller. In fact, in the dual of $J_{14}$ the

*Elongated and gyroelongated dipyramids (bipyramids): Johnson solids 14-17.*

lateral sides of the isosceles trapezoids are ~0.5064, and the two parallel sides are ~1.2404 and exactly 2, while in the dual of $J_{16}$ the lateral sides of the isosceles trapezoids are ~0.8720, and the two parallel sides are ~0.6335 and ~1.2361, so the lateral sides are longer than the shorter base of the trapezoid.

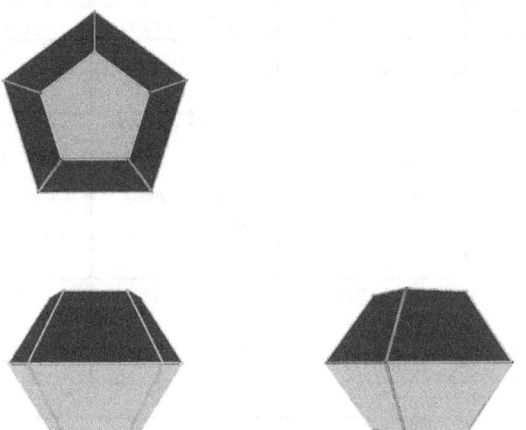

*Figure 32: The dual of Johnson solid 16 (the elongated pentagonal bipyramid): a bifrustum of a pentagonal bipyramid.*

## The Johnson Solids and Their Duals

| Johnson solid number | 14 | 15 | 16 |
|---|---|---|---|
| Name | Elongated triangular bipyramid (dipyramid) | Elongated square bipyramid (dipyramid) | Elongated pentagonal bipyramid (dipyramid) |
| Symmetry* | $D_{3h}$ | $D_{4h}$ | $D_{5h}$ |
| Faces | [6](3) + [3](4) | [8](3) + [4](4) | [10](3) + [5](4) |
| Edges* | 15 | 20 | 25 |
| Vertices | 8 | 10 | 12 |
| Dihedral angles | 3-3: 70.5288° (70° 31′ 44″), 3-4: 160.5288° (160° 31′ 44″), 4-4: 60° | 3-3: 109.4712° (109° 28′ 16″), 3-4: 144.7356° (144° 44′ 8″), 4-4: 90° | 3-3: 138.1897° (138° 11′ 23″), 3-4: 127.3774° (127° 22′ 39″), 4-4: 108° |
| Area | 5.59807621 = 3 + (3/2)√3 | 7.46410162 = 4 + 2√3 | 9.33012702 = 5 + (5/2)√3 |
| Volume | 0.66871496 | 1.47140452 | 2.32348307 |
| **Dual** | | | |
| Name | Bifrustum of triangular bipyramid (dipyramid) | Bifrustum of square bipyramid (dipyramid) | Bifrustum of pentagonal bipyramid (dipyramid) |
| Faces | [2](3) + [6](4) | [2+8](4) | [10](4) + [2](5) |
| Vertices | 9 | 12 | 15 |
| Face angles† | 3: [3]60°; 4: [2]41.4096° (41° 24′ 35″), [2]138.5904° (138° 35′ 25″) | 4i: [2]60°, [2]120°; 4s: [4]90° | 4: [2]110.2118° (110° 12′ 42″), [2]69.7882° (69° 47′ 18″); 5: [5]108° |
| Dihedral angles | 3-4: 130.8934° (130° 53′ 36″), 4-4: 98.2132° (98° 12′ 48″)♦ | 4i-4i: 109.4712° (109° 28′ 16″), 4i-4s: 125.2644° (125° 15′ 52″)¤ | 4-4: 119.1072° (119° 6′ 26″), 4-5: 120.4464° (120° 26′ 47″)♦ |
| Area | 4.58858711 | 5.92340274 | 9.03053606 |
| Volume | 0.58615413 | 1.06531983 | 2.19467052 |

Table 9: Properties of Johnson solids 14, 15, and 16 and their duals.

*The symmetry and number of edges are always the same for each solid and its dual, so are not repeated in the "dual"

*Elongated and gyroelongated dipyramids (bipyramids): Johnson solids 14-17.*

section.

†Face angles are only given for the duals, as the Johnson solids have regular faces, whose angles can be found in Table 1 on p. 3.

✦Although there are two kinds of 4-4 edges, the dihedral angles at both of them are the same 98.2132° (98° 12′ 48″) in $J_{14}$ and the same 119.1072° (119° 6′ 26″) in $J_{16}$.

⁎The designation 4i refers to the eight lateral isosceles trapezoids; 4s refers to the two bases. Although there are two kinds of 4i-4i edges, the dihedral angles at both of them are the same 109.4712° (109° 28′ 16″).

There is only one gyroelongated bipyramid, the gyroelongated square bipyramid, $J_{17}$, illustrated in Figure 33 below, among the Johnson solids. There is no gyroelongated triangular or pentagonal bipyramid in Johnson's list; the reasons for these are quite different. If one tried to construct a gyroelongated triangular bipyramid, a triangle originating from one of the two pyramids and the adjacent triangle originating from the inserted antiprism would have a 180° dihedral angle at the edge where they meet, and fall into the same plane. On the other hand, a gyroelongated pentagonal bipyramid, constructed with all equilateral triangles, is simply a regular icosahedron.

The gyroelongated square bipyramid is also, as its faces are all equilateral triangles, a member of the set of polyhedra called *deltahedra*. (See also p. 61 in Chapter 5.) Since there are sixteen faces, it has been called (e. g., in Pugh's book, *Polyhedra: a Visual Approach*) a *heccaideltahedron*.

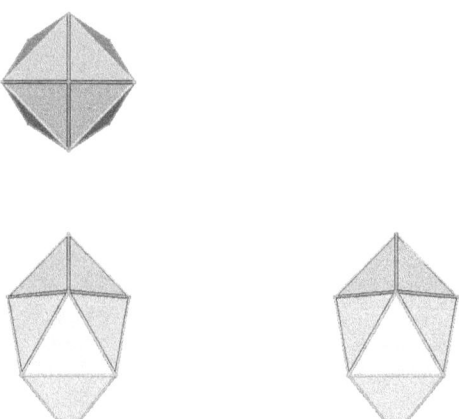

Figure 33: Johnson solid 17: the gyroelongated square bipyramid.

The dual of the gyroelongated square bipyramid, illustrated in Figure 34 below, can obviously be seen to have the same relationship to a regular dodecahedron that the gyroelongated square bipyramid itself has to the regular icosahedron; namely, having a fourfold axis rather than a fivefold. As $J_{17}$ can be derived from the regular icosahedron by replacing one pair of opposite quinquevalent vertices by quadrivalent ones, making the

*Elongated and gyroelongated dipyramids (bipyramids): Johnson solids 14-17.*

axis through those two vertices a fourfold axis (and simultaneously destroying all the other threefold and fivefold axes, so that the symmetry becomes $C_{4v}$), its dual can be derived from the *dual* of the regular icosahedron, the regular dodecahedron, by replacing one pair of opposite pentagonal faces by squares, making the axis through the centers of those two faces a fourfold axis (and simultaneously destroying all the other threefold and fivefold axes, so that the symmetry becomes $C_{4v}$). It is exactly a dual process, as demonstrated by the similarity in the words used here to describe the two processes.

The resulting polyhedron has ten faces: two square and eight isosceles pentagonal ("isosceles," once more, defined as in my earlier book, *Polyhedra: a New Approach*). While there is no name known to me for a polyhedron with two *n*-gonal faces as bases and a band between them consisting of 2*n* isosceles pentagons alternating in orientation, if a name were devised for this family, this polyhedron would be the square member and the regular dodecahedron would be the pentagonal member.

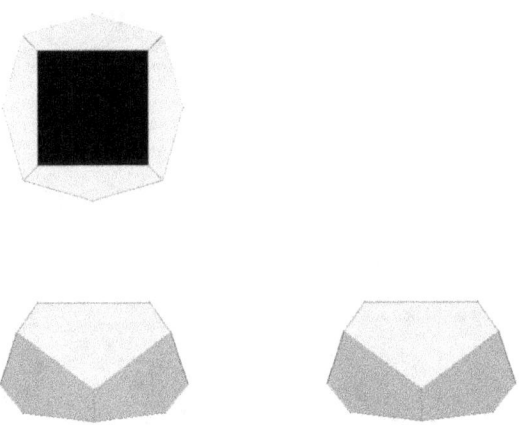

Figure 34: *The dual of Johnson solid 17 (the gyroelongated square bipyramid)*

In the dodecahedron, of course, the twelve faces are all identical; in this figure, two have become squares, but the remaining ten have simply been reduced in number to eight in keeping with the fourfold symmetry, and become isosceles rather than regular. (In fact, while an isosceles pentagon need have only two sets of two equal angles, in this case the fifth angle is in fact exactly equal to the two on either side of it.)

In another way, too, this solid shows its kinship to the regular dodecahedron. While there are two kinds of 5-5 edge, it can be seen that they both have exactly the same dihedral angle.

*Elongated and gyroelongated dipyramids (bipyramids): Johnson solids 14-17.*

| Johnson solid number | 17 |
|---|---|
| Name | Gyroelongated square bipyramid (dipyramid) |
| Symmetry* | $S_{8v}$ |
| Faces | [8+8](3) |
| Edges* | 24 |
| Vertices | 10 |
| Dihedral angles | 3a-3a: 109.4712° (109° 28′ 16″),<br>3a-3b: 158.5718° (158° 34′ 18″),<br>3b-3b: 127.5516° (127° 33′ 6″)♣ |
| Area | 6.92820323 = $4\sqrt{3}$ |
| Volume | 1.42840450 |
| **Dual** | |
| Name | Unnamed |
| Faces | [2](4) + [8](5) |
| Vertices | 16 |
| Face angles† | 4: [4]90°;<br>5: [3]101.9529° (101° 57′ 10″),<br>[2]117.0707° (117° 4′ 14″) |
| Dihedral angles | 4-5: 120.7359° (120° 44′ 9″),<br>5-5: 105.1415° (105° 8′ 29″) |
| Area | 4.43781962 |
| Volume | 0.71381871 |

Table 10: Properties of Johnson solid 17 and its dual.

*The symmetry and number of edges are always the same for each solid and its dual, so are not repeated in the "dual" section.

†Face angles are only given for the duals, as the Johnson solids have regular faces, whose angles can be found in Table 1 on p. 3.

♣The designation 3a refers to the triangles around either apex; 3b refers to the triangles forming the belt around the center.

## Chapter 7: Elongated and gyroelongated cupolae and rotundae: Johnson solids 18-25.

As has been stated earlier, the terms "elongated" and "gyroelongated" are used by Johnson to designate polyhedra with a prism (for the elongated forms) or antiprism (for the gyroelongated forms) fused to a face (or in the case of diapical polyhedra, between the two pieces of a polyhedron obtained by dissecting a polyhedron along a plane).

As was noted for the pyramids, the 3-4-5 trend rule (introduced on p. 12 in Chapter 2) applies to the elongated and gyroelongated cupolae, a fact which will be apparent on looking at the figures.

The elongated triangular cupola, Johnson solid 18, illustrated in Figure 35 below, is the first of the three elongated cupolae to be described in this chapter.

Since there is so much in common among the descriptions of the elongated cupolae, to avoid repetition there will be given here a descruption that is shared by all three.

All the elongated cupolae have square faces falling into three transitivity classes common to them all: one from the cupola and two different ones from the prismatic "elongation." (In the case of the elongated square cupola, the roof of the cupola constitutes a face that forms a fourth transitivity class.) The elongated triangular cupola has triangles that constitute two different transitivity classes (one triangle that forms the roof of the cupola, and three in a second class, consisting of all the other triangles). In the other two elongated cupolae, only this second transitivity class exists. And the base, a hexagon in the case of the elongated triangular cupola, is a transitivity class in itself.

So, for all the elongated cupolae, the faces fall into six transitivity classes: three common sets of squares; a common set of triangles; the fifth class being a second set, consisting of only one member, of triangles in this case, a fourth set, consisting of only one member, of squares in the case of the elongated square cupola, and a single pentagon in the case of the elongated pentagonal cupola; and the sixth set also being a single-member set, a polygon of twice as many sides as the one constituting the fifth set.

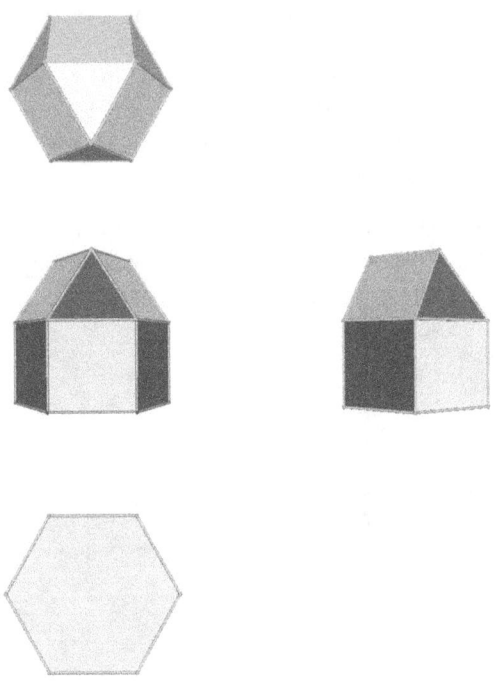

Figure 35: Johnson solid 18: the elongated triangular cupola.

The dual of the elongated triangular cupola, illustrated in Figure 36 below, shows the typical features of the duals of the elongated cupolae. At the apex are three kite-quadrilaterals (constituting an *antibipyramid-like apex*, as defined on p. 20, Chapter 3) and at the antiapex, six isosceles triangles making a pyramid-like antiapex, as we have defined it. Between those two is a belt of asymmetric quadrilaterals (though two angles of each quadrilateral are equal, they are adjacent, not opposite as needed to form a kite or parallelogram) alternating in chirality.

The faces of this polyhedron can be classified into three transitivity classes; the same will be seen to be the case for the other duals of elongated cupolae. One transitivity class consists of the kite-quadrilaterals that form the antibipyramid-like apex, which will be designated "4k" in Table 11, one transitivity class consists of the isosceles triangles that form the pyramid-like antiapex, which can simply be designated as "3" since they are the only triangular faces, and the third transitivity class consists of the remaining quadrilateral faces along the belt, which are asymmetric and appear in an alternating-chirality arrangement, and which will be designated as "4a" in Table 11. If the order of the principal axis is equal to $n$, there are $n$ kite-quadrilaterals, $2n$ isosceles triangles, and $2n$ asymmetric quadrilaterals in all three of these elongated cupola duals.

One thing that seems a little surprising is that while the 4a-4a edges alternate in *length*, the dihedral angles between these faces are the same at the longer edges as at the shorter ones, as can be seen in Table 11 below.

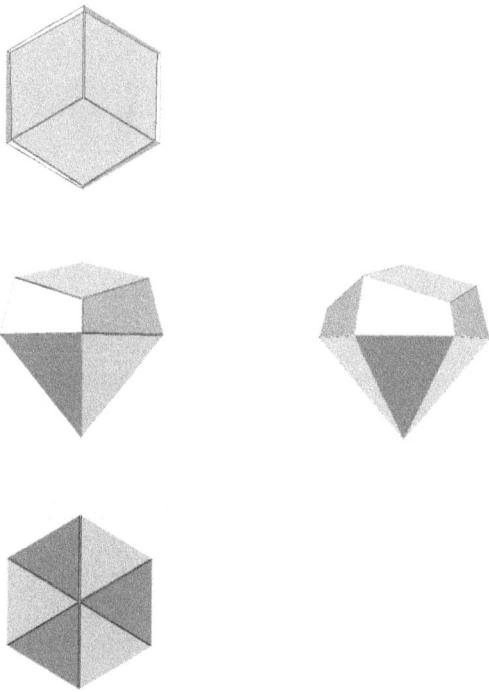

*Figure 36: The dual of Johnson solid 18 (the elongated triangular cupola).*

*Elongated and gyroelongated cupolae and rotundae: Johnson solids 18-25.*

There is no particular name that I know of for a quadrilateral which has two adjacent angles equal; if the two equal angles were opposite, it would be a kite-quadrilateral, while if the remaining two angles were also equal, it would be an isosceles trapezoid. The quadrilaterals designated "4a" in Table 11, however, have this structure: the angles are 115.1806° (115° 10′ 50″), 79.9694° (79° 58′ 10″), and two angles of 82.4250° (82° 25′ 30″). And the duals of the elongated cupolae are not the only Johnson duals that will be found to have such quadrilaterals; the duals of other Johnson solids, like $J_{49}$ and $J_{57}$, have them as well.

Although the duals of the elongated cupolae have not been given names, one might note that every one is a polyhedron which can be obtained by fusion of a pyramid (of $2n$-gonal base, not $n$-gonal) to a figure that *itself* has no name. If this latter polyhedron (whose faces, other than the $2n$-gon that is lost in the fusion) are $n$ kite-quadrilaterals and $2n$ irregular quadrilaterals of the type discussed in the previous paragraph) had a name, and we put "X" for that name, then the duals of the elongated cupolae could simply be described as (basally) augmented X's. Interestingly, if one starts with a *basally augmented* cupola, its dual is essentially the same X that we have just discussed. (The proportions are not identical, but the arrangements of triangles and quadrilaterals are the same, as is the $n$-fold symmetry.)

This is really a special case of a general phenomenon which might be worth discussing at this point. Suppose one takes a starting polyhedron which we call *A*. Construct a polyhedron *B* by fusing a pyramid to a face of *A*, and a different polyhedron *C* by fusing a prism to the same face of *A*. (In this case, one can consider *A* to be the cupola $J_3$, *B* to be the basally augmented cupola mentioned in the previous paragraph, and *C* to be the elongated triangular cupola $J_{18}$.) If one generates the dual of *C*, and calls it *D* (and since we are taking as our example $J_{18}$ for *C*, *D* is the $J_{18}$ dual illustrated in Figure 36), it will include a pyramid whose apex corresponds to the face of the prism *opposite* the face that was fused to *A* in constructing *C*. The dual of *B* — which we will call *E* — will by contrast have a face corresponding to the apex of the pyramid that was fused to *A* in order to construct *B*. Now what we see is that not only does

*D* include a pyramid as mentioned earlier, but in fact this pyramid is fused to a polyhedron which is essentially *E*. (I say "essentially" because it will not necessarily have the same edge-lengths and angles as *E*, but it will have the same structure in terms of what kinds of polygons for faces — where "kind" means how many sides/angles they have — and what valencies of vertices are present.) We have seen that here for *A* being a cupola, and another easy case to visualize is where *A* is a pyramid. *B* is then a bipyramid and *C* an elongated pyramid. The dual of *B* is a prism; the dual of *C* another elongated pyramid, which is what one gets by fusing a pyramid to the base of a prism such as *B*.

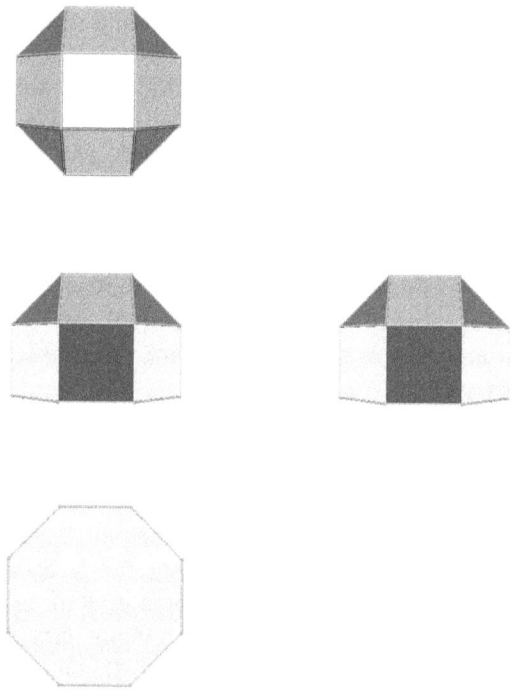

Figure 37: Johnson solid 19: the elongated square cupola.

In Chapter 3 it was seen that there were three cupolae which satisfied the criteria for a Johnson solid; namely, they could be constructed with all faces as regular polygons. As a result, in most cases where a derivative of one cupola is found to be a Johnson solid with the same $C_{nv}$ ($n$ = 3, 4, or 5) symmetry as the cupola itself, there is a set of three with the other two values of $n$. The elongated square cupola, Johnson solid 19, illustrated in Figure 37 below, is the $n$ = 4 member of the same family as $J_{18}$ above, with the appropriate $C_{4v}$ symmetry, and the next member of the Johnson solid list, $J_{20}$, is the $n$ = 5 member.

In examining Table 11 below, it might seem an amazing coincidence that all the dihedral angles between squares are equal, and that all the dihedral angles between triangles and squares are equal. However, this is *not*, actually, mere coincidence. If one takes a rhombicuboctahedron, an Archimedean solid, and slices it into two pieces by cutting it with a plane (oriented so it is parallel to a square face of the rhombicuboctahedron and passes through all the eight vertices that are one edge away from the vertices of that square face), the two pieces are a square cupola ($J_4$) and an elongated square cupola. Since the rhombicuboctahedron is an Archimedean solid, symmetry requires all 3-4 edges to be equivalent, and all 4-4 edges similarly. The measure of the dihedral angles at those edges is retained when the rhombicuboctahedron is cut into two pieces, so that all the 3-4 and 4-4 dihedral angles in both pieces (the square cupola and the elongated square cupola) retain the values they had in the rhombicuboctahedron, respectively 144.7356° (144° 44′ 8″) and 135°.

The relationship of the dual of the elongated square cupola, illustrated in Figure 38 below, to the dual of the elongated triangular cupola is the same as the relationship that the elongated cupolae themselves have to each other, and so for details the reader is referred back to the discussion about the dual of the elongated triangular cupola on p. 79: it is a similar shape, but with a fourfold axis instead of a threefold axis, so the same antibipyramid-like apex, pyramid-like antiapex, and belt of asymmetric quadrilaterals of alternating chirality, noted for the dual of the elongated triangular cupola, are found, except that the number of kite-quadrilaterals is now four rather than three, and the number of asymmetric quadrilaterals, as well as the number of isosceles triangles at the antiapex, is now eight rather than six.

The differences in the proportions of the edges and faces

of different type can be laid down to consequences of the 3-4-5 trend rule, making this polyhedron longer and narrower than the dual of the elongated triangular cupola, as expected, but these differences do not constitute major differences in the appearance of the the dual of the elongated square cupola as opposed to the the dual of the elongated triangular cupola. The reader is invited to compare Figure 38 with Figure 36 above, and to consult Table 11 below as well.

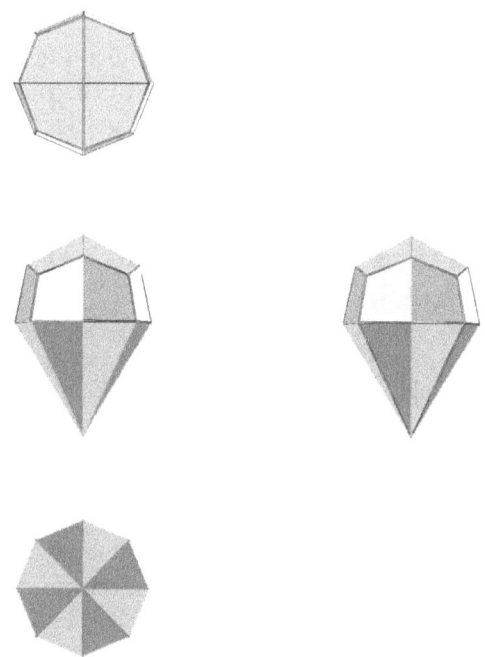

*Figure 38: The dual of Johnson solid 19 (the elongated square cupola).*

The elongated pentagonal cupola, Johnson solid 20, is illustrated in Figure 39 below. Examination of Table 11 below shows some interesting points, consequences of the 3-4-5 trend rule. In Johnson solid 18, the triangular faces have a smaller dihedral angle with the squares that form the cupola than with the squares that form the elongation; in Johnson solid 19, the two dihedral angles are equal, so that no distinction is made between 3-4c and 3-4L. In Johnson solid 20, they have almost exactly interchanged values; the 3-4c dihedral angle of $J_{18}$ is very close to the 3-4L dihedral angle of $J_{20}$, and vice versa. Exactly the same can be noticed comparing the 4c-4L and 4L-4L dihedral angles; they differ in one sense in $J_{18}$, are equal in $J_{19}$, and differ in the opposite sense in $J_{20}$, with the 4c-4L dihedral angle of $J_{18}$ very close to the 4L-4L dihedral angle of $J_{20}$, and vice versa. Of course, the 3-4L and 4c-4L dihedral angles measure the "squatness" of the polyhedron, and while the 3-4c and 4L-4L angles are only dependent on the fact that as the order of the principal axis of rotation increases, these angles increase as well, the interchange of values makes for a neat demonstration of the 3-4-5 trend rule.

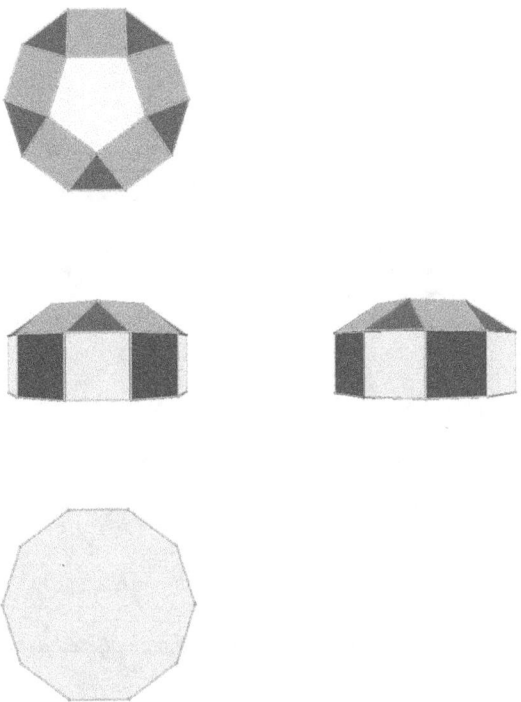

Figure 39: Johnson solid 20: the elongated pentagonal cupola.

As was noted for the other duals of the elongated cupolae, the dual of the elongated pentagonal cupola, illustrated in Figure 40 below, shows an *n*-gonal antibipyramid-like apex and a 2*n*-gonal pyramid-like antiapex, with a belt of 2*n* asymmetric quadrilaterals, alternating in chirality, between the two; in this case, of course, with *n* = 5. Figure 40 can be compared with Figures 36 and 38 above, and details can be found for comparison in Table 11 below. Except for the principal axis being fivefold, rather than three- or fourfold, whatever can be said about this polyhedron has already been said about the other duals of the elongated cupolae, and the reader is referred back to the previous discussion concerning the elongated triangular cupola on p. 79.

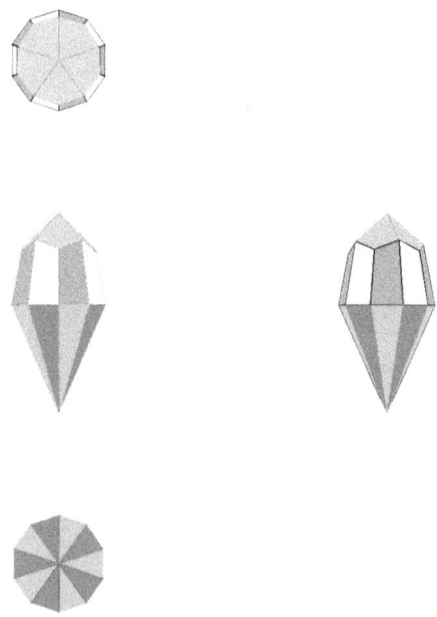

*Figure 40: The dual of Johnson solid 20 (the elongated pentagonal cupola).*

In Chapter 4 there was a major discussion regarding the nature of polyhedra obtained by fusion, and it would perhaps be a good idea to review it at this point. The Johnson solids treated in this chapter, like those in Chapter 4, are obtainable by fusion, and in Chapter 3 (Tables 4 and 5) it was noted that

*Elongated and gyroelongated cupolae and rotundae: Johnson solids 18-25.*

the 3-4 dihedral angles of the cupolae are 125.2644° (125° 15′ 52″), 144.7356° (144° 44′ 8″), and 159.0948° (159° 5′ 41″) respectively for n = 3, 4, and 5, and the 3-5 angles of the pentagonal rotunda are 142.6226° (142° 37′ 21″), regardless of which faces of those numbers of sides are used. (In part, this comes from the fact that $J_3$ is a portion of a cuboctahedron, $J_4$ a portion of a rhombicuboctahedron, $J_5$ is a portion of a rhombicosidodecahedron, and $J_6$ is a portion of an icosidodecahedron, and these latter polyhedra are Archimedean, thus with all edges alike, so since the Johnson solids produce Archimedean solids by fusion, they must share the values of the dihedral angles of the Archimedean solids.) As a result, these particular values of the dihedral angles will be found to recur in many more tables, not merely in Table 11, but throughout this book.

Table 11 below summarizes the important properties of the elongated cupolae and their duals.

## The Johnson Solids and Their Duals

| Johnson solid number | 18 | 19 | 20 |
|---|---|---|---|
| Name | Elongated triangular cupola | Elongated square cupola | Elongated pentagonal cupola |
| Symmetry* | $C_{3v}$ | $C_{4v}$ | $C_{5v}$ |
| Faces | [1+3](3) + [3+3+3](4) + [1](6) | [4](3) + [1+4+4+4](4) + [1](8) | [5](3) + [5+5+5](4) + [1](5) + [1](10) |
| Edges* | 27 | 36 | 45 |
| Vertices | 15 | 20 | 25 |
| Dihedral angles | 3-4c: 125.2644° (125° 15′ 52″), 3-4L: 160.5288° (160° 31′ 44″), 4c-4L: 144.7356° (144° 44′ 8″), 4L-4L: 120°, 4L-6: 90°♦ | 3-4: 144.7356° (144° 44′ 8″), 4-4: 135°, 4-8: 90°* | 3-4c: 159.0948° (159° 5′ 41″), 3-4L: 127.3774° (127° 22′ 39″), 4c-4L: 121.7175° (121° 43′ 3″), 4L-4L: 144°, 4c-5: 148.2825° (148° 16′ 57″), 4L-10: 90°♦ |
| Area | 13.33012702 = 9 + (5/2)√3 | 19.56047793 = 15 + 2√2 + √3 | 26.57974975 |
| Volume | 3.77658751 | 6.77123617 | 10.01825416 |
| **Dual** | | | |
| Name | Unnamed | Unnamed | Unnamed |
| Faces | [6](3) + [6](4a) + [3](4k) | [8](3) + [8](4a) + [4](4k) | [10](3) + [10](4a) + [5](4k) |
| Vertices | 14 | 18 | 22 |

*Elongated and gyroelongated cupolae and rotundae: Johnson solids 18-25.*

| Johnson solid number | 18 | 19 | 20 |
|---|---|---|---|
| Face angles[†] | 3: [2]70.6944° (70° 41′ 40″), 38.6111° (38° 36′ 40″); 4a: [2]82.4250° (82° 25′ 30″), 79.9694° (79° 58′ 10″), 115.1806° (115° 10′ 50″); 4k: [2]71.1171° (71° 7′ 2″), 104.4884° (104° 29′ 18″), 113.2774° (113° 16′ 39″)[□] | 3: [2]78.4463° (78° 26′ 47″), 23.1074° (23° 6′ 27″); 4a: [2]85.4024° (85° 24′ 9″), 76.3449° (76° 20′ 42″), 112.8504° (112° 51′ 1″); 4k: [2]82.4394° (86° 26′ 22″), 77.7951° (77° 47′ 42″), 117.3262° (117° 19′ 34″)[□] | 3: [2]82.6038° (82° 36′ 14″), 14.7924° (14° 47′ 33″); 4a: [2]86.6662° (86° 39′ 58″), 71.6843° (71° 41′ 3″), 114.9832° (114° 59′ 0″); 4k: [2]93.5176° (93° 31′ 3″), 53.5345° (53° 32′ 4″), 119.4303° (119° 25′ 49″)[□] |
| Dihedral angles | 3-3: 133.1625° (133° 9′ 45″), 3-4a: 129.3287° (129° 19′ 43″), 4a-4a: 121.7711° (121° 46′ 16″), 4a-4k: 127.3788° (127° 22′ 44″), 4k-4k: 130.7979° (130° 47′ 53″)[□] | 3-3: 141.1187° (141° 7′ 7″), 3-4a: 139.2322° (139° 13′ 56″), 4a-4a: 135.9016° (135° 54′ 6″), 4a-4k: 133.7942° (133° 47′ 39″), 4k-4k: 130.6149° (130° 36′ 54″)[□] | 3-3: 147.0874° (147° 5′ 14″), 3-4a: 146.1246° (146° 7′ 28″), 4a-4a: 144.6027° (144° 36′ 10″), 4a-4k: 142.9266° (142° 55′ 36″), 4k-4k: 129.9472° (129° 56′ 50″)[□] |
| Area | 7.97458921 | 11.88805416 | 9.07106599 |
| Volume | 1.83331557 | 3.41405632 | 2.17450819 |

Table 11: Properties of Johnson solids 18, 19, and 20 and their duals.

\*The symmetry and number of edges are always the same for each solid and its dual, so are not repeated in the "dual" section.

†Face angles are only given for the duals, as the Johnson solids have regular faces, whose angles can be found in Table 1 on p. 3.

\*Although there are squares in four different transitivity

classes, all 4-4 edges have the same 135° dihedral angle and all 4-8 edges have the same 90° dihedral angle.

♦The designation 4c refers to the squares forming the cupola; 4L to the squares forming the lateral belt (the elongation). Although there are two types of triangle in the case of $J_{18}$, both the roof triangle and the triangles forming the lateral part of the cupola make the same dihedral angle of 125.2644° (125° 15′ 52″) with the cupola squares.

¤See the text beginning on p. 79 for an explanation of the difference between the "4a" and "4k" quadrilaterals. Note that the 4a quadrilaterals are *not* kite-quadrilaterals, even though they have two equal angles, because the equal angles are not opposite each other.

*Elongated and gyroelongated cupolae and rotundae: Johnson solids 18-25.*

The elongated pentagonal rotunda, Johnson solid 21, illustrated in Figure 41 below, has properties that can be inferred from its being a fusion between a pentagonal prism and a rotunda. As from the pentagonal cupola ($J_5$), with 12 faces, 15 vertices, and 25 edges, by fusion with a pentagonal prism, one gets the elongated pentagonal cupola ($J_{20}$), with 22 faces, 25 vertices, and 45 edges, similarly from the pentagonal rotunda ($J_6$), with 17 faces, 20 vertices, and 35 edges, by fusion with a pentagonal prism, one gets the elongated pentagonal rotunda ($J_{21}$), with 27 faces, 30 vertices, and 55 edges.

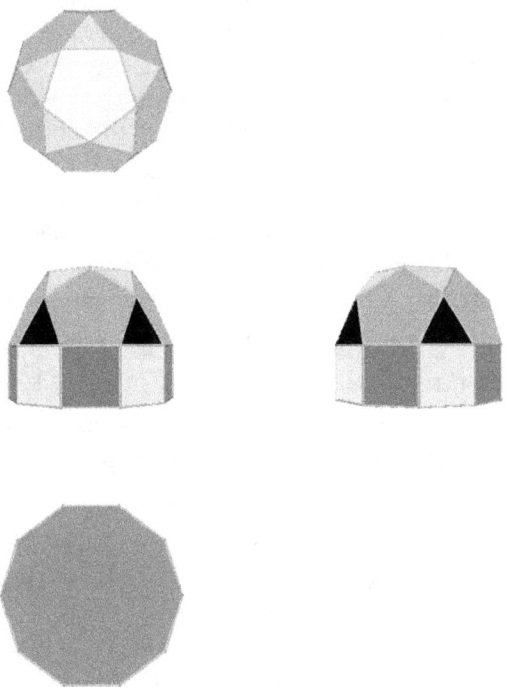

*Figure 41: Johnson solid 21: the elongated pentagonal rotunda.*

And because the elongated pentagonal rotunda is a fusion of a pentagonal rotunda and a pentagonal prism, all the dihedral angles among the triangular and pentagonal faces are identical to those between corresponding faces in the pentagonal rotunda; the dihedral angles between the triangular and pentagonal faces, on the one hand, and the square faces, on the other, are exactly 90° more than the dihedral angles between the triangular and pentagonal faces and the decagonal base of the rotunda; the dihedral angles between square faces are equal to the 144° angle at a vertex of a decagon; and the dihedral angles between the square faces and the decagonal base are all 90° (as are all dihedral angles between a lateral face of a prism and its bases).

The dual of the elongated pentagonal rotunda, illustrated in Figure 42 below, has three different types of quadrilateral faces. The irregular quadrilaterals are in a similar arrangement, alternating in chirality, to what has been seen before, for example in the duals of $J_{18}$ through $J_{20}$. However, because the equal angles are so close to being right angles, in this case they are *almost* trapezoids. The other quadrilaterals are kite-quadrilaterals, falling under the definitions of apical and lateral kites, as defined on p. 30. And it can be seen that the apical/lateral kite rule, discussed on p. 30, applies, relating the acute and obtuse angles of the apical and lateral kites and noting that both are close to rhombi in shape. The two sets of kite-quadrilaterals are *almost* identical to each other and *almost* rhombi; the ones designated as 4k in Table 12 have edges equal to ~1.0321 and ~1.0438, with the two equal angles being 116.5699° (116° 34′ 12″) and the other two 63.0328° (63° 1′ 58″) and 63.8274° (63° 49′ 388″), while those designated as 4t have edges equal to ~1.0277 and ~1.0321, with the two equal angles being 63.4475° (63° 26′ 51″) and the other two 116.1582° (116° 9′ 29″) and 116.9468° (116° 56′ 48″). (The number of "almosts" in this paragraph are amazing!) Note that the two exactly equal angles of one set are the obtuse angles, while they are the acute angles of the other, but in both cases the two angles that are almost equal are less than a degree apart in size. Additionally, the 4-4 dihedral angles (even those involving the 4a faces!) are all within a degree of each other.

The near-congruence of the two sets of kite-quadrilaterals in the dual of $J_{21}$ extends to other duals of Johnson solids related to the rotunda. In each case, both are near-rhombi, but the acute angles of one set remain equal, while the obtuse angles are the ones in the other set that remain equal. It is not as clearly seen in the dual of $J_6$ itself, but this phenomenon will be apparent, for example, when we get to the duals of $J_{40}$ and $J_{41}$.

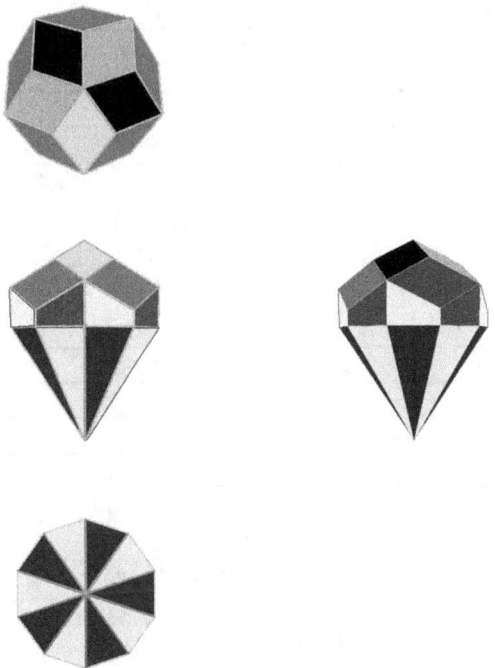

Figure 42: The dual of Johnson solid 21 (the elongated pentagonal rotunda).

Exactly like the duals of $J_{18}$ through $J_{20}$, the dual of $J_{21}$ can be viewed as a fusion of a pyramid (decagonal in this case) with another, as-yet-unnamed solid; however in this case the solid has *three* types of quadrilateral faces as mentioned above.

Table 12 below summarizes the properties of Johnson solid 21 and its dual.

| | |
|---|---|
| Johnson solid number | 21 |
| Name | Elongated pentagonal rotunda |
| Symmetry* | $C_{5v}$ |
| Faces | [5+5](3) + [5+5](4) + [1+5](5) + [1](10) |
| Edges* | 55 |
| Vertices | 30 |
| Dihedral angles | 3-4: 169.1877° (169° 11′ 16″), 3-5: 142.6226° (142° 37′ 21″), 4-4: 144°, 4-5: 153.4349° (153° 26′ 6″), 4-10: 90°♣ |
| Area | 32.34720027 |
| Volume | 14.61197181 |
| **Dual** | |
| Name | Unnamed |
| Faces | [10](3) + [5+5+10](4) |
| Vertices | 27 |
| Face angles† | 3: [2]80.0502° (80° 3′ 1″), 19.8997° (19° 53′ 59″); 4a: [2]90.4352° (90° 26′ 7″), 62.7974° (62° 47′ 51″), 116.3321° (116° 19′ 56″); 4k: [2]116.5699° (116° 34′ 12″), 63.0328° (63° 1′ 58″), 63.8274° (63° 49′ 38″); 4t: [2]63.4475° (63° 26′ 51″), 116.1582° (116° 9′ 29″), 116.9468° (116° 56′ 48″)‡ |
| Dihedral angles | 3-3: 149.8470° (149° 50′ 49″), 3-4a: 148.6627° (148° 39′ 46″), 4a-4a: 144.0102° (144° 0′ 37″), 4a-4t: 143.7853° (143° 47′ 7″), 4k-4k: 143.2465° (143° 14′ 47″), 4k-4t: 143.3985° (143° 28′ 54″)‡ |
| Area | 26.05017934 |
| Volume | 11.35772957 |

Table 12: Properties of Johnson solid 21 and its dual.

°The symmetry and number of edges are always the same for each solid and its dual, so are not repeated in the "dual" section.

†Face angles are only given for the duals, as the Johnson solids have regular faces, whose angles can be found in Table 1 on p. 3.

♦Although there are triangles, squares, and pentagons in two different transitivity classes each, all 3-5 edges have the same 142.6226° dihedral angle, all 4-4 edges have the same 144° dihedral angle, and all 4-10 edges have the same 90° dihedral angle.

⁺The designation 4a refers to the ten irregular quadrilaterals adjacent to the pyramidal faces; 4k and 4t to the two sets of kite-quadrilateral faces, where the 4k faces are adjacent to each other around the other apex and the 4t faces are the belt between the 4k and 4a faces. Within the rotunda dual part, the 4k quadrilaterals are apical kites and the 4t quadrilaterals are lateral kites, as defined on p. 30.

The gyroelongated triangular cupola, Johnson solid 22, illustrated in Figure 43 below, is obtained by fusing a triangular cupola ($J_3$) to a hexagonal antiprism. Its properties, therefore, can mainly be determined by considering those of the two polyhedra whose fusion produces it.

*Figure 43: Johnson solid 22: the gyroelongated triangular cupola.*

The cupola originally has a triangular roof, a lateral ring of three triangles and three squares, and a hexagonal base. This makes eight faces: four triangular, three square, and one hexagonal. The antiprism has a lateral ring of twelve triangles between two hexagonal bases, thus fourteen faces: twelve triangular and two hexagonal. Fusing them eliminates one hexagonal face from each polyhedron, so the resulting polyhedron has twenty faces: sixteen triangular, three square, and one hexagonal. The original cupola has nine vertices: three on the roof and six on the base. The antiprism has twelve vertices: six on each base. Six vertices of one polyhedron are merged with six of the other in the fusion process, making a total of fifteen vertices

— 96 —

(nine plus twelve minus six). And either a similar analysis or the use of the Euler formula $F + V = E + 2$ will suffice to demonstrate that there are thirty-three edges. The symmetry is $C_{3v}$, with three mirror planes (each bisecting one side of the triangular roof of the cupola and passing through the opposite vertex of that roof, bisecting a square face on the side where it had bisected a side of the roof, a triangular face deriving from the cupola on the other side, and passing through two opposite vertices of the hexagonal base) all passing through a threefold axis.

The dual of the gyroelongated triangular cupola, illustrated in Figure 44 below, shares the $C_{3v}$ symmetry of the original Johnson solid, as every polyhedron and its dual must share symmetry. There are three different classes of kite-quadrilaterals; three of the nine kite-quadrilaterals (here designated 4a) meet at one apex, and the remaining six, at the opposite one. To describe these kites, it is useful to refer to the vertex of each kite which is opposite the one at an apex of the polyhedron as the "furthest vertex." The six that come together at one apex really divide into two transitivity classes, because three of them (here designated 4b) have their furthest vertices aligned with the furthest vertices of the 4a kites, while the other three kites (here designated 4c) have their furthest vertices aligned with the edges between the 4a kites. For most purposes, however, the 4b and 4c kites are identical; it is only this rather remote relationship with the 4a kites that defines the difference, and although this leads to designating the 4b and 4c kites with different symbols, this does not show itself in the geometric properties with which we are concerned.

This is another case where there is a greater degree of symmetry of some parts of a Johnson dual than might be expected. The $C_{3v}$ symmetry only requires three mirror planes and a threefold axis. Those three mirror planes bisect the three 4a kites, and guarantees that they will in fact be kites and not irregular quadrilaterals. And if one follows one of the mirror planes through the furthest vertex of the 4a kite that it bisects, it continues onward to bisect a 4b kite, again making the 4b faces certain to be kites. On the opposite side of the solid, the same plane runs along the edge between the two 4a kites that were not bisected, and continues onward to bisect a 4c kite. So it is forced by symmetry that all three types of quadrilateral face are kites. What is *not* forced is that the 4b and 4c kites are identical. The edges between 4b and 4c kites are *not* along mirror planes.

The fact that every one of the edges passing through the apex where the 4b and 4c kites cluster is a 4b-4c edge, however, means that the pair of equal sides of the 4b kites that extend to that apex and the pair of equal sides of the 4c kites that extend to that apex must be the same length, since each side of one is included in the other set. This does not guarantee congruence, however, because there is no constraint that would require the angle between them in a 4b face to be equal to the angle between them in a 4c face. And there is also no constraint that would require the remaining sides of the 4b kites to equal the corresponding sides of the 4c kites. Yet in fact these equalities are obtained in the actual solid, as constructed and illustrated in Figure 44.

We will see the same phenomenon with regard to the duals of $J_{23}$ and $J_{24}$, but the discussion will not be repeated there because the only difference is the fact that the symmetry of $J_{23}$ and its dual is fourfold and that of $J_{24}$ and its dual is fivefold, with four or five kites in each transitivity class rather than three, and eight or ten pentagons, alternating in chirality, instead of six, respectively. (It is true that in $J_{22}$ and $J_{24}$, each mirror plane passes through an edge between two 4a kites as well as bisecting one, while in $J_{23}$, half the mirror planes bisect 4a kites and the others pass through 4a-4a edges. This is simply a difference that always occurs between odd-order and even-order $C_{nv}$ rotational symmetry, and makes no significant difference in the discussion.)

Earlier, in Chapter 3, the apex of a cupola dual or similar polyhedron, where a number of kite-quadrilaterals came together equal to the order of the principal axis of rotation, was described as an antibipyramid-like apex, and there is such an apex in all of the duals of the gyroelongated cupolae. The opposite apex (antiapex, if you will) differs only in one respect: the number of kite-quadrilaterals is *twice* the order of the principal axis of rotation. It would seem reasonable, and there will not be any reason to avoid doing so, to describe the antiapex of this polyhedron as antibipyramid-like as well, therefore.

*Elongated and gyroelongated cupolae and rotundae: Johnson solids 18-25.*

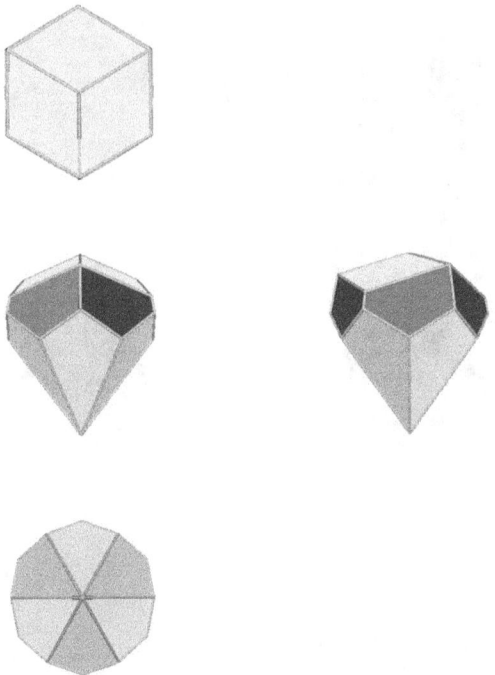

*Figure 44: The dual of Johnson solid 22 (the gyroelongated triangular cupola).*

All the duals of the gyroelongated cupolae have a ring of irregular pentagons, alternating in chirality, separating the two antibipyramid-like apices, and this is the basic structure of a gyroelongated cupola dual. The precise values of the angles of all three of the duals of the Johnson gyroelongated cupolae are given in Table 13 below.

The gyroelongated square cupola, Johnson solid 23, illustrated in Figure 45 below, can be described by simply consulting the descriptions above for the gyroelongated triangular cupola, and making the appropriate changes to accommodate the fourfold symmetry, replacing "three" by "four," "six" by "eight," "triangle" (insofar as it refers to the base, rather than a lateral face) by "square," "threefold" by "fourfold," etc. Because of that, it is unnecessary to provide a detailed summary of the properties of this solid, and the reader is referred back to the discussion of the gyroelongated triangular cupola. The symmetry is $C_{4v}$.

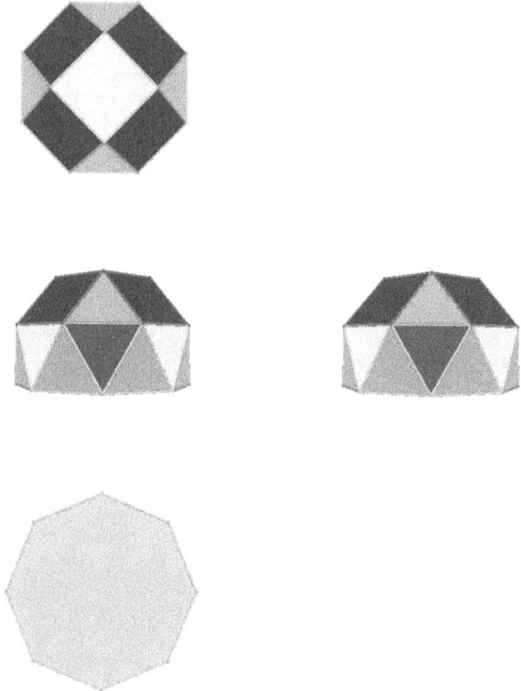

*Figure 45: Johnson solid 23: the gyroelongated square cupola.*

The dual of the gyroelongated square cupola, illustrated in Figure 46 below, also possesses $C_{4v}$ symmetry, and can be best described by simply consulting the rather long description above for the dual of the gyroelongated triangular cupola and making the appropriate changes to accommodate the fourfold symmetry, as described for $J_{23}$ itself. It is not particularly to repeat the description with those changes here. Even the division of the quadrilateral faces into three transitivity classes is the same; only the number of each, and the number of irregular pentagons, must be changed to accommodate the fourfold symmetry. (While in the case of $J_{23}$ itself, triangles are replaced by squares and the hexagon by an octagon, in the dual it is only the valencies of vertices that show similar changes. The types of polygons do not.) An overview of the differences between the dual of the gyroelongated triangular cupola and the dual of the gyroelongated square cupola can best be obtained by comparing Figures 44 and 46; a more detailed treatment can be obtained by consulting Table 13 below.

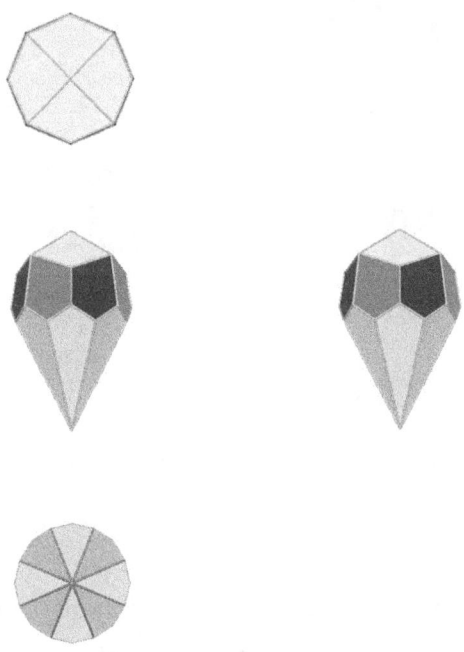

Figure 46: The dual of Johnson solid 23 (the gyroelongated square cupola).

The gyroelongated pentagonal cupola, Johnson solid 24, illustrated in Figure 47 below, can be described by simply consulting the descriptions above for the gyroelongated triangular cupola and the gyroelongated square cupola, and making the appropriate changes to accommodate the fivefold symmetry, replacing "three" or "four" by "five," "six" or "eight" by "ten," "triangle" (insofar as it refers to the base, rather than a lateral face) or "square" by "regular pentagon," "threefold" or "fourfold," by "fivefold," etc. The symmetry is $C_{5v}$. Compare Figure 47 with Figures 43 and 45 for the similarities and differences among these three polyhedra. In addition, a more detailed comparison can be obtained by consulting Table 13 below.

*The Johnson Solids and Their Duals*

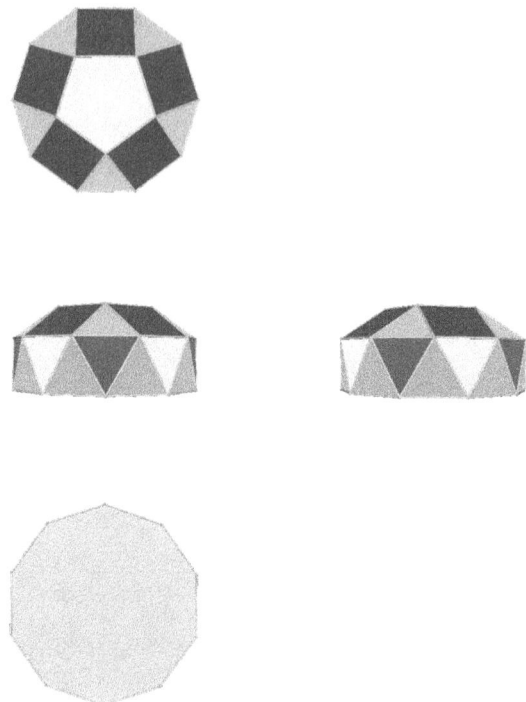

*Figure 47: Johnson solid 24: the gyroelongated pentagonal cupola.*

As was mentioned for the dual of the gyroelongated square cupola, the dual of the gyroelongated pentagonal cupola, illustrated in Figure 48 below, can similarly be described by simply consulting the description above for the dual of the gyroelongated triangular cupola and making the appropriate changes to accommodate the fivefold symmetry, as the description of the gyroelongated pentagonal cupola itself can be obtained by making the changes appropriate to the fivefold symmetry in the description of the gyroelongated triangular cupola. It is unnecessary, of course, to repeat the words of the earlier discussion with those changes. The symmetry, as for its dual, is $C_{5v}$. The reader is advised to compare Figure 48 with Figures 44 and 46 for the similarities and differences among these three polyhedra, and it will be noticed that the effects of the 3-4-5 trend rule are clearly visible. And in addition, a more detailed comparison can be obtained by consulting Table 13 below.

*Elongated and gyroelongated cupolae and rotundae: Johnson solids 18-25.*

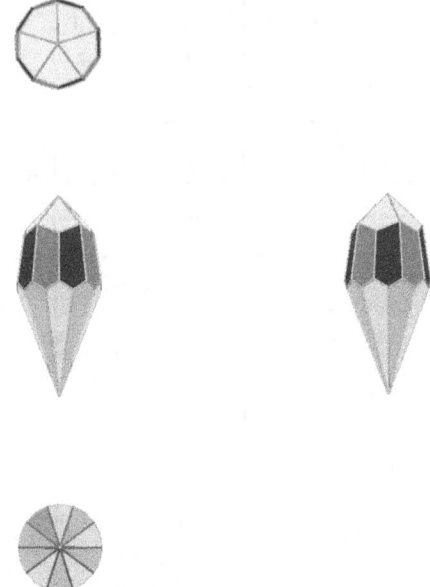

*Figure 48: The dual of Johnson solid 24 (the gyroelongated pentagonal cupola).*

| Johnson solid number | 22 | 23 | 24 |
|---|---|---|---|
| Name | Gyroelongated triangular cupola | Gyroelongated square cupola | Gyroelongated pentagonal cupola |
| Symmetry* | $C_{3v}$ | $C_{4v}$ | $C_{5v}$ |
| Faces | [1+3+3+6](3) + [3](4) + [1](6) | [4+4+4+8](3) + [1+4](4) + [1](8) | [5+5+5+10](3) + [5](4) + [1](5) + [1](10) |
| Edges* | 33 | 44 | 55 |
| Vertices | 15 | 20 | 25 |
| Dihedral angles | 3a-3b: 145.2219° (145° 13′ 19″), 3a-3c: 169.4282° (169° 25′ 42″), 3a-4: 153.6350° (153° 38′ 6″), 3c-4: 125.2644° (125° 15′ 52″), 3b-6: 98.8994° (98° 53′ 58″)♣ | 3a-3b: 153.9624° (153° 57′ 45″), 3a-3c: 151.3301° (151° 19′ 48″), 3a-4: 141.5945° (141° 35′ 40″), 3c-4: 144.7356° (144° 44′ 8″), 3b-8: 96.5945° (96° 35′ 40″), 4-4: 135°□ | 3a-3b: 159.1865° (159° 11′ 11″), 3a-3c: 132.6240° (132° 37′ 26″), 3a-4: 126.9641° (126° 57′ 51″), 3c-4: 159.0948° (159° 5′ 41″), 3b-10: 95.2466° (95° 14′ 48″), 4-5: 148.2825° (148° 16′ 57″)® |
| Area | 12.52627944 = 3 + (11/2)√3 | 18.48868116 7 + 2√2 + 5√3 | 25.24000379 |
| Volume | 3.51605309 | 6.21076579 | 9.07333319 |
| Dual | | | |
| Name | Unnamed | Unnamed | Unnamed |
| Faces | [3+3+3](4) + [6](5) | [4+4+4](4) + [8](5) | [5+5+5](4) + [10](5) |
| Vertices | 20 | 26 | 32 |

*Elongated and gyroelongated cupolae and rotundae: Johnson solids 18-25.*

| Johnson solid number | 22 | 23 | 24 |
|---|---|---|---|
| Face angles[†] | 4a: [2]70.8937° (70° 53′ 37″), 105.6533° (105° 39′ 12″), 112.5592° (112° 33′ 33″); 4b, 4c: [2]110.3519° (110° 21′ 7″), 35.8587° (35° 51′ 31″), 103.4375° (103° 26′ 15″); 5: [2]115.6090° (115° 36′ 32″), 77.8701° (77° 52′ 12″), 113.7435° (113° 44′ 37″), 117.1684° (117° 10′ 6″)[△] | 4a: [2]82.8111° (82° 48′ 40″), 76.5871° (76° 35′ 14″), 117.7908° (117° 47′ 27″); 4b, 4c: [2]114.0901° (114° 5′ 24″), 21.0975° (21° 5′ 51″), 110.7223° (110° 43′ 20″); 5: [2]117.2835° (117° 17′ 1″), 74.9215° (74° 55′ 17″), 112.3454° (112° 20′ 43″), 118.1661° (118° 9′ 58″)[△] | 4a: [2]94.5114° (94° 30′ 41″), 51.8410° (51° 50′ 28″), 119.1362° (119° 8′ 10″); 4b, 4c: [2]116.1251° (116° 7′ 30″), 13.3106° (13° 18′ 38″), 114.4392° (114° 26′ 21″); 5: [2]117.9915° (117° 59′ 29″), 70.3558° (70° 21′ 21″), 115.0988° (115° 5′ 56″), 118.5624° (118° 33′ 45″)[△] |
| Dihedral angles | 4a-4a: 128.4932° (128° 29′ 35″), 4b-4c: 131.0740 (131° 4′ 26″), 4a-5: 125.4558° (125° 27′ 21″), 4b-5, 4c-5: 127.3963° (127° 23′ 47″), 5-5[b]: 121.0318° (121° 1′ 54″)[△] | 4a-4a: 128.5664° (128° 33′ 59″), 4b-4c: 140.0232° (140° 1′ 24″), 4a-5: 132.9446° (132° 56′ 40″), 4b-5, 4c-5: 138.2944° (138° 17′ 40″), 5-5[b]: 135.5605° (135° 33′ 38″)[△] | 4a-4a: 128.1865° (128° 11′ 11″), 4b-4c: 146.4752° (146° 28′ 31″), 4a-5: 142.8795° (142° 56′ 46″), 4b-5, 4c-5: 145.6265° (145° 37′ 35″), 5-5[b]: 144.4021° (144° 24′ 8″)[△] |
| Area | 7.96517233 | 10.62948066 | 8.45614592 |
| Volume | 1.83060940 | 2.84489415 | 1.90682816 |

Table 13: *Properties of Johnson solids 22, 23, and 24 and their duals.*

[\*]The symmetry and number of edges are always the same for each solid and its dual, so are not repeated in the "dual" section.

†Face angles are only given for the duals, as the Johnson solids have regular faces, whose angles can be found in Table 1 on p. 3.

♦Although the triangles fall into *five* transitivity classes, only three distinctions need be made: 3a designates the triangular faces from the antiprismatic elongation that are adjacent to the cupola (those adjacent to square faces form a different transitivity class from those adjacent only to triangular faces, but both have the same dihedral angles with 3b faces, and only one set participates in 3a-4 edges while the other participates in 3a-3c edges); 3b designates the triangular faces from the antiprismatic elongation that are adjacent to the hexagonal base, and 3c represents the triangular faces that derive from the cupola (again falling into two transitivity classes, but both having identical dihedral angles with adjacent squares, while no ambiguity is otherwise introduced).

□Although the triangles fall into *four* transitivity classes, and the squares into *two*, no distinctions need be made among the squares and only three distinctions need be made among the triangles: 3a designates the triangular faces from the antiprismatic elongation that are adjacent to the cupola (those adjacent to square faces form a different transitivity class from those adjacent only to triangular faces, but both have the same dihedral angles with 3b faces, and only one set participates in 3a-4 edges while the other participates in 3a-3c edges); 3b designates the triangular faces from the antiprismatic elongation that are adjacent to the octagonal base, and 3c represents the triangular faces that derive from the cupola.

⊛Although the triangles fall into *four* transitivity classes, only three distinctions need be made: 3a designates the triangular faces from the antiprismatic elongation that are adjacent to the cupola (those adjacent to square faces form a different transitivity class from those adjacent only to triangular faces, but both have the same dihedral angles with 3b faces, and only one set participates in 3a-4 edges while the other participates in 3a-3c edges); 3b designates the triangular faces from the antiprismatic elongation that are adjacent to the decagonal base, and 3c represents the triangular faces that derive from the cupola.

ᐃFor the distinction between the 4a, 4b, and 4c kite-quadrilaterals, see the discussion on p. 97.

ᵇAlthough there are two different types of 5-5 edges, the dihedral angles are the same at both.

The gyroelongated pentagonal rotunda, Johnson solid 25, illustrated in Figure 49 below, has properties that can be inferred from its being a fusion between a pentagonal antiprism and a rotunda. As from the pentagonal cupola ($J_5$), with 12 faces, 15 vertices, and 25 edges, by fusion with a pentagonal antiprism, one gets the gyroelongated pentagonal cupola ($J_{24}$), with 32 faces, 25 vertices, and 55 edges, similarly from the pentagonal rotunda ($J_6$), with 17 faces, 20 vertices, and 35 edges, by fusion with a pentagonal antiprism, one gets the gyroelongated pentagonal rotunda ($J_{25}$), with 37 faces, 30 vertices, and 65 edges.

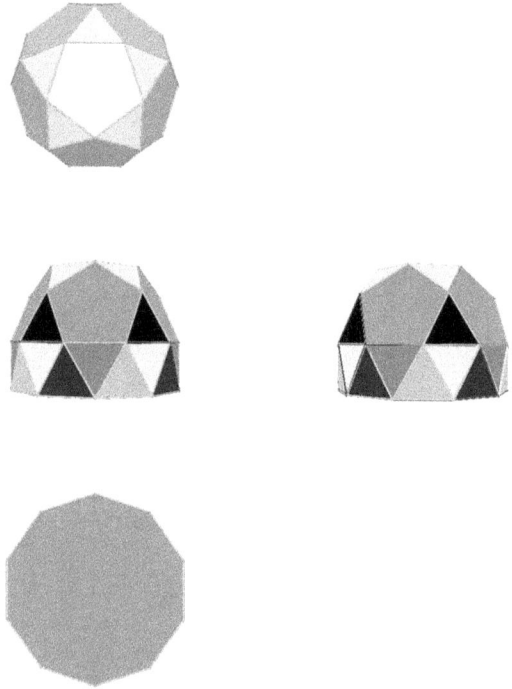

Figure 49: Johnson solid 25: the gyroelongated pentagonal rotunda.

The dual of the gyroelongated pentagonal rotunda, illustrated in Figure 50 below, has quadrilateral faces which fall into four different transitivity classes, but although the ten faces that meet at the decavalent vertex are members of two different transitivity classes, they are all congruent, much as was

*Elongated and gyroelongated cupolae and rotundae: Johnson solids 18-25.*

found in the cases of the duals of $J_{22-24}$. Although in the discussion of the duals of $J_{22-24}$ it was useful to use two different designations for those two sets of kites and term them 4b and 4c, in discussing the dual of $J_{25}$ we will group them together as 4a, since the distinctions between them were discussed previously and will not be gone into here.

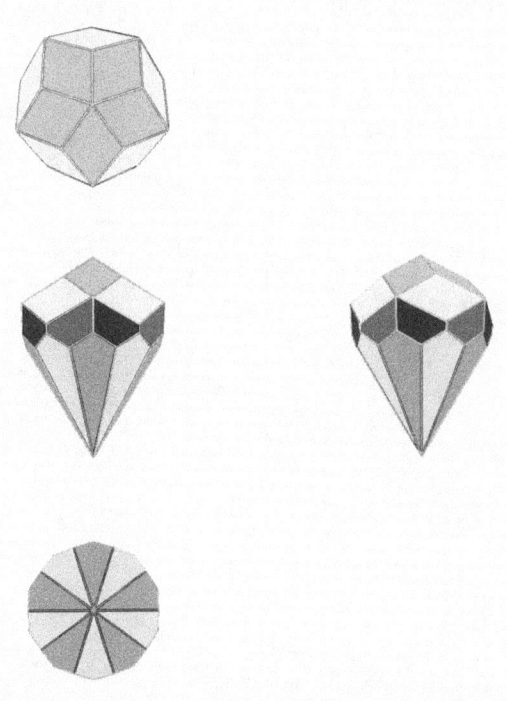

Figure 50: The dual of Johnson solid 25 (the gyroelongated pentagonal rotunda).

The other two classes of kite-quadrilateral will be termed 4b and 4c, and although they are both kite-quadrilaterals, both are *almost* rhombi. (And like the similarly sited kite-quadrilaterals in the dual of $J_{21}$, they are *almost* identical to each other, but one set has the obtuse angles exactly equal and the acute angles close, while the other set reverses this situation; the ones designated as 4b in Table 14 have edges equal to ~0.9856 and ~0.9950, with the two equal angles being 63.4984° [63° 29′

54″] and the other two 115.6156° [115° 36′ 56″] and 117.3876° [117° 23′ 15″], while those designated as 4c have edges equal to ~0.9950 and ~1.0207, with the two equal angles being 116.5902° [116° 35′ 25″] and the other two 62.5089° [62° 30′ 32″] and 64.3108° [64° 18′ 39″].) Within the rotunda dual part, the 4b quadrilaterals are apical kites and the 4c quadrilaterals are lateral kites, as defined on p. 30.

| Johnson solid number | 25 |
|---|---|
| Name | Gyroelongated pentagonal rotunda |
| Symmetry* | $C_{5v}$ |
| Faces | [5+5+5+5+10](3) + [1+5](5) + [1](10) |
| Edges* | 65 |
| Vertices | 30 |
| Dihedral angles | 3a-3b: 159.1865° (159° 11′ 11″), 3a-3r: 174.4343° (174° 26′ 4″), 3a-5: 158.6816° (158° 40′ 54″), 3r-5: 142.6226° (142° 37′ 21″), 3b-10: 95.2466°.(95° 14′ 48″)♦ |
| Area | 31.00745430 |
| Volume | 13.66705084 |
| Dual ||
| Name | Unnamed |
| Faces | [5+5+5+5](4) + [10](5) |
| Vertices | 37 |
| Face angles† | 4a: [2]116.0503 (116° 3′ 1″), 18.6282° (18° 37′ 41″), 109.2713° (109° 16′ 17″); 4b: [2]63.4984 (63° 29′ 54″), 115.6156° (115° 36′ 56″), 117.3876° (117° 23′ 15″); 4c: [2]116.5902° (116° 35′ 25″), 62.5089° (62° 30′ 32″), 64.3108° (64° 18′ 39″); 5: [2]120.9815° (120° 58′ 53″), 62.0050° (62° 0′ 18″), 116.0716° (116° 4′ 18″), 119.9605° (119° 57′ 38″)⊛ |
| Dihedral angles | 4a-4a: 149.0568° (149° 3′ 24″), 4b-4c: 142.6643° (142° 39′ 51″), 4c-4c: 142.2976° (142° 17′ 51″), 4a-5: 147.7779° (147° 46′ 40″), 4b-5: 143.5648° (143° 33′ 53″), 5-5: 144.0497° (144° 2′ 59″)⊛ |

| Johnson solid number | 25 |
|---|---|
| Area | 24.49260875 |
| Volume | 10.27897757 |

Table 14: Properties of Johnson solid 25 and its dual.

*The symmetry and number of edges are always the same for each solid and its dual, so are not repeated in the "dual" section.

†Face angles are only given for the duals, as the Johnson solids have regular faces, whose angles can be found in Table 1 on p. 3.

♦Although the triangles fall into *five* transitivity classes, and the pentagons into two, no distinctions need be made among the pentagons and only three distinctions need be made among the triangles: 3a designates the triangular faces from the antiprismatic elongation that are adjacent to the rotunda (those adjacent to pentagonal faces form a different transitivity class from those adjacent only to triangular faces, but both have the same dihedral angles with 3b faces, and only one set participates in 3a-5 edges while the other participates in 3a-3r edges); 3b designates the triangular faces from the antiprismatic elongation that are adjacent to the decagonal base, and 3r represents the triangular faces that derive from the rotunda (again falling into two transitivity classes, but both having identical dihedral angles with adjacent triangles, while no ambiguity is otherwise introduced).

⊗For the distinctions between 4a, 4b, and 4c kite-quadrilaterals, see the discussion on p. 109.

# Chapter 8: Bicupolae, cupolarotundae, and birotundae: Johnson solids 26-34.

Both the cupolae and the rotundae have bases with twice the number of sides of their order of rotational symmetry. Thus, for any species of these polyhedra, they can be fused to copies of themselves or to each other. In every case there are two possibilities: the roofs of the two can be aligned or opposite. Johnson's terminology is systematic: if the two are both cupolae, the resulting solid is a "bicupola"; if the two are both rotundae, the resulting solid is a "birotunda"; and if one is a cupola and the other a rotunda, the resulting solid is a "cupolarotunda." If the two roofs of the pieces are aligned, the prefix "ortho-" is used; if they are oriented oppositely, so that a side of one is oriented over a vertex of the other, the prefix "gyro-" is used. Since (when regular polygonal faces are required) there are three kinds of cupola (triangular, square, and pentagonal), there would seem to be three kinds of orthobicupola and three kinds of gyrobicupola. However, a triangular gyrobicupola is not a Johnson solid, because the construction would produce an Archimedean solid, the cuboctahedron. Because there is only (when regular polygonal faces are required) a pentagonal rotunda, only pentagonal forms of the other four kinds (orthocupolarotunda, gyrocupolarotunda, orthobirotunda, and gyrobirotunda) can be constructed. But again the pentagonal gyrobirotunda is not a Johnson solid, because the construction would produce an Archimedean solid, the icosidodecahedron. While these would make eight members of the Johnson solids list, there is a ninth, though Johnson did not name it as such, though its placement as $J_{26}$ in the list makes it clear that he recognized that it belongs in this family, just before the triangular member of the bicupolae. If one considers a triangular prism as a digonal cupola (which, in fact it is, if one of its square faces is thought of as the cupola base), one could not form an orthobicupola because the triangles adjacent to the two bases would have a 180° dihedral angle at the edge where they meet and fall into the same plane. But a digonal gyrobicupola can be constructed, although Johnson called it a "gyrobifastigium." It is illustrated in Figure 51 below. All of the gyrobicupolae (*including* the gyrobifastigium) and the pentagonal gyrobirotunda have $S_{2nv}$ symmetry, where $n$ is the order of the principal axis of rotation. The orthobicupolae and the pentagonal orthobirotunda have $D_{nh}$ symmetry, and because the cupolarotundae have two different pieces flanking the fusion plane, their symmetry is only $C_{5v}$. The first member of this set, Johnson solid 26, the gyrobifastigium or digonal gyrobicupola, is illustrated in Figure 51 below. Although its $S_{4v}$ symmetry is exactly what would be expected based on its membership in the family of gyrobicupolae as just indicated, this symmetry will not be met again among the Johnson solids until the last of the chapters describing them, and in fact it will be seen that only three of the ninety-two Johnson solids have this symmetry.

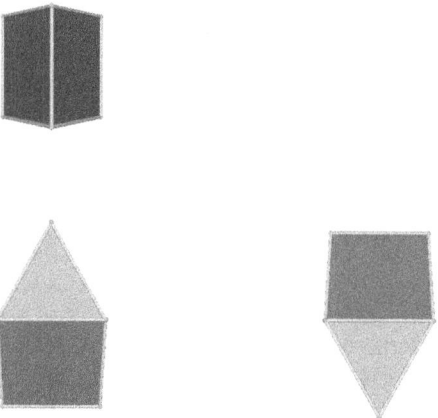

*Figure 51: Johnson solid 26, called by him the gyrobifastigium, the digonal member of the gyrobicupolae.*

The dual of $J_{26}$ has no name, but I think it ought to; it is such a simple polyhedron to describe and, in this author's opinion, is a rather attractive figure, and is illustrated in Figure 52 below. As is the case of all duals, the symmetry is the same as that of the figure whose dual it is, in this case $S_{4v}$. The lateral faces are all parallelograms (four of them) with one set of parallel sides all parallel to the $S_4$ axis (which can be termed "axial edges"). They alternate in direction, but are all congruent, and there is an edge connecting the two highest vertices (designating as "height" the distance along the $S_4$ axis) and an edge

*Bicupolae, cupolarotundae, and birotundae: Johnson solids 26-34.*

connecting the two lowest. Each edge bounds two isosceles triangles, and the remaining sides of those triangles are among the sides of the lateral faces that are not axial edges.

While many of the duals of the Johnson solids are not, to my eyes, very esthetic, this one is different. It resembles a prism in having a family of edges that are all equal and parallel to an axis of symmetry; they alternate in position instead of all being directly related by an *n*-fold rotation, but this alternation is regular. If one took a 2*n*-gonal prism, translated every other vertex in either of the two base planes by the same amount in a direction parallel to the lateral edges, and then performed the same translation on the remaining vertices so as to keep all the lateral edges equal, the $n = 4$ case would be the $J_{26}$ dual.

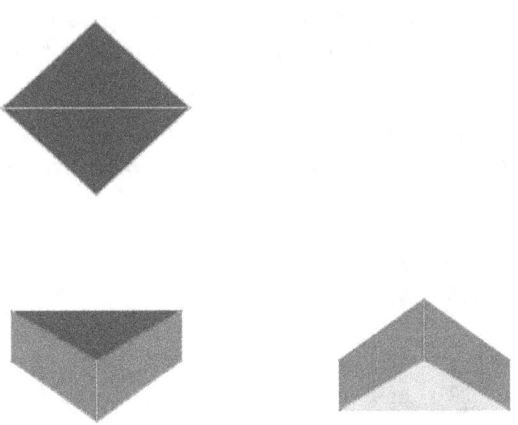

Figure 52: *The dual of Johnson solid 26 (the gyrobifastigium), an unnamed but very simple polyhedron.*

The four triangular faces are isosceles, and the four quadrilateral faces are parallelograms.

As was noted for the pyramids, the 3-4-5 trend rule (introduced on p. 12 in Chapter 2) applies to the bicupolae, a fact which will be apparent on looking at the figures.

It is interesting to look at the duals of the bicupolae ($J_{27-31}$) and the duals of the cupolae ($J_{3-5}$), because the bicupolae are obtained by fusing two cupolae together, and this comparison gives insight as to what, in general, the effects of fusion and dualization are when combined. For ease in discussion, note that in a dual, a vertex corresponds to a face of the original polyhedron. The fusion operation eliminates two faces, one on each polyhedron, and the corresponding vertices of the duals can be termed the "disappearing vertices." The vertices that are connected to the disappearing vertices will here be termed the "adjacent vertices" — specifically, adjacent *to the disappearing vertices*, but this will not be expressed in the following discussion. If one takes two copies of the dual of $J_3$ and removes the disappearing vertices, then takes the adjacent vertices in one cupola and the adjacent vertices of the other and joins them in corresponding pairs, one gets the dual of $J_{27}$ — at least in shape; the proportions will not be alike, because of differences in the reciprocation center used. The same is true if one takes two copies of the dual of $J_4$ and does the same operation: depending on their relative orientation, one gets something that looks like the dual of $J_{28}$ or $J_{29}$.

The triangular orthobicupola, Johnson solid 27, illustrated in Figure 53 below, is the first of the bicupolae just discussed.

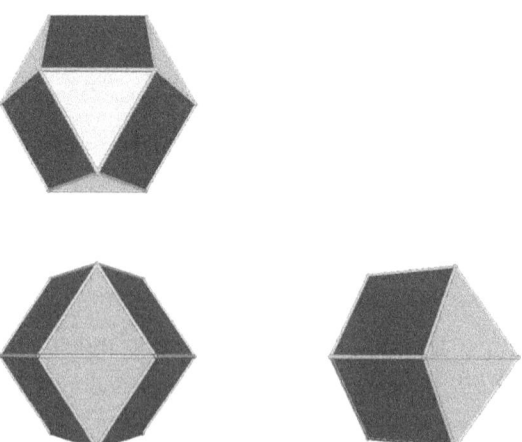

*Figure 53: Johnson solid 27: the triangular orthobicupola.*

As the name suggests, the triangular orthobicupola can be formed by fusing two triangular cupolae at their bases, taking care to orient their roof triangles alike. The triangular orthobicupola, of course, preserves the threefold axis of rotation of the original cupolae, as can be seen, but introduces a new mirror plane where they were fused, so the resulting symmetry is $D_{3h}$.

Because, as in previous chapters, the orthobicupolae can be related to each other by simply consulting the description of the triangular member of the class and altering it as appropriate for the square and pentagonal members, a relatively long description of the properties of the triangular orthobicupola will be inserted here. There are two triangular bases (derived from the roofs of the cupolae that were fused), which will be square or pentagonal in the case of the other orthobicupolae, and twelve lateral faces, six triangles and six squares, in two bands between the bases, each band consisting of three squares and three triangles. (In the description of the square and pentagonal orthobicupolae, "three," "six," and "twelve" would be replaced by "four" or "five," "eight" or "ten," and "sixteen" or "twenty," but otherwise the description of the lateral faces is unchanged.) The only distinction between the orthobicupolae and the gyrobicupolae that are also discussed in this chapter is that the two lateral bands of the *orthobicupolae* are oriented so that a triangular face in one band is adjacent to a triangular face in the other, while a square face in one band is adjacent to a square face in the other, while the *gyrobicupolae* show the opposite orientation, with a triangular face of either band adjacent to a square face of the other band. Altogether, the number of faces is fourteen for the triangular member of the family, eighteen for the square, and twenty-two for the pentagonal member, as can be seen by totaling the numbers just given. Since each base of the triangular orthobicupola has three vertices and the central row of vertices where the two cupolae were fused has six, the total number of vertices is twelve (and for the square and pentagonal members it is sixteen or twenty). The number of edges can be calculated by using the Euler formula, $V + F = E + 2$, without needing to go into detail, which can be seen to give twenty-four for the triangular, thirty-two for the square, and forty for the pentagonal member of the family. All these figures do not depend on orientation, so that the gyrobicupolae have the same face/vertex/edge counts. All this is included in Tables 15 and 16, below.

The dual of the triangular orthobicupola, illustrated in

Figure 54 below, is a figure with twelve faces, exactly half of which are rhombi and the remainder of which are isosceles trapezoids. Since any polyhedron with twelve faces can be termed a dodecahedron, although the most common such figure is the Platonic solid with five regular pentagons as faces (the *pentagon-dodecahedron* or *regular dodecahedron*) and the next most commonly encountered such figure is the Catalan solid (or Archimedean dual solid) with twelve rhombi as faces, it is necessary to distinguish *this* dodecahedron from these, and the term "trapezo-rhombic dodecahedron" has been used to describe this polyhedron.

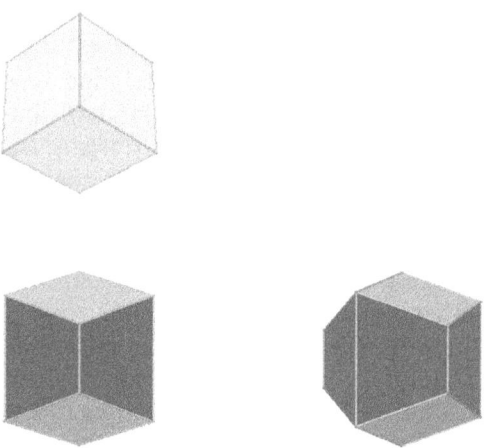

Figure 54: The dual of Johnson solid 27 (the triangular orthobicupola): a trapezo-rhombic dodecahedron.

The trapezo-rhombic dodecahedron obtained by dualizing the Johnson triangular orthobicupola has some interesting special properties: all the edges of the rhombi, of course, are equal, but in addition the longest edges of the isosceles trapezoids are exactly twice their shortest edges, with the other edges (which are also edges of the rhombi) as well as the edges between the rhombi, being halfway between those lengths. In other words, if the shortest edges of the isosceles trapezoids are equal to 2 units, all edges of the rhombi will be 3 and the longest edges of the isosceles trapezoids will be 4 units. Also, the angles of both kinds of quadrilaterals are the same; the acute angles are the same 70.5288° (70° 31′ 44″) as the dihedral

*Bicupolae, cupolarotundae, and birotundae: Johnson solids 26-34.*

angle in a regular tetrahedron, and the obtuse angles are the same 109.4712° (109° 28' 16") as the dihedral angle in a regular octahedron. The dihedral angles between any two faces are exactly 120°.

It will be seen that, following the 3-4-5 trend rule, as one goes from $J_{27}$ to $J_{31}$, the duals start off approximately spherical (in $J_{27}$) and end up very long and thin, resembling a pencil sharpened at both ends.

There is no "triangular gyrobicupola" among the Johnson solids, because it turns out that fusing two triangular cupolae in this orientation gives a cuboctahedron, an Archimedean solid. (The dual of that cuboctahedron, a rhombic dodecahedron, is exactly what one would expect by this procedure.) Therefore, the next Johnson solid in the list is the square orthobicupola, Johnson solid 28, illustrated in Figure 55 below.

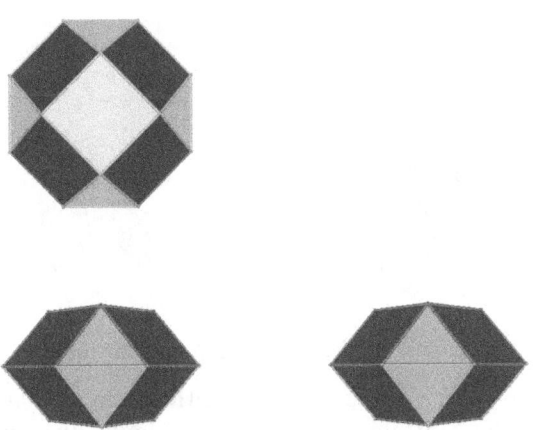

*Figure 55: Johnson solid 28: the square orthobicupola.*

As the name suggests, the square orthobicupola can be formed by fusing two square cupolae at their bases, taking care to orient their roof squares alike. The square orthobicupola, of course, preserves the fourfold axis of rotation of the original cupolae, as can be seen, but introduces a new mirror plane where they were fused, so the resulting symmetry is $D_{4h}$. The properties can be inferred from those of the triangular orthobicupola on p. 117 by simply allowing for the change from threefold to fourfold symmetry.

The dual of the square orthobicupola, illustrated in Figure 56 below, has sixteen faces, half of which are kite-quadrilaterals and half of which are isosceles trapezoids. As for the square orthobicupola itself, the properties of its dual can be inferred from those of the dual of the triangular orthobicupola on p. 117 by simply allowing for the change from threefold to fourfold symmetry. However, because of symmetry effects, the rhombi in the dual of the triangular orthobicupola become kite-quadrilaterals in the dual of the square orthobicupola.

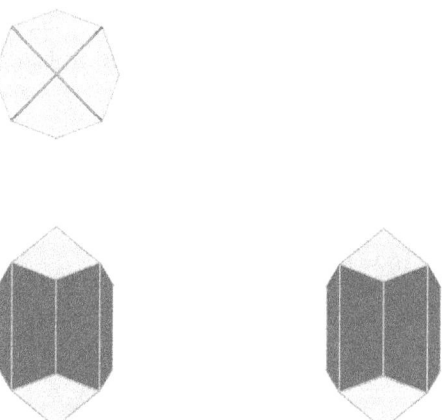

*Figure 56: The dual of Johnson solid 28 (the square orthobicupola).*

Because the symmetry is $D_{4h}$, with a mirror plane perpendicular to the fourfold axis, the bottom looks just like the top, and the lateral faces are all isosceles trapezoids, giving the polyhedron an appearance somewhat like an eraserless pencil, sharpened at both ends, as was noted earlier.

The square gyrobicupola, Johnson solid 29, is illustrated in Figure 57 below. When two cupolae (or two rotundae, as will be seen later) are fused in the "gyro" orientation, a mirror plane is not introduced, because the *n*-gonal vertices of the two roofs are not directly over each other; however, it can be seen that instead an *alternating axis* is introduced by the fusion, which will be 2*n*-fold. So, in the case of a *square* gyrobicupola, the symmetry will be $S_{8v}$. Except for the 45° rotation of the bottom cupola, the polyhedron is identical to the square orthobicupola, illustrated in Figure 55 above. The face/vertex/edge count is the same, but because the parameters of the square orthobicupola were only implied by referring to the triangular orthobicupola, with the assumed change from threefold to fourfold symmetry, it should be mentioned here that each of the two square cupolae contributes 10 faces, 12 vertices, and 20 edges. The two octagonal bases are eliminated in the fusion operation, so the total number of faces is 20 − 2 = 18. The eight vertices in each base are merged, so the 24 vertices of the two cupolae before fusion become 16, and the same happens to the eight edges that define each octagonal base, so the forty edges of the two cupolae become 32. (The same figures, 18 faces, 16 vertices, and 32 edges, would apply to the square orthobicupola, of course.)

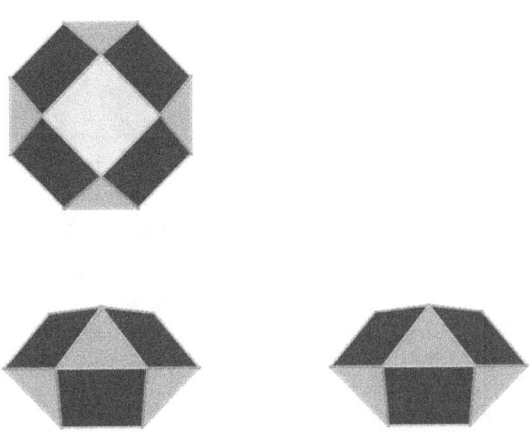

Figure 57: *Johnson solid 29: the square gyrobicupola.*

The dual of the square gyrobicupola, illustrated in Figure 58 below, is much like the dual of the square orthobicupola, except that the twist of the lower half by 45° compared to the upper causes the lateral faces, which were isosceles trapezoids in the case of the dual of the square orthobicupola, to be parallelograms. All the angles (face and dihedral), however are the same for both polyhedra; and the only difference is that in the isosceles trapezoidal faces of the dual of the square orthobicupola, equal angles are adjacent, while in the parallelogram faces of the dual of the square gyrobicupola, they are opposite. Of course, since any polyhedron and its dual have the same symmetry, the dual of the square gyrobicupola has $S_{8v}$ symmetry.

Comparing Figure 58 with Figure 56, the great similarity of these two polyhedra is very clear.

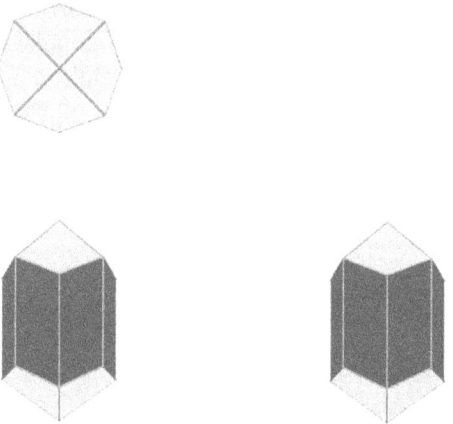

Figure 58: The dual of Johnson solid 29 (the square gyrobicupola).

Bicupolae, cupolarotundae, and birotundae: Johnson solids 26-34.

| Johnson solid number | 26 | 27 | 28, 29 |
|---|---|---|---|
| Name | Gyrobifastigium (digonal gyrobicupola) | Triangular orthobicupola | 28: Square orthobicupola<br>29: Square gyrobicupola |
| Symmetry* | $S_{4v}$ | $D_{3h}$ | 28: $D_{4h}$<br>29: $S_{8v}$ |
| Faces | [4](3) + [4](4) | [2+6](3) + [6](4) | [8](3) + [2+8](4) |
| Edges* | 14 | 24 | 32 |
| Vertices | 8 | 12 | 16 |
| Dihedral angles | 3a-4a: 90°,<br>3a-4b: 150°,<br>4a-4a: 60°♣ | 3-3: 141.0576°<br>(141° 3' 27"),<br>3-4: 125.2644°<br>(125° 15' 52"),<br>4-4: 109.4712°<br>(109° 28' 16")✱ | 3a-3b: 109.4712°<br>(109° 28' 16"),<br>3a-4a, 3b-4b: 144.7356°<br>(144° 44' 8"),<br>3a-4b: 99.7356°<br>(99° 44' 8"),<br>4a-4b: 90°,<br>4r-4: 135°♭ |
| Area | 5.73205081 = 4 + √3 | 9.46410162 = 6 + 2√3 | 13.46410162 = 10 + 2√3 |
| Volume | 0.86602540 = ½√3 | 2.35702260 | 3.88561808 |
| Dual | | | |
| Name | Unnamed | Unnamed | Unnamed |
| Faces | [4](3) + [4](4) | [6+6](4) | [8+8](4) |
| Vertices | 8 | 14 | 18 |
| Face angles† | 3: [2]49.1066°<br>(49° 6' 24"),<br>81.7868°<br>(81° 47' 12");<br>4: [2]112.2077°<br>(112° 12' 28"),<br>[2]67.7923°<br>(67° 47' 32") | 4i, 4r: [2]70.5288°<br>(70° 31' 44"),<br>[2]109.4712°<br>(109° 28' 16")✱ | 4i/4p: [2]69.0590°<br>(69° 3' 32"),<br>[2]110.9410°<br>(110° 56' 28");<br>4k: [2]85.0968°<br>(85° 5' 48"),<br>70.5288°<br>(70° 31' 44"),<br>119.2776°<br>(119° 16' 39")⊛ |

## The Johnson Solids and Their Duals

| Johnson solid number | 26 | 27 | 28, 29 |
|---|---|---|---|
| Dihedral angles | 3-3: 120°, 3-4: 110.7048° (110° 42′ 17″), 4-4: 90° | 4-4: 120°* | 4i-4i/4p-4p: 135°, 4i-4k/4k-4p: 130.7895° (130° 47′ 22″), 4k-4k: 120°⊗ |
| Area | 1.97119712 | 9.54594155 | 7.12338552 |
| Volume | 0.19245009 | 2.38648539 | 1.54415588 |

Table 15: Properties of Johnson solids 26, 27, 28, and 29 and their duals.

*The symmetry and number of edges are always the same for each solid and its dual, so are not repeated in the "dual" section.

†Face angles are only given for the duals, as the Johnson solids have regular faces, whose angles can be found in Table 1 on p. 3.

♦The designations 3a and 4a refer to the triangles and squares around one of the two edges that are considered digonal faces; 3b and 4b refer to the opposite one. (There is no need to say which is which, because of symmetry; the actual distinction is simply between *pairs* of faces at a common digon and pairs at opposite digons.)

⊗No distinction need be made between different types of triangles, or between different types of squares. All 3-3 dihedral angles are equal, as are all 3-4 and all 4-4 dihedral angles.

♭The designations 3a and 4a refer to the triangles and squares forming the lateral faces of one of the two cupolae involved in the fusion; 3b and 4b refer to the opposite one. (There is no need to say which is which, because of symmetry; the actual distinction is simply between *pairs* of faces in a common cupola and pairs in opposite cupolae.) The designation 4r refers to the two squares forming the roofs of the cupolae, and the 4a/4b distinction is unnecessary in this case. Only $J_{28}$ has 3a-3b and 4a-4b edges, and only $J_{29}$ has 3a-4b edges.

‡There are two kinds of quadrilateral face in this solid, an isosceles trapezoid and a rhombus. However, both have the same angles, only that equal angles are opposite each other in

the rhombic faces and in adjacent pairs in the isosceles trapezoids. Additionally, all dihedral angles are equal.

⁸The duals of Johnson solids 28 and 29 possess identical kite-quadrilateral-shaped faces, designated 4k. The remaining faces are isosceles trapezoids (designated 4i) in the dual of $J_{28}$ and parallelograms (designated 4p) in the dual of $J_{29}$, but the angles are the same in both, only differing in that pairs of equal angles are adjacent in the dual of $J_{28}$ and opposite in the dual of $J_{29}$. The dihedral angles between these faces are the same in both, as are the dihedral angles between the kite-quadrilaterals and these 4i/4p faces.

The pentagonal orthobicupola, Johnson solid 30, is illustrated in Figure 59 below. As the name suggests, it can be formed by fusing two pentagonal cupolae at their bases, taking care to orient their roof pentagons alike. The properties can be inferred from those of the triangular orthobicupola on p. 117 by simply allowing for the change from threefold to fivefold symmetry, as as the properties of the the square orthobicupola can be inferred from those of the the triangular orthobicupola on p. 117 by simply allowing for the change from threefold to fourfold symmetry. Comparisons among all three can be made by consulting Tables 15 (above) and 16 (below).

The pentagonal orthobicupola, of course, preserves the fivefold axis of rotation of the original cupolae, as can be seen, but introduces a new mirror plane where they were fused, so the resulting symmetry is $D_{5h}$.

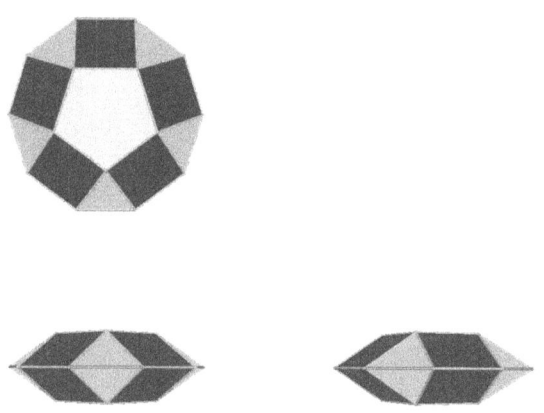

Figure 59: Johnson solid 30: the pentagonal orthobicupola.

The dual of the pentagonal orthobicupola is illustrated in Figure 60 below. As is the case for the pentagonal orthobicupola itself, the properties of its dual can be inferred from those of the dual of the triangular orthobicupola on p. 117 by simply allowing for the change from threefold to fivefold symmetry in the same way as the properties of the dual of the square orthobicupola can be inferred from those of the dual of the triangular orthobicupola, described on p. 117, by allowing for the change from threefold to fourfold symmetry, changing the appropriate numbers as has been described in other discussions of this type.

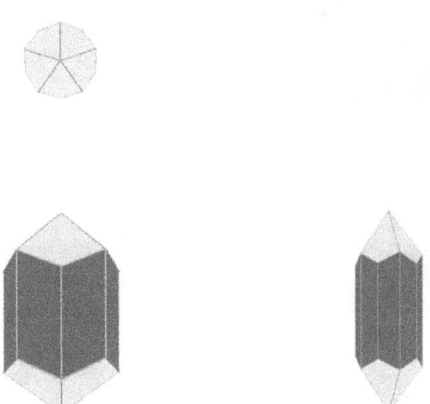

Figure 60: The dual of Johnson solid 30 (the pentagonal orthobicupola).

Comparisons among all three of the duals of the orthobicupolae can be made by consulting Tables 15 (above) and 16 (below).

The pentagonal gyrobicupola, Johnson solid 31, is illustrated in Figure 61 below. Except for the 36° rotation of the bottom cupola relative to the top one, the polyhedron is identical to the pentagonal orthobicupola, illustrated in Figure 59 above. The face/vertex/edge count is the same: 22 faces, 20 vertices, and 40 edges. As was noted in the discussion of the square gyrobicupola previously, the fusion process, generating this polyhedron by combining two pentagonal cupolae ($J_5$), introduces an alternating axis, in this case tenfold, so the symmetry of this polyhedron is $S_{10v}$.

As had been noted earlier, there is no triangular gyrobicupola, because a triangular gyrobicupola with all regular faces would be an Archimedean solid (a cuboctahedron), but comparisons of all the properties which are tabulated in this book among the two gyrobicupolae (square and pentagonal), as well as comparisons between the gyrobicupolae and the orthobicupolae, can easily be made by consulting Tables 15 (above) and 16 (below).

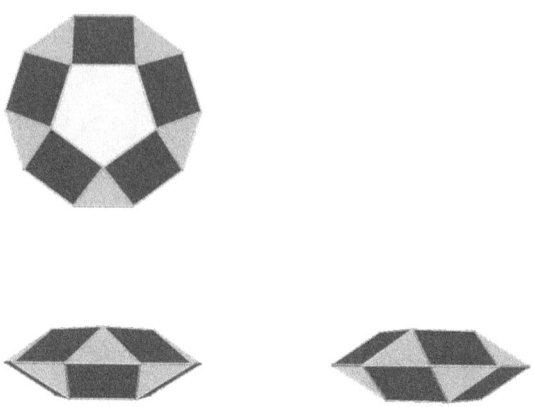

*Figure 61: Johnson solid 31: the pentagonal gyrobicupola.*

The dual of the pentagonal gyrobicupola, illustrated in Figure 62 below, is the same as the dual of the square gyrobicupola with the appropriate changes in the description arising from the change from fourfold to fivefold symmetry. It has the same $S_{10v}$ symmetry as the pentagonal gyrobicupola itself, of course, and most of the numerical values of the properties tabulated in this book are the same as those of the dual of the pentagonal orthobicupola, as can be seen in Table 16. Comparison with the dual of the square gyrobicupola described in Table 15 (and with the rhombic dodecahedron which is the dual of the cuboctahedron) illustrates the 3-4-5 trend rule, as is normal.

All of the bicupolae, it can be seen, have duals that can be considered to consist of three pieces, joined together: a central belt and two apical caps. (The joining is not a *fusion*, as I defined it in my earlier book, *Polyhedra: a New Approach*, because the pieces do not terminate in planes. Rather, one should simply see two zigzag sets of vertices and edges as dividing each polyhedron into these three sets of faces.) While the central belt of the duals of all the orthobicupolae consists of isosceles trapezoids, the central belt of the duals of all the gyrobicupolae (*including* the gyrobifastigium) consists of parallelograms. The apical caps of all (with the *exception* of the gyrobifastigium) are formed of kite-quadrilaterals (thus are antibipyramid-like apices), but the gyrobifastigium instead has pairs of isosceles triangles as caps.

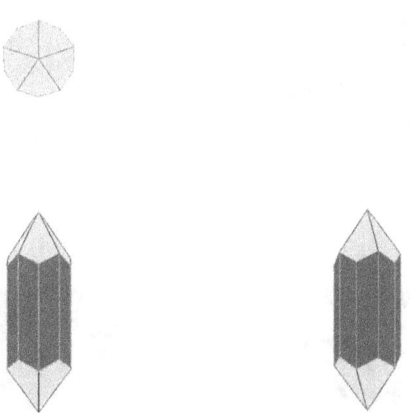

Figure 62: The dual of Johnson solid 31 (the pentagonal gyrobicupola).

Oddly, it should be noted that the dihedral angles between the kite-quadrilaterals at the apices remain a constant 120°, regardless of the number of those kite-quadrilaterals, as can be seen in Tables 15 and 16.

This concludes the discussion of the Johnson solids that can be obtained by fusion of two identical cupolae (including the triangular prism considered as a digonal cupola). As there is some similarity between the cupolae and the pentagonal rotunda (which is why they were treated together in Chapter 3), this chapter will also consider the Johnson solids in which either a cupola and a rotunda or two rotundae are fused. When a cupola and a rotunda are fused, of course, it is not correct to name the polyhedron as a "bi-" anything, so the two words are combined as "cupolarotunda" in the names of the next two Johnson solids.

The pentagonal orthocupolarotunda, Johnson solid 32, illustrated in Figure 63 below, is one of two polyhedra obtainable by the fusion of a pentagonal cupola ($J_5$) and a pentagonal rotunda

($J_6$) at their decagonal faces. Since it is not possible to form a triangular or square rotunda within the Johnson requirement of all faces being regular polygons, it is only the pentagonal cupola, not the triangular or square cupolae, that can be fused to a corresponding rotunda in this way.

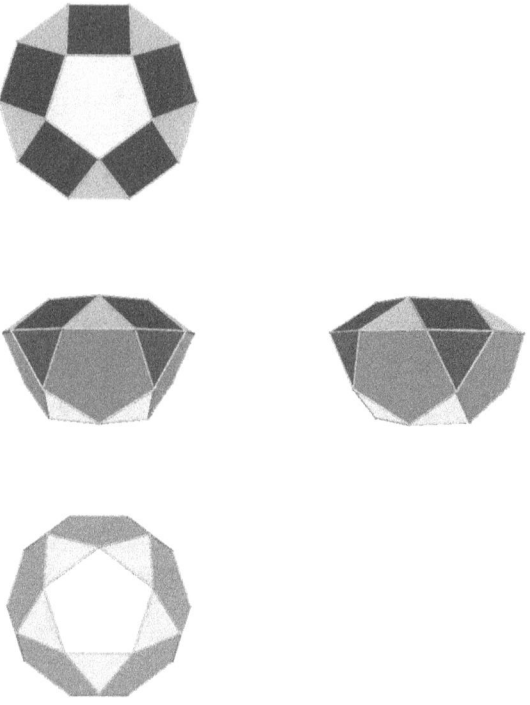

*Figure 63: Johnson solid 32: the pentagonal orthocupolarotunda.*

The cupola contributes twelve faces, fifteen vertices, and twenty-five edges; the rotunda, seventeen faces, twenty vertices, and thirty-five edges. As was done in previous discussions involving Johnson solids obtainable by fusion, one can inventory the faces/vertices/edges in this way: the two polyhedra combined contribute 29 faces, 35 vertices, and 60 edges. Two faces are lost in the fusion, leaving a total of 27 remaining. Ten vertices at the base of the cupola merge with ten at the base of the rotunda, so the number of vertices after the fusion is 25. And a similar merger of ten edges at each base occurs, so the resulting polyhedron has 50 edges. (Note that this does not de-

*Bicupolae, cupolarotundae, and birotundae: Johnson solids 26-34.*

pend on whether the cupola and rotunda are in an "ortho" or a "gyro" orientation, so the same count will apply to the pentagonal gyrocupolarotunda, next in the Johnson solid list.) One can also note that the dihedral angles between all faces of the cupola and also between all faces of the rotunda (with trhe exception of those angles involving the decagonal bases) remain unchanged by the fusion, and the angles at the base are simply added, so one can compare the numbers in Table 16 with those in Tables 4 and 5.

The dual of the pentagonal orthocupolarotunda is illustrated in Figure 64 below. It is to be noted that unlike the duals of the bicupolae, which have central sections with all parallel edges, which are also parallel to the principal axis of rotation (of course this is required by the $S_{2nv}$ or $D_{nh}$ symmetry!), the duals of the cupolarotundae are rather pear-shaped, with the part coming from the rotunda distinctly wider than the part coming from the cupola. This is an artifact of the larger number of vertices which the rotunda contributes as compared with the cupola, so when one determines the centroid of the cupolarotunda for use as the center of polar reciprocation, it is not in the fusion plane but slightly below (using terminology that describes the cupola side as "top" and the rotunda side as "bottom," as the orientation of Figure 63 implies.) If the polar reciprocation is done using a center that is actually in the center of the fusion plane, as opposed to the convention followed in this book where the centroid of the vertices is used as the center of the polar reciprocation process, this lateral band *does* have its edges parallel to the principal axis of rotation, and the polygons become parallelograms (or in the case of $J_{33}$, isosceles trapezoids), while the antibipyramid-like apex on the cupola side becomes taller than the one which is shown in Figure 64. In such a case, the dual becomes much more like a combination of the dual of a pentagonal bicupola and the dual of a pentagonal birotunda (choosing, as appropriate, the "ortho" or "gyro" orientation), and the reader is invited to look at those illustrations, Figures 60, 62, and 68 (There is no "pentagonal gyrobirotunda" because it is in fact an Archimedean solid, the icosidodecahedron, but one can imagine what its dual looks like by considering the dual of the pentagonal orthobirotunda and applying the 36° twist as usual) to see what those look like. In this book, however, there are enough illustrations of duals which were generated by polar reciprocation using the centroid of the vertices as center that it would be confusing to include additional illustrations where some other center is used.

Figure 64: The dual of Johnson solid 32 (the pentagonal orthocupolarotunda).

It might also be noted that the primary difference, other than this pear-shape, between this figure and the bicupola duals is that the end that derives from the rotunda has an extra set of kite-quadrilaterals at the rotunda end, forming the lateral kite/apical kite arrangement first discussed when discussing the dual of $J_6$ earlier. And it can be seen that the apical/lateral kite rule, discussed on p. 30, applies, relating the acute and obtuse angles of the apical and lateral kites and noting that both are close to rhombi in shape.

The pentagonal gyrocupolarotunda, Johnson solid 33, illustrated in Figure 65 below, is identical to the pentagonal orthocupolarotunda previously discussed, except for the fact that the rotunda and the cupola have a relative orientation rotated by 36° from that of the orthocupolarotunda.

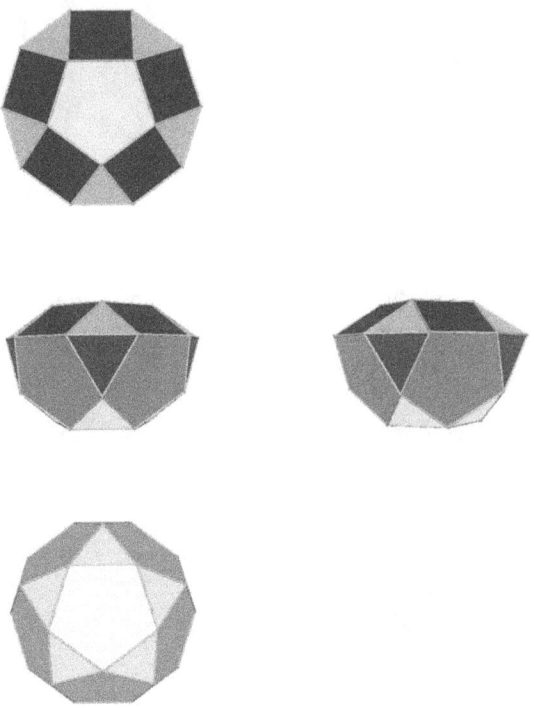

Figure 65: Johnson solid 33: the pentagonal gyrocupolarotunda.

The dual of the pentagonal gyrocupolarotunda, illustrated in Figure 66 below, has the same description as the dual of the pentagonal orthocupolarotunda, except for the same twisting of one end relative to the other that has been seen in other "ortho"/"gyro" pairs. Each of the irregular quadrilaterals comprising the belt around the middle has exactly the same angles as any of the irregular quadrilaterals comprising the belt around the middle of $J_{32}$, but they are arranged differently, so that the shapes of the two differ; but while, in the case of the duals of the bicupolae, this makes the difference between parallelograms and isosceles trapezoids, in the case of the duals of the cupolarotundae, both sets of belt quadrilaterals are irregular, neither parallelograms nor isosceles trapezoids. (When we get to the duals of the birotundae, they would be parallelograms and isosceles trapezoids again, except that the pentagonal gyrobirotunda is not a Johnson solid.)

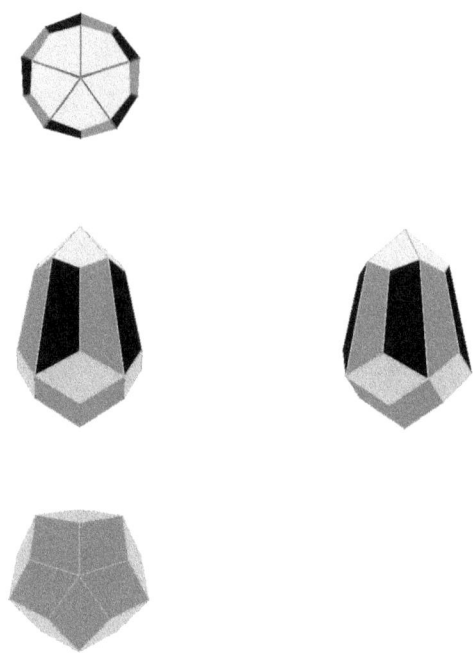

Figure 66: The dual of Johnson solid 33 (the pentagonal gyrocupolarotunda).

*Bicupolae, cupolarotundae, and birotundae: Johnson solids 26-34.*

| Johnson solid number | 30, 31 | 32, 33 |
|---|---|---|
| Name | 30: Pentagonal orthobicupola<br>31: Pentagonal gyrobicupola | 32: Pentagonal orthocupolarotunda<br>33: Pentagonal gyrocupolarotunda |
| Symmetry* | 30: $D_{5h}$<br>31: $S_{10v}$ | $C_{5v}$ |
| Faces | [10](3) + [10](4) + [2](5) | [5+5+5](3) + [5](4) + [1+1+5](5) |
| Edges* | 40 | 50 |
| Vertices | 20 | 25 |
| Dihedral angles | 3a-3b: 74.7547° (74° 45' 17"),<br>3a-4a, 3b-4b: 159.0948° (159° 5' 41"),<br>3a-4b, 3b-4a: 69.0948° (69° 5' 41"),<br>4a-4b: 63.4349° (63° 26' 6"),<br>4-5: 148.2825° (148° 16' 57")♣ | 3c-3r: 116.5651° (116° 33' 54"),<br>3r-5r: 142.6226° (142° 37' 21"),<br>3c-4: 159.0948° (159° 5' 41"),<br>3r-4: 110.9052° (110° 54' 19"),<br>3c-5r: 100.8123° (100° 48' 44"),<br>4-5c: 148.2825° (148° 16' 57"),<br>4-5r: 95.1524° (95° 9' 9")* |
| Area | 17.77108182 = 10 + $\sqrt{[25 + (5/2)\sqrt{5} + (5/2)\sqrt{(75 + 30\sqrt{5})}]}$ | 23.53853233 |
| Volume | 4.64809064 | 9.24180829 |
| **Dual** | | |
| Name | Unnamed | Unnamed |
| Faces | [10+10](4) | [5+5+5+10](4) |
| Vertices | 22 | 27 |

## The Johnson Solids and Their Duals

| Johnson solid number | 30, 31 | 32, 33 |
|---|---|---|
| Face angles[†] | 4i/4p: [2]63.4349° (63° 26' 6"), [2]116.5651° (116° 33' 54"); 4k: [2]100.8123° (100° 48' 44"), 41.8103° (41° 48' 37"), 116.5651° (116° 33' 54")[⊗] | 4a: 57.2234° (57° 13' 24"), 72.7179° (72° 43' 5"), 114.9329° (114° 55' 58"), 115.1258° (115° 7' 33"); 4b: [2]92.7936° (92° 47' 37"), 54.7936° (54° 47' 48"), 119.6162° (119° 36' 58"); 4c: [2]117.1208° (117° 7' 15"), 58.6317° (58° 37' 54"), 67.1266° (67° 7' 36"); 4d: [2]64.6742° (64° 40' 27"), 111.3441° (111° 20' 39"), 119.3075° (119° 18' 27")[‡] |
| Dihedral angles | 4i-4i/4p-4p: 144°, 4i-4k/4k-4p: 144°, 4k-4k: 120°[⊗] | 4a-4a: 144.7960° (144° 47' 46"), 4a-4b: 143.0355° (143° 2' 8"), 4a-4d: 143.2338° (143° 14' 2"), 4b-4b: 131.3520° (131° 21' 7"), 4c-4c: 136.2018° (136° 12' 7"), 4c-4d: 138.5946° (138° 35' 40")[‡] |
| Area | 6.95288237 | 10.71415555 |
| Volume | 1.32788237 | 2.94738245 |

Table 16: Properties of Johnson solids 30, 31, 32, and 33 and their duals.

[*]The symmetry and number of edges are always the same for each solid and its dual, so are not repeated in the "dual" section.

[†]Face angles are only given for the duals, as the Johnson solids have regular faces, whose angles can be found in Table 1

## Bicupolae, cupolarotundae, and birotundae: Johnson solids 26-34.

on p. 3.

♦The designations 3a and 4a refer to the triangles and squares forming the lateral faces of one of the two cupolae involved in the fusion; 3b and 4b refer to the opposite one. (There is no need to say which is which, because of symmetry; the actual distinction is simply between *pairs* of faces in a common cupola and pairs in opposite cupolae.) The 4a/4b distinction is unnecessary in the case of the 4-5 edge. Only $J_{30}$ has 3a-3b and 4a-4b edges, and only $J_{31}$ has 3a-4b edges.

°Although the triangular faces fall into three different transitivity classes, and so do the pentagonal faces, both kinds of triangle from the rotunda make the same dihedral angle with the pentagonal faces, and both kinds of pentagon from the rotunda make the same dihedral angle with the triangular faces; only one type of rotunda pentagon, and only one kind of rotunda triangle, shares an edge with a cupola triangle or with a square face. It is therefore sufficient to designate the triangular faces as 3c and 3r, and the pentagonal faces as 5c and 5r, depending on whether they derive from the cupola or the rotunda. It is also to be noted that only $J_{32}$ has 3r-4 or 3c-5r edges, while only $J_{33}$ has any 4-5r or 3c-3r edges.

⊛The duals of Johnson solids 30 and 31 possess identical kite-quadrilateral-shaped faces, designated 4k. The remaining faces are isosceles trapezoids (designated 4i) in the dual of $J_{30}$ and parallelograms (designated 4p) in the dual of $J_{31}$, but the angles are the same in both, only differing in that pairs of equal angles are adjacent in the dual of $J_{30}$ and opposite in the dual of $J_{31}$. The dihedral angles between these faces, and the dihedral angles between the kite-quadrilaterals and these 4i/4p faces, are the same in both.

*The designation 4a refers to the band of ten irregular quadrilaterals around the middle of the polyhedron. The designations 4b, 4c, and 4d refer to three different sets of kite-quadrilaterals: 4b to the five around the apex derived from the cupola roof; 4c to the five around the apex derived from the rotunda roof (apical kites, as defined on p. 30); 4d to the other five (lateral kites, as defined on p. 30), which are derived from the rotunda and separate the 4a faces from the 4c faces.

The pentagonal orthobirotunda, Johnson solid 34, illustrated in Figure 67 below, is the only birotunda among the Johnson solids. As was noted in Chapter 3 above, the only rotunda that can be constructed with all regular polygons as faces is the pentagonal rotunda, Johnson solid 6. And fusing two pentagonal rotundae in the "gyro" orientation does not produce a Johnson solid, because such a fusion is in fact an icosidodecahedron, an *Archimedean* solid. The pentagonal orthobirotunda, of course, preserves the fivefold axis of rotation of the original rotundae, as can be seen, but introduces a new mirror plane where they were fused, so the resulting symmetry is $D_{5h}$.

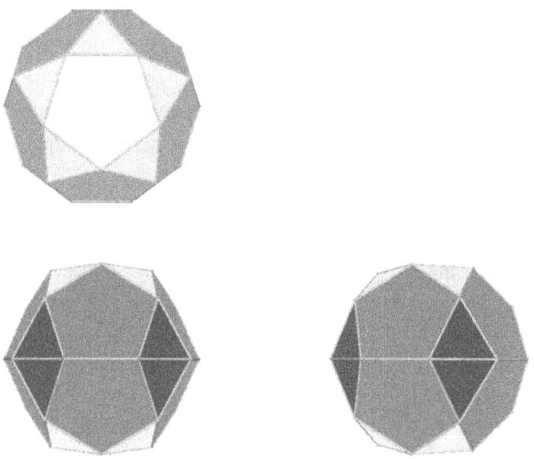

*Figure 67: Johnson solid 34: the pentagonal orthobirotunda.*

Because the icosidodecahedron is an Archimedean solid, with all its edges identical, slicing it into two congruent halves, rotating one by 36°, and fusing the pieces together again leaves all the 3-5 dihedral angles which remain equal to each other. In addition, while it creates new 3-3 and 5-5 edges along the fusion plane, it can be seen that in fact the dihedral angle at the 3-3 edge must be *greater* than the 3-5 dihedral angle by exactly the same amount by which the 5-5 dihedral angle is *less* than the 3-5 dihedral angle. (In fact, each of the pieces is a pentagonal rotunda, $J_6$, with all its 3-5 angles equal to 142.6226° [142° 37' 21"], the same as the dihedral angle of the icosidodecahedron. The 3-10 and 5-10 dihedral angles must sum to the same amount, since when the pieces are fused to form an icosidodecahedron, one 3-10 and one 5-10 edge fuse to become an edge of the icosidodecahedron. These values are given as 79.1877° [79° 11′ 16″] and 63.4349° [63° 26′ 6″] in Table 5 in Chapter 3. When the pieces are rotated by 36°, two 3-10 edges are fused and two 5-10 edges are, which leads to the values being equally spaced above and below the value of the 3-5 dihedral angle.)

The dual of the pentagonal orthobirotunda, illustrated in Figure 68 below, has the same $D_{5h}$ symmetry as $J_{34}$, since dualization preserves symmetry. All of its faces are quadrilaterals; all except for the ten forming the "belt" are rhombi.

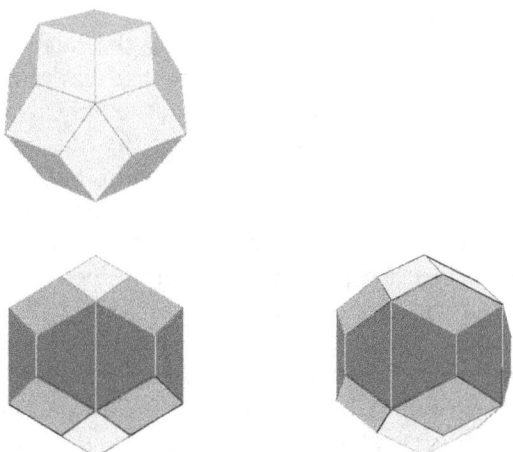

*Figure 68: The dual of Johnson solid 34 (the pentagonal orthobirotunda).*

The "belt" polygons are isosceles trapezoids, whose obtuse angles are equal to the obtuse angles of the rhombi, at 116.5651° (116° 33′ 54″), and whose acute angles are equal to the acute angles of the rhombi at 63.4349° (63° 26′ 6″).

And at this point we can now explain the origin of the apical/lateral kite rule, first enunciated on p. 30. As noted, there is no pentagonal gyrobirotunda among the Johnson solids, because fitting two pentagonal rotundae together in the "gyro" orientation produces an icosidodecahedron, an Archimedean solid. The dual of the icosidodecahedron is a Catalan solid, the rhombic triacontahedron. All Catalan solids have identical faces, and in the case of the rhombic triacontahedron, these are thirty rhombi with angles of 63.4349° (63° 26′ 6″) and 116.5651° (116° 33′ 54″). Forming $J_{34}$, one simply slices the icosidodecahedron in half and rotates one half relative to the other. As we have seen in the cases of other "ortho"/"gyro" pairs, this does not change any face angles in the dual, because the centroid, used as the center of polar reciprocation in forming the dual, does not change. Only the rhombi along the slice are turned into isosceles trapezoids because one half of each becomes reversed in orientation. If we take only one of these halves, however, either by itself to make $J_6$, or fusing it to something else which makes some of the other Johnson solids with a rotunda as part of its faces, the centroid, which, as was just noted, is used as the center of polar reciprocation in forming the dual, has moved, but remains along the one fivefold axis that was retained when the icosidodecahedron was cut in half. (The only symmetry elements retained are this fivefold axis and the five mirror planes through that axis.) This distorts the rhombi, but only a small amount, so the five nearest to the fivefold axis become apical kites and the remaining five become lateral kites. (They will be *kites*, rather than *irregular* quadrilaterals, because a mirror plane bisects them through two vertices. They will not remain rhombi, because the centroid no longer has the same relationship to the vertices of these quadrilaterals that it had in the icosidodecahedron.) Each will have two angles that will be close to the 63.4349° (63° 26′ 6″) angles of the original rhombic faces of the triacontahedron and two that will be close to the 116.5651° (116° 33′ 54″) angles. The two that are constrained to be equal because they are on opposite sides of a mirror plane (the acute angles of the lateral kites, the obtuse angles of the apical kites) will remain equal; the other two have no such constraint, so they can differ.

Since the departure of all these faces from precise rhombi and the departure of the apical and the lateral ones from exact congruence from each other is related to the departure of the centroid of the polyhedron whose dual it is from where it was in the original icosidodecahedron or birotunda, the more that centroid is diverted from the fusion plane, the further the apical and lateral kites will be from congruence with each other, and from being rhombi. The greatest move of the centroid, of course, is if *nothing at all* is fused to the rotunda, which is why $J_6$ has the greatest departure from having the apical and lateral kites congruent with each other, and from their being rhombi.

# The Johnson Solids and Their Duals

| Johnson solid number | 34 |
|---|---|
| Name | Pentagonal orthobirotunda |
| Symmetry* | $D_{5h}$ |
| Faces | [10+10](3) + [2+10](5) |
| Edges* | 60 |
| Vertices | 30 |
| Dihedral angles | 3-3: 158.3754° (158° 22' 31"),<br>3-5♣: 142.6226° (142° 37' 21"),<br>5-5: 126.8699° (126° 52' 12") |
| Area | 29.30598284 |
| Volume | 13.83552594 |
| **Dual** | |
| Name | Unnamed |
| Faces | [10+10+10](4) |
| Vertices | 32 |
| Face angles† | 4i, 4r: [2]63.4349 (63° 26' 6"),<br>[2]116.5651° (116° 33' 54")° |
| Dihedral angles | 4-4: 144°△ |
| Area | 30.33813729 |
| Volume | 14.80021243 |

Table 17: Properties of Johnson solid 34 and its dual.

*The symmetry and number of edges are always the same for each solid and its dual, so are not repeated in the "dual" section.

†Face angles are only given for the duals, as the Johnson solids have regular faces, whose angles can be found in Table 1 on p. 3.

♣Though there are three different types of 3-5 edges, all the 3-5 dihedral angles are equal.

°There are two different transitivity classes of rhombi which are congruent to each other, but oriented differently: one consists of two sets of five arranged around each apex, with their long axes in the five mirror planes that contain the fivefold axis; the other consists of two sets of five immediately adjacent to them, with their short axes in those five mirror planes. Additionally, there are ten isosceles trapezoids, all of them bisecting the remaining mirror plane (the one which is perpendicular to the fivefold axis). All thirty quadrilaterals, however, have the same angles; the acute angles are adjacent and so are the obtuse angles in the isosceles trapezoids, while equal angles are opposite in the rhombi.

△All the dihedral angles in this solid, regardless of what kind of edges they are in terms of transitivity classes of the adjacent faces, are identical at 144°.

## Chapter 9: Elongated bicupolae, cupolarotundae, and birotundae: Johnson solids 35-43.

After seeing Chapter 6, which describes the elongated and gyroelongated bipyramids, it is rather obvious what is meant by elongated and gyroelongated bicupolae and birotundae. For an elongated bicupola or birotunda, you put two identical cupolae or rotundae with their bases facing each other, and (unlike the bicupolae and birotundae in Chapter 8) between them, before fusing them, insert a prism, which may be termed the "belt." (The gyroelongated bicupolae and birotundae, formed by using an antiprism rather than a prism as the "belt," will be treated in Chapter 10.)

With cupolae and rotundae, however, unlike the case of pyramids, it is possible to fit these three pieces together in two different ways. While the bipyramids have belts that are *n*-gonal prisms, since their bases are *n*-gons, the cupolae and rotundae have 2*n*-gonal bases, and half the edges at those bases are different from the other edges. In the case of the bicupolae, half those edges are between *triangles* and the 2*n*-gonal base, while the other half are between *squares* and the base. If the two cupolae are oriented so that half of the faces of the prism forming the belt are adjacent to triangles on each cupola, while the other half are adjacent to squares on each cupola, the fused polyhedron is an *elongated orthobicupola*; if they are oriented so that each face of the prism forming the belt is adjacent to a triangle on one cupola and a square on the other cupola, the fused polyhedron is an *elongated gyrobicupola*. (In Chapter 8, a somewhat different definition was given, involving the orientation of the *roofs* of the cupolae. In fact, both definitions are really equivalent as regards the elongated bicupolae, but it is easier to visualize the orientations in this case by looking at the belt and the adjacent faces of the cupolae.) A glance at Figures 69 and 71 below will make this difference clear.

The first of the Johnson solids to be treated in this chapter is the elongated triangular orthobicupola, Johnson solid 35. It is illustrated in Figure 69 below. The symmetry is $D_{3h}$, and most of what is necessary to characterize it has already been stated in the previous paragraph.

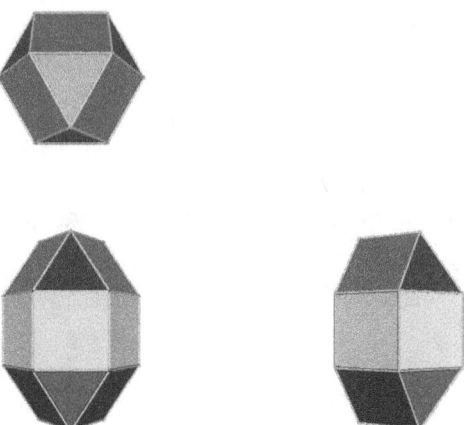

Figure 69: Johnson solid 35: the elongated triangular orthobicupola.

The dual of the elongated triangular orthobicupola is illustrated in Figure 70 below. It may be noted that all of its faces are quadrilaterals, and this will be seen to be true of every elongated bicupola dual in this chapter.

This dual of the elongated triangular orthobicupola illustrates a phenomenon that is true of every one of the Johnson solid duals in this chapter. As is easy to see, it consists of two pieces fused together. In this case, the two pieces are identical, as they will be whenever the two pieces fused to the prism in the original Johnson solid are identical; only the two elongated pentagonal cupolarotundae ($J_{42}$ and $J_{43}$) differ here, but it is still true that the duals of $J_{42}$ and $J_{43}$ are describable as the fusion of two pieces; the pieces simply differ. In the specific case of $J_{35}$, if one were to take either of the pieces of the dual separately, and dualize it, the figure that would be obtained would be a *triangular cupolapyramid* (or *basally augmented triangular cupola*), a figure that cannot be constructed as a Johnson solid (because it has six triangles all meeting at its apex) but can easily be visualized as a fusion of a triangular cupola and a hexagonal pyramid. In the case of every one of the Johnson solids that are described in this chapter, in fact the same thing can be done, and every dual of a Johnson solid described in this chapter is recognizable as a fusion of two parts at a median plane; and even for polyhedra that are not Johnson solids this is the case:

If a polyhedron is obtainable by elongation of a polyhedron which can be described as the fusion of two *component polyhedra* (i. e., insertion of a prism between the two components that were fused), its dual will be itself a fusion of two components; each of those components will be itself the dual of a polyhedron which can be described as the fusion of a pyramid with one of the component polyhedra of the original (unelongated) fusion. (One could describe those pyramids which were fused to the component polyhedra as lost in the process.)

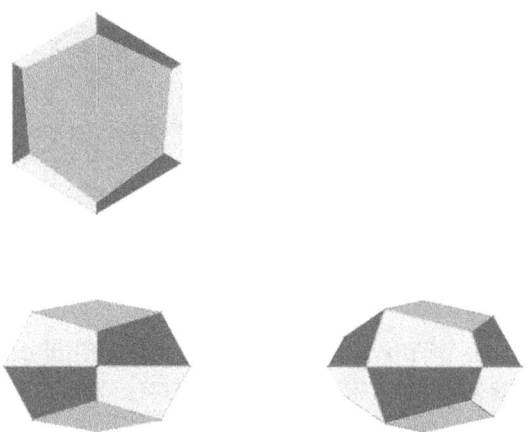

Figure 70: The dual of Johnson solid 35 (the elongated triangular orthobicupola).

Alternatively, this can be stated:

If two polyhedra each consist of a fusion of a pyramid with some other polyhedron, and the two pyramids are identical, then considering the remainder of each as a component, a polyhedron containing both components fused to the bases of a prism has a dual which is itself the fusion of the duals of the two original polyhedra.

Either of these two statements can be termed the *fusion-*

*elongation-dual rule*, and this will be referred to as needed in the remainder of this book.

In Chapter 6, this phenomenon was observed in the case of the elongated bipyramids, but some of the details could not be easily observed. First of all, the component polyhedra are *pyramids*, which cannot be distinguished from the pyramids that get lost in the fusion. Second of all, combination of those pyramids with the pyramids that get lost produces *bipyramids*, whose duals are prisms, and those prisms become distorted to frusta of pyramids in the process of fusing them, so the relationships described in these observations are obscured.

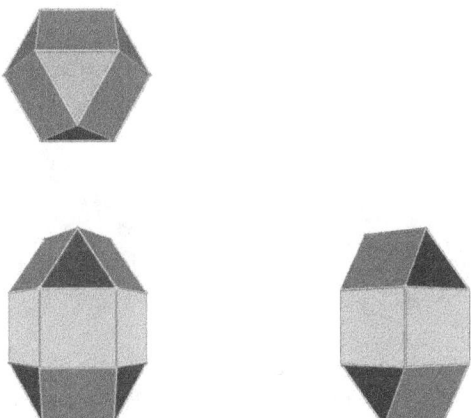

Figure 71: Johnson solid 36: the elongated triangular gyrobicupola.

It might be noted that all the parts of $J_{35}$ and those of $J_{36}$ are in fact identical, except for the twisted orientation of the cupolae compared to each other, so all the statistics (angles, area, volume, etc.) of the two are identical. However, the *symmetries* of the two are quite different. There is a mirror plane through the belt of squares in $J_{35}$, as well as the three mirror planes through the threefold axis perpendicular to the one through the belt. Thus the symmetry of $J_{35}$ is $D_{3h}$. However, the threefold axis of $J_{36}$ is actually an alternating axis (a *sixfold* alternating axis) so that $J_{36}$ in fact has $S_{6v}$ symmetry. (In fact, it is the *only* one of the ninety-two Johnson solids with $S_{6v}$ symmetry.)

The dual of the elongated triangular gyrobicupola is illustrated in Figure 72 below. All the comments about the symmetry of the elongated triangular ortho- and gyrobicupola given above, of course, apply to the duals. Of course, the fusion-elongation-dual rule, stated on p. 146 above, applies here.

Comparison of Figure 72 with Figure 70 is instructive. The entire difference is just a 60° twist of the lower half relative to the upper; all the pieces — the kite-quadrilateral caps and the irregular quadrilaterals that form the central belt — are shaped exactly the same in both polyhedra.

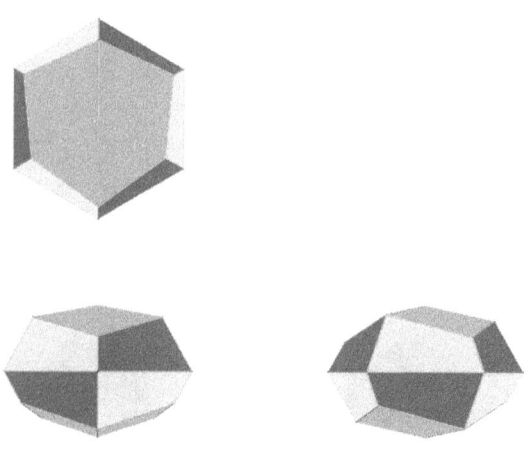

*Figure 72: The dual of Johnson solid 36 (the elongated triangular gyrobicupola).*

Johnson solid 37, the elongated square gyrobicupola, has been the occasion of some controversy. It is also referred to as Miller's solid after J. C. P. Miller, who accidentally constructed it, or as the pseudorhombicuboctahedron for reasons to be described immediately. An elongated square *ortho*bicupola is an Archimedean solid, the (small) rhombicuboctahedron. In fact it was in an attempt to make a model of that solid that Miller discovered the one that Johnson ultimately listed as $J_{37}$. (All three names, elongated square gyrobicupola, Miller's solid, and pseudorhombicuboctahedron, are given on p. 91 in Cromwell's book, *Polyhedra*, one of the best books on the subject.) There are people who would class $J_{37}$ as an Archimedean solid as well, because all its faces are regular polygons and all its vertices are identical 3•4•3•4, but others insist that, since the vertices are not all interconvertible by a symmetry transformation of the whole polyhedron, it should not be considered as an Archimedean solid. (As a result, some claim that the number of Archimedean solids is 13, and others give the number as 14.) Since Johnson included this polyhedron in his list, he obviously did not count it among the Archimedean solids. The solid is illustrated in Figure 73 below.

As was noted in the discussion of $J_{36}$, the fourfold axis of each of the cupolae involved in the fusion that produces $J_{37}$ becomes, in $J_{37}$, an alternating axis (an *eightfold* alternating axis) so that $J_{37}$ in fact has $S_{8v}$ symmetry.

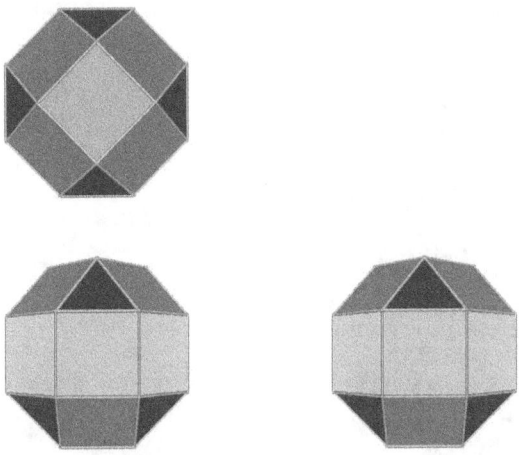

Figure 73: *Johnson solid 37: the elongated square gyrobicupola, also known as Miller's solid or the pseudorhombicuboctahedron.*

In my earlier book, *Polyhedra: a New Approach*, I stated just this: that there were 13 undisputed Archimedean solids and a fourteenth that some did, and some did not, include, and a reviewer criticized me, saying in effect that I was accepting Miller's solid as an Archimedean solid unjustifiedly, but I was simply asserting something that is an indisputable fact: that some do, and some do not, include it among the Archimedean solids. I am not claiming one way or the other: it just depends on how one wants to define the term "Archimedean solid."

The dual of the elongated square gyrobicupola is illustra-

*The Johnson Solids and Their Duals*

ted below in Figure 74. As is the case for $J_{37}$ itself, its dual has a lot of symmetry, more than most of the Johnson duals. (Of course, any dual has the same symmetry as the polyhedron whose dual it is.) As Johnson solid 37 is *almost* an Archimedean solid, its dual is *almost* a Catalan solid. Like the Catalan solids, it has all faces identical, and it also has the dihedral angles at all of its edges equal. (However, there are two types of edge because the edges have two different *lengths*.) The faces can be described as kite-quadrilaterals, as they have each two opposite angles equal, but beyond the typical kite-quadrilateral, one of the other angles of each is equal to those two. There is, however, no name I know of that specifically refers to a quadrilateral with three equal angles. As $J_{37}$ is sometimes called a pseudorhombicuboctahedron, its dual may be called a pseudodeltoidal icositetrahedron.

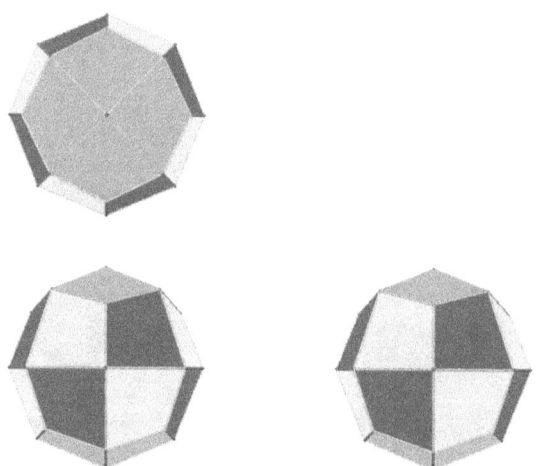

*Figure 74: The dual of Johnson solid 37 (the elongated square gyrobicupola).*

So just as in the case of $J_{37}$, its dual has $S_{8v}$ symmetry, not higher, but there are a lot more equal parts than the $S_{8v}$ symmetry requires. Of course, it was noted in the discussions of the duals of $J_{35}$ and $J_{36}$ that all the faces of the one were similarly shaped to those of the other, with just a 60° twist of the lower half. So it perhaps should not be a total surprise that the dual of $J_{37}$ should be obtained in the same way by simply taking the Catalan solid, the deltoidal icositetrahedron (which is the dual of the rhombicuboctahedron), cutting it in half, and rotating the lower half by 45°. In addition, one should note that the fusion-elongation-dual rule, stated on p. 146 above, applies here, providing another way of decomposing this polyhedron.

The elongated pentagonal orthobicupola, Johnson solid 38, illustrated in Figure 75 below, is built analogously to the elongated triangular orthobicupola, except for those differences associated with the change from threefold to fivefold symmetry. The symmetry is $D_{5h}$ rather than $D_{3h}$, and the two bases are pentagonal instead of triangular, with the appropriate changes in the counts of the triangular and square faces between the two bases.

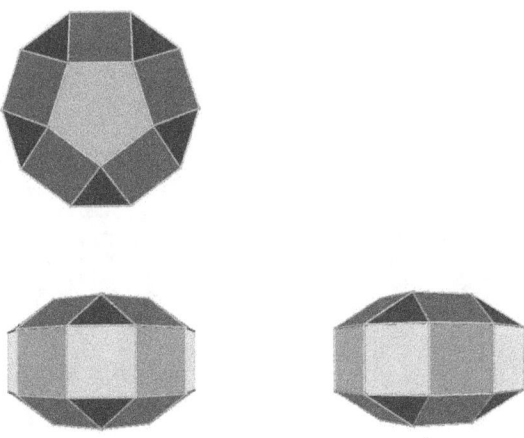

Figure 75: Johnson solid 37: the elongated pentagonal orthobicupola.

The dual of the elongated pentagonal orthobicupola, illustrated in Figure 76 below, can similarly be characterized by considering the dual of the elongated triangular orthobicupola and allowing for the change from threefold to fivefold symmetry. Like the elongated pentagonal orthobicupola itself, it has $D_{5h}$ symmetry. Of course, the fusion-elongation-dual rule, stated on p. 146 above, applies here; like the other elongated bicupola duals, it consists of two identical parts, each composed of an antibipyramid-like apical cap and a belt of irregular quadrilaterals, fused together. Like the dual of the elongated triangular orthobicupola, and *unlike* the duals of the elongated triangular and square gyrobicupolae, it has the two belts of irregular quadrilaterals oriented the same way, so as to provide a mirror plane where the two halves can be considered to be fused. Comparing Figure 76 with Figure 70 above, the effects of the 3-4-5 trend rule are clearly visible; it might be noted that the missing fourfold member of the family has all faces identical, being the dual of an Archimedean solid. (And the illustrations of the corresponding elongated *gyrobicupola* duals, in Figures 72, 74, and 78, which *do* include a fourfold member, may aid in visualizing the effects of the 3-4-5 trend rule.)

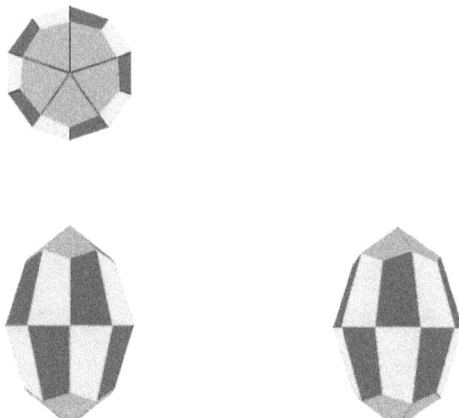

Figure 76: The dual of Johnson solid 38 (the elongated pentagonal orthobicupola).

As was the case for all the previous elongated bicupola duals, the faces consist of two classes of quadrilaterals, kite-quadrilaterals at the apices and irregular quadrilaterals of alternating chirality around the central portion which can be considered a fusion plane of two halves. (The halves, of course, would have octagonal bases, lost in the fusion process.)

The elongated pentagonal gyrobicupola, Johnson solid 39, illustrated in Figure 77 below, can simply be described by referring to the description of the elongated pentagonal orthobicupola and allowing for the twisting of one half by 36° relative to the other, or by referring to the description of the elongated triangular and square gyrobicupolae and allowing for the change from threefold or fourfold to fivefold symmetry. Because of this, there is little reason to provide a detailed description. The face/vertex/edge count will be identical to that of the elongated pentagonal orthobicupola, namely 32 faces, 30 vertices, and 60 edges; and the symmetry is $S_{10v}$, as expected.

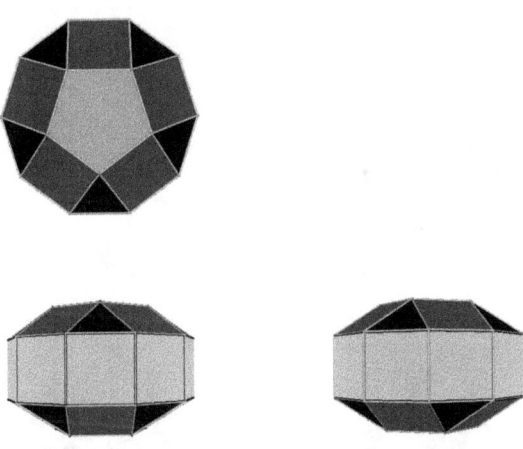

*Figure 77: Johnson solid 39: the elongated pentagonal gyrobicupola.*

Comparing it with the elongated triangular and square gyrobicupolae, the 3-4-5 trend rule can be seen to apply; as the cupolae themselves were seen to embody this rule, as the pentagonal cupola is much squatter than the triangular cupola and the square cupola is intermediate, it should be no surprise, as the elongated bicupolae are simply fusions involving those cupolae to a decagonal prism. (One should really combine the elongated triangular ortho- and gyrobicupolae, and the elongated pentagonal ortho- and gyrobicupolae, in looking at the application of the 3-4-5 trend rule here, as the twists of the bottom relative to the top do not affect this comparison.)

The dual of the elongated pentagonal gyrobicupola, illustrated in Figure 78 below, does not need a very detailed description, because most of its properties can be inferred either from the properties of the duals of the elongated triangular and square gyrobicupolae (allowing for the change from threefold or fourfold to fivefold symmetry) or from those of the dual of the elongated pentagonal orthobicupola (allowing for the 36° twist of the lower part relative to the top). As was noted for the pyramids, the 3-4-5 trend rule (introduced on p. 12 in Chapter 2) applies to the elongated bicupolae. The fusion-elongation-dual rule, stated on p. 146 above, applies here as well, and these two rules too provide ways to figure out the properties expected for this polyhedron.

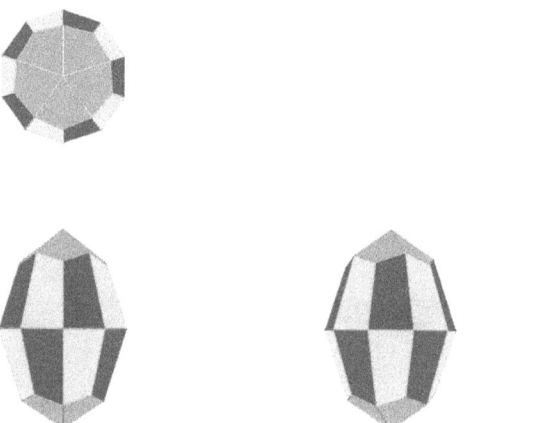

*Figure 78: The dual of Johnson solid 39 (the elongated pentagonal gyrobicupola).*

The properties of all the elongated bicupolae are summarized in Table 18 below in order to enable the reader to compare them easily.

*Elongated bicupolae, cupolarotundae, and birotundae: Johnson solids 35-43.*

| Johnson solid number | 35, 36 | 37 | 38, 39 |
|---|---|---|---|
| Name | 35: Elongated triangular orthobicupola 36: Elongated triangular gyrobicupola | Elongated square gyrobicupola (pseudorhombicuboctahedron, Miller's solid) | 38: Elongated pentagonal orthobicupola 39: Elongated pentagonal gyrobicupola |
| Symmetry* | 35: $D_{3h}$ 36: $S_{6v}$ | $S_{8v}$ | 38: $D_{5h}$ 39: $S_{10v}$ |
| Faces | [2+6](3) + [6+6](4)* | [8](3) + [2+8+8](4) | [10](3) + [10+10](4)* + [2](5) |
| Edges* | 36 | 48 | 60 |
| Vertices | 18 | 24 | 30 |
| Dihedral angles | 3-4c: 125.2644° (125° 15' 52"), 3-4L: 160.5288° (160° 31' 44"), 4c-4L: 144.7356° (144° 44' 8"), 4L-4L: 120°♣ | 3-4: 144.7356° (144° 44' 8"), 4-4: 135°△ | 3-4c: 159.0948° (159° 5' 41"), 3-4L: 127.3774° (127° 22' 39"), 4c-4L: 121.7175° (121° 43' 3"), 4L-4L: 144°, 4c-5: 148.2825° (148° 16' 57")* |
| Area | 15.46410162 = 12 + 2√3 | 21.46410162 = 18 + 2√3 | 27.77108182 = 20 + √[25 + (5/2)√5 + (5/2)√(75 + 30√5)] |
| Volume | 4.95509882 = (5√2)/3 + (3√3)/2 | 8.71404521 | 12.34229948 = [10 + 8√5 + 15√(5 + 2√5)]/6 |
| Dual ||||
| Name | Unnamed | Pseudodeltoidal icositetrahedron | Unnamed |
| Faces | [6+12](4) | [8+16](4) | [10+20](4) |
| Vertices | 20 | 26 | 32 |

*The Johnson Solids and Their Duals*

| Johnson solid number | 35, 36 | 37 | 38, 39 |
|---|---|---|---|
| Face angles[†] | 4a: [2]72.3987° (72° 23′ 55″), 99.6597° (99° 39′ 15″), 115.5429° (115° 32′ 34″); 4k: [2]75.5225° (75° 31′ 21″), 87.6404° (87° 38′ 25″), 121.3146° (121° 18′ 53″)[⊛] | 4: [3]81.5789° (81° 34′ 44″), 115.2632° (115° 15′ 47″)[□] | 4a: [2]84.5204° (84° 31′ 13″), 75.9514° (75° 57′ 5″), 115.0078° (115° 0′ 28″); 4k: [2]90.8038° (90° 48′ 14″), 58.4322° (58° 25′ 56″), 119.9603° (119° 57′ 37″)[⊛] |
| Dihedral angles | 4a-4a: 126.8699° (126° 52′ 12″), 4a-4k: 136.1088° (136° 6′ 32″), 4k-4k: 139.2921° (139° 17′ 32″)[⊛] | 4-4: 138.1180° (138° 7′ 5″)[□] | 4a-4a: 145.6559° (145° 39′ 21″), 4a-4k: 143.8338° (143° 50′ 2″), 4k-4k: 135.9247° (135° 55′ 29″)[⊛] |
| Area | 10.81731849 | 21.51345465 | 14.02317899 |
| Volume | 2.92947603 | 8.75069057 | 4.49204377 |

Table 18: *Properties of Johnson solids 35 through 39 and their duals.*

\*The symmetry and number of edges are always the same for each solid and its dual, so are not repeated in the "dual" section.

†Face angles are only given for the duals, as the Johnson solids have regular faces, whose angles can be found in Table 1 on p. 3.

□The twelve square faces of $J_{35}$ divide into three transitivity classes of 3, 3, and 6; but in the case of $J_{36}$ there are only two transitivity classes, containing 6 squares each. Similarly, the twenty square faces of $J_{38}$ divide into three transitivity classes of 5, 5, and 10; but in the case of $J_{39}$ there are only two transitivity classes, containing 10 squares each.

⊛Although the triangular faces of both $J_{35}$ and $J_{36}$ divide into *two* transitivity classes and the square faces of $J_{35}$ into *three*, it is unnecessary for this tabulation to make any distinction except between the square faces derived from the cupolae (designated 4c) and those derived from the prismatic elongation (designated 4L). In particular, all the triangular faces derived from the cupolae make the *same* dihedral angle with the cupola squares.

△Although the square faces of $J_{37}$ divide into *three* transitivity classes, all make the same dihedral angles with the triangular faces as well as the same dihedral angles with each other, so no distinctions are made in this table.

⁎Although the square faces of $J_{38}$ (but *not* of $J_{39}$) divide into *three* transitivity classes, it is only necessary to distinguish between the square faces derived from the cupolae (designated 4c) and those derived from the prismatic elongation (designated 4L).

⊛The symbol 4a designates the irregular quadrilaterals (twelve in the dual of $J_{35\text{-}36}$, twenty in the dual of $J_{38\text{-}39}$) that are among the faces of this polyhedron, 4k designates the remaining faces (six in the dual of $J_{35\text{-}36}$, ten in the dual of $J_{38\text{-}39}$) that are kite-quadrilaterals. Note that the 4a quadrilaterals are *not* kite-quadrilaterals, even though they have two equal angles, because the equal angles are not opposite each other.

□Although the kite-quadrilateral faces of $J_{37}$ divide into *two* transitivity classes, all have the same face angles as well as making the same dihedral angles with each other, so no distinctions are made in this table. The symbol 4 will be used for all.

*The Johnson Solids and Their Duals*

In Chapter 8, in addition to the bicupolae and birotunda, the cupolarotundae were introduced. In exactly the same way as the bicupolae and birotunda, the assembly of the cupolarotundae can be combined with the insertion of a decagonal prism as a belt between the pieces, giving Johnson solids 40 and 41, the elongated pentagonal ortho- and gyrocupolarotunda. Since only a pentagonal rotunda and not a triangular or square rotunda can be constructed within Johnson's limitation to regular-polygonal faces, only the pentagonal cupolarotundae are to be found in Johnson's list. Again, two possible orientations can be set, giving the two examples.

The elongated pentagonal orthocupolarotunda, Johnson solid 40, is illustrated in Figure 79 below.

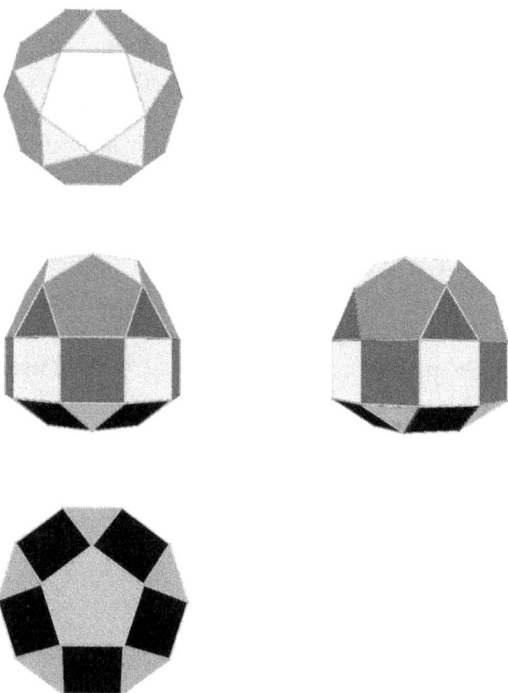

*Figure 79: Johnson solid 40: the elongated pentagonal orthocupolarotunda.*

After the descriptions of the elongated bicupolae above, the properties of the elongated pentagonal orthocupolarotunda should be easy to work out. Instead of two cupolae, there are one cupola and one rotunda fused to the prism, but the fact that these are two different objects means that there is no mirror plane halfway through the prism, so that the symmetry is $C_{5v}$ rather than $D_{5h}$. All the most important properties of this polyhedron can be inferred from the properties of the cupola, rotunda, and prism, as has been done for the other Johnson solids described in this chapter.

The dual of the elongated pentagonal orthocupolarotunda, illustrated in Figure 80 below, contains *three different sets* of kite-quadrilaterals among its faces. Both of the sets of quadrilaterals that are immediately adjacent to the fusion plane are irregular, not kite-quadrilaterals at all.

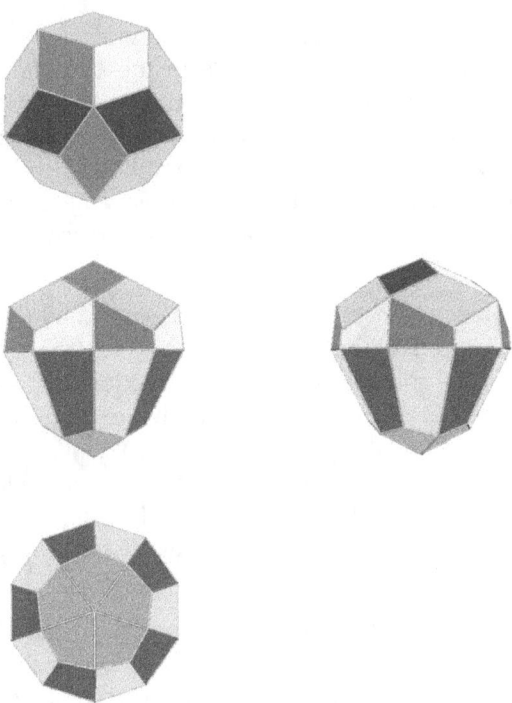

Figure 80: *The dual of Johnson solid 40 (the elongated pentagonal orthocupolarotunda).*

But it is worthwhile to look at the two sets of kite-quadrilaterals that are adjacent to *each other*, on the side that derives from the rotunda. In Table 19 below, these faces are designated 4c and 4d, with the latter being the ones at the apex. The sides of the 4c quadrilaterals are ~0.9722 and ~0.9534, and the angles include two angles of 63.6969° (63° 41′ 49″), with the remaining, not quite equal, angles being 114.4980° (114° 29′ 53″) and 118.1082° (118° 6′ 30″). The sides of the 4d quadrilaterals are ~0.9534 and ~0.9076, and the angles include two angles of 116.6585° (116° 39′ 31″), with the remaining, not quite equal, angles being 61.6031° (61° 36′ 11″) and 65.0800° (65° 4′ 48″). As one can see, both are nearly rhombi; the *short* sides of the 4c quadrilaterals are equal to the *long* sides of the 4d quadrilaterals because they are the same edges, but in each case all the sides are nearly equal. In addition, the acute angle in the 4c quadrilaterals is approximately halfway between the two values of the acute angles in the 4d quadrilaterals (all three being almost equal), and the obtuse angle in the 4d quadrilaterals is approximately halfway between the two values of the obtuse angles in the 4c quadrilaterals (all three being almost equal). Thus it can be seen that the apical/lateral kite rule, discussed on p. 30, applies, relating the cute and obtuse angles of the apical and lateral kites and noting that both are close to rhombi in shape.

Of course it can be noted that the fusion-elongation-dual rule, stated on p. 146 above, applies here. Thus there is clearly a division of this polyhedron into two sections separated by a plane; one (the top in Figure 80) deriving from the rotunda, with the apical and lateral kites expected when a rotunda is part of the figure whose dual is being studied, and the other (the bottom) deriving from the cupola, looking much like the halves of the elongated bicupolae discussed earlier in this chapter.

The elongated pentagonal gyrocupolarotunda, Johnson solid 41, illustrated in Figure 81 below, is exactly like the elongated pentagonal orthocupolarotunda, except for the 36° twist of the rotunda relative to the cupola. All the comments about the elongated pentagonal gyrocupolarotunda apply here as well, so no detailed description is necessary. Comparing Figure 81 with Figure 80 above, they are almost indistinguishable; it takes a close examination to note the 36° twist. And the face/vertex/edge counts, the dihedral angles, and such are identical, as can be seen in Table 19 below.

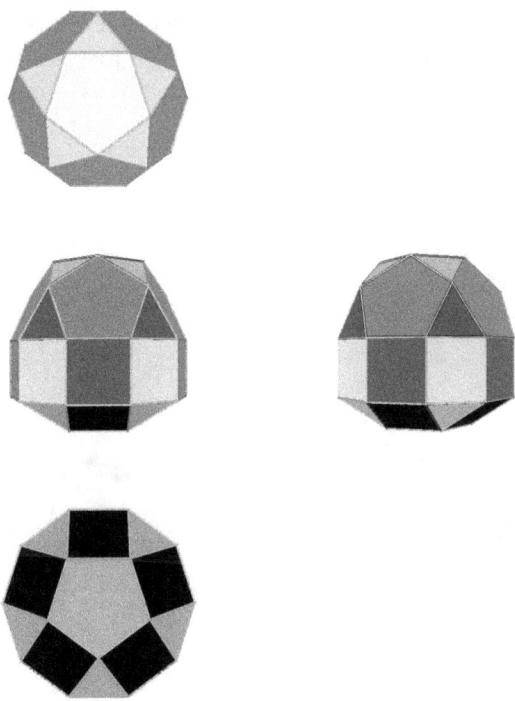

Figure 81: Johnson solid 41: the elongated pentagonal gyrocupolarotunda.

Because the two objects fused to the prismatic belt are different (a cupola and a rotunda), unlike the elongated bicupolae discussed earlier in this chapter, where the "ortho" member of the pair has $D_{nh}$ symmetry and the "gyro" member has the distinctly different $S_{2nv}$ symmetry, the elongated cupolarotundae have neither a mirror plane cutting through the prism parallel to the original bases (lost in the fusion) nor an alternating axis, and both of them therefore have the same $C_{5v}$ symmetry.

The dual of the elongated pentagonal gyrocupolarotunda, illustrated in Figure 82 below, looks very much like the polyhedron illustrated in Figure 80. As was so for all the other duals in this chapter, the fusion-elongation-dual rule, stated on p. 146 above, applies here, so that the polyhedron can be seen to be composed of the two sections that are easily seen in Figure 82, each of which is like the corresponding section in Figure 80, with the edges that separate these sections all lying in a horizontal plane. However, careful examination will show that the bottom section has been rotated 36° (one tenth of the way around) as compared with Figure 80 – just as the bottom section of Figure 81 has been rotated 36° as compared with Figure 79.

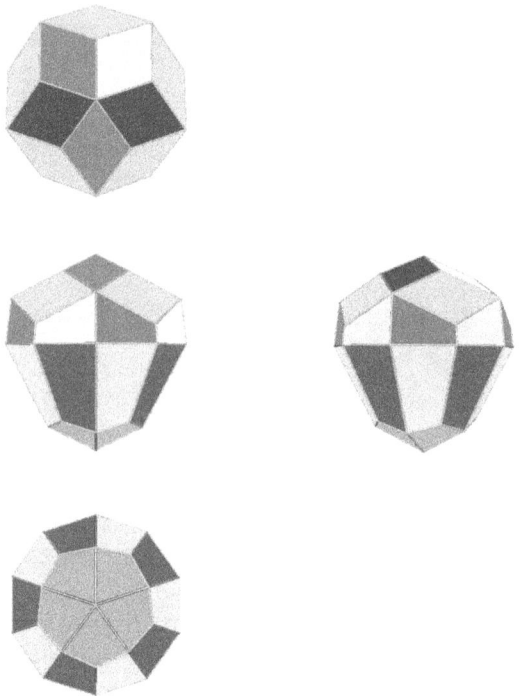

*Figure 82: The dual of Johnson solid 41 (the elongated pentagonal gyrocupolarotunda).*

The elongated pentagonal orthobirotunda, Johnson solid 42, is illustrated in Figure 83 below. Unlike the elongated pentagonal orthocupolarotunda, but like the elongated pentagonal orthobicupola, this is formed by the fusion of two *identical* polyhedra to the bases of a prism, and therefore it has $D_{5h}$ symmetry. If one starts with two pentagonal rotundae (each with 17 faces, 20 vertices, and 35 edges) and a decagonal prism (with 12 faces, 20 vertices, and 30 edges) and fuses them, the decagonal bases of the rotundae and the prism (4 altogether) are lost in the fusion, giving 2(17) + 12 − 4 = 42 faces. Ten vertices of each rotunda are merged with vertices of the prism, giving a total number of vertices of 2(20) + 20 − 2(10) = 40 vertices. And similarly the ten edges around the bases of each rotunda are merged with those of the bases of the prism, so that there are altogether 2(35) + 30 − 2(10) = 80 edges.

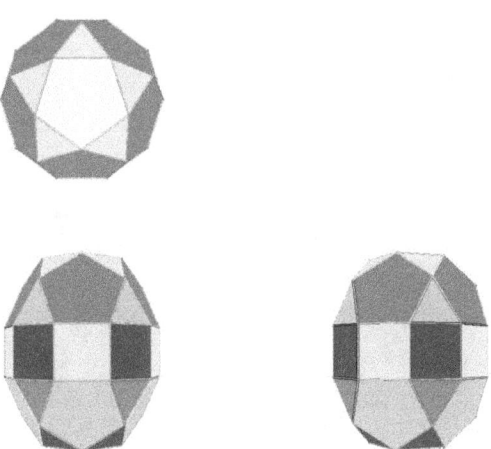

*Figure 83: Johnson solid 42: the elongated pentagonal orthobirotunda.*

(When we get to the elongated pentagonal gyrobirotunda, later in this chapter, the same face/vertex/edge counts will apply, since the same three component polyhedra are fused on the same faces, so this derivation will not be repeated there.)

The dual of the elongated pentagonal orthobirotunda, illustrated in Figure 84 below, like the duals of all the other elongated polyhedra discussed in this chapter, is a polyhedron which consists of two sections fused together, in accordance with the fusion-elongation-dual rule, stated on p. 146 above. Each half is much like the "rotunda" part of the duals of the elongated cupolarotundae, with a set of apical kites, a set of lateral kites, and a ring of alternating-chirality irregular quadrilaterals separating the lateral kites from the fusion plane. Unlike the two parts of the duals of the elongated cupolarotundae, but like the parts of the duals of the elongated bicupolae, the two halves are identical, since they both come from the dualization process applied to identical rotundae, placed the same distance from the center of the polar reciprocation process. The symmetry is $D_{5h}$, of course, since every polyhedron has the same symmetry as its dual.

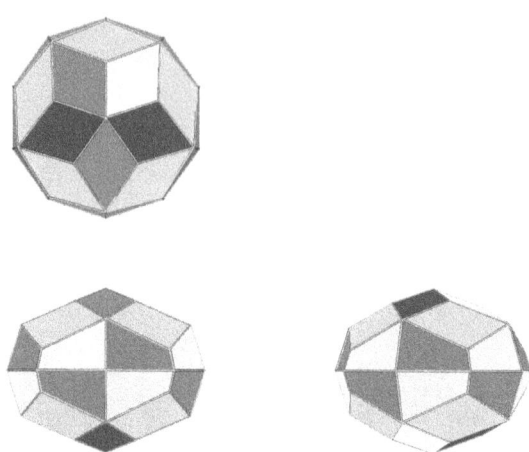

*Figure 84: The dual of Johnson solid 42 (the elongated pentagonal orthobirotunda).*

It might be noted that, although the elongated pentagonal orthobirotunda has triangular, square, and pentagonal faces, all the faces of the dual are quadrilaterals (whether kites or irregular quadrilaterals).

The elongated pentagonal gyrobirotunda, Johnson solid 43, illustrated in Figure 85 below, is the last of the Johnson solids that will be discussed in this chapter; although it might be possible to consider the next chapter as a continuation of this one, I think it better to separate the two. Because it can simply be described by referring to the description of the elongated pentagonal orthobirotunda and allowing for the twisting of one half by 36° relative to the other, no detailed description is required. The reader can see how similar these two polyhedra are by comparing Figure 85 with Figure 83. In keeping with what has been said before about the differences between "ortho" and "gyro" versions of solids, the main difference is the one that has

*Elongated bicupolae, cupolarotundae, and birotundae: Johnson solids 35-43.*

been seen in earlier such pairs: that the elongated pentagonal orthobirotunda has a horizontal mirror plane and $D_{5h}$ symmetry, and the elongated pentagonal gyrobirotunda has an alternating axis and $S_{10v}$ symmetry. All the other statistics, including the face/vertex/edge counts, the area, and the volume, are identical, so the two are listed together in a single column of Table 19 below, in the same way as other "ortho" and "gyro" versions have been in earlier tables.

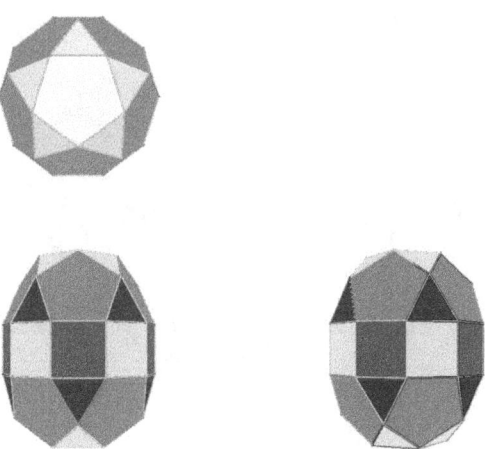

*Figure 85: Johnson solid 43: the elongated pentagonal gyrobirotunda.*

The chapter will be concluded with the discussion of the dual of the elongated pentagonal gyrobirotunda, which is illustrated in Figure 86 below.

In the same way as the properties of the two elongated pentagonal birotundae are extremely similar, so that the description of one can be taken with only slight modifications to apply to the other, with the major change to be taken into account being the difference in symmetry ($D_{5h}$, with a horizontal mirror plane, versus $S_{10v}$, with an alternating axis), the same applies to their duals. As a result, most of what was written, earlier in this chapter, about the dual of the elongated pentagonal orthobirotunda applies to this polyhedron, and the reader is referred to that earlier discussion. If the reader will compare Figure 86 with the earlier Figure 84, it will clearly show just how similar the two polyhedra are, in the same manner as other comparisons of the "ortho" and "gyro" versions of polyhedra treated in this book.

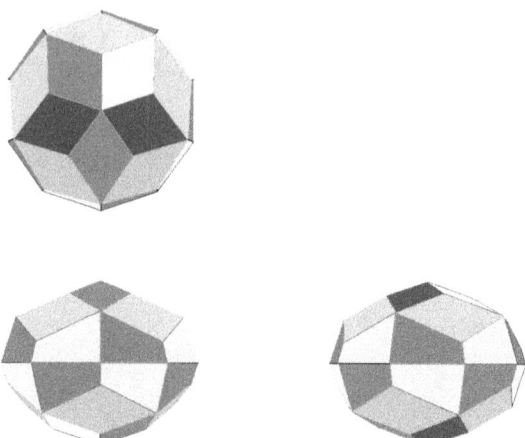

*Figure 86: The dual of Johnson solid 43 (the elongated pentagonal gyrobirotunda).*

*Elongated bicupolae, cupolarotundae, and birotundae: Johnson solids 35-43.*

| Johnson solid number | 40, 41 | 42, 43 |
|---|---|---|
| Name | 40: Elongated pentagonal orthocupolarotunda<br>41: Elongated pentagonal gyrocupolarotunda | 42: Elongated pentagonal orthobirotunda<br>43: Elongated pentagonal gyrobirotunda |
| Symmetry* | $C_{5v}$ | 42: $D_{5h}$<br>43: $S_{10v}$ |
| Faces | [5+5+5](3) + [5+5+5](4) + [1+1+5](5) | [10+10](3) + [10](4)* + [2+10](5) |
| Edges* | 70 | 80 |
| Vertices | 35 | 40 |
| Dihedral angles | 3c-4c: 159.0948° (159° 5' 41"),<br>3c-4L: 127.3774° (127° 22' 39"),<br>3r-4L: 169.1877° (169° 11' 16"),<br>3r-5r: 142.6226° (142° 37' 21"),<br>4c-4L: 121.7175° (121° 43' 3"),<br>4L-4L: 144°,<br>4c-5c: 148.2825° (148° 16' 57"),<br>4L-5r: 153.4349° (153° 26' 6") | 3-4: 169.1877° (169° 11' 16"),<br>3-5: 142.6226° (142° 37' 21"),<br>4-4: 144°,<br>4-5: 153.4349° (153° 26' 6")* |
| Area | 33.53853233 | 39.30598284 |
| Volume | 16.93601713 | 21.52973478 |
| **Dual** | | |
| Name | Unnamed | Unnamed |
| Faces | [5+5+5+10+10](4) | [10+10+20](4) |
| Vertices | 37 | 42 |

| Johnson solid number | 40, 41 | 42, 43 |
|---|---|---|
| Face angles[†] | 4a: [2]81.6884° (81° 41' 18"), 80.8045° (80° 48' 16"), 115.8187° (115° 49' 7"); 4b: [2]87.8802° (87° 52' 49"), 66.3322° (66° 19' 56"), 117.9073° (117° 54' 26"); 4c: [2]63.6969° (63° 41' 49"), 114.4980° (114° 29' 53"), 118.1082° (118° 6' 30"); 4d: [2]116.6585° (116° 39' 31"), 61.6031° (61° 36' 11"), 65.0800° (65° 4' 48"); 4e: [2]88.5682° (88° 34' 5"), 63.0510° (63° 3' 3"), 119.8127° (119° 48' 46")[♣] | 4a: [2]84.5204° (84° 31' 13"), 70.2821° (70° 16' 56"), 120.6771° (120° 40' 38"); 4b: [2]64.9962° (64° 59' 46"), 110.5141° (110° 30' 51"), 119.4934° (119° 29' 36"); 4c: [2]117.0729° (117° 4' 23"), 58.8674° (58° 52' 3"), 66.9867° (66° 59' 12")[◻] |
| Dihedral angles | 4a-4a: 147.9544° (147° 57' 16"), 4a-4b: 146.7395° (146° 44' 22"), 4a-4e: 146.5999° (146° 36' 0"), 4b-4b: 144.2423° (144° 14' 32"), 4b-4c: 145.4247° (145° 25' 29"), 4c-4d: 146.8588° (146° 51' 32"), 4d-4d: 147.3446° (147° 20' 40"), 4e-4e: 143.2801° (143° 16' 48")[♣] | 4a-4a: 145.6559° (145° 39' 21"), 4a-4b: 148.7976° (148° 47' 51"), 4b-4c: 151.2031° (151° 12' 11"), 4c-4c: 151.9087° (151° 54' 31")[◻] |
| Area | 24.50893121 | 28.26225240 |
| Volume | 10.70326405 | 13.24937311 |

Table 19: Properties of Johnson solids 40, 41, 42, and 43 and their duals.

°The symmetry and number of edges are always the same for each solid and its dual, so are not repeated in the "dual" section.

†Face angles are only given for the duals, as the Johnson solids have regular faces, whose angles can be found in Table 1 on p. 3.

°The ten square faces in $J_{42}$ divide into two transitivity classes of five squares each; in the case of $J_{43}$ they are all equivalent.

⁺Although there are triangles and pentagons (and squares, in the case of $J_{42}$) in two different transitivity classes each, all 3-5 edges have the same 142.6226° dihedral angle, and all 4-4 edges have the same 144° dihedral angle.

♦The symbols 4a and 4b designate the two bands of ten irregular quadrilaterals around the center of the polyhedron, with 4a referring to the set on the side of the fusion plane which come from the cupola, and 4b the other set, which come from the rotunda. 4c refers to the set of five quadrilaterals immediately adjacent to the 4b faces (on the other side from where the 4a faces are). 4d and 4e refer to the two sets of five quadrilaterals at the apices, 4d being the ones derived from the rotunda and 4e the ones derived from the cupola. Note that the 4a and 4b quadrilaterals are *not* kite-quadrilaterals, even though they have two equal angles, because the equal angles are not opposite each other. The 4c, 4d, and 4e quadrilaterals *are* kite-quadrilaterals, on the other hand. Within the rotunda dual part, the 4c quadrilaterals are lateral kites and the 4d quadrilaterals are apical kites, as defined on p. 30.

"The designation 4a refers to the twenty irregular quadrilaterals (*not* kite-quadrilaterals, even though they have two equal angles, because the equal angles are not opposite each other) around the middle of the polyhedron, 4b and 4c to the two sets of ten kite-quadrilaterals: according to the definition on p. 30, the 4b quadrilaterals are lateral kites and the 4c quadrilaterals are apical kites from the two rotundae that were fused.

# Chapter 10: Gyroelongated bicupolae, cupolarotundae, and birotundae: Johnson solids 44-48.

The gyroelongated bicupolae, Johnson solids 44-46, are interesting for an unusual reason. Although they can in each case be obtained by fusion of a pair of cupolae and an antiprism, none of which is a chiral polyhedron, the resulting polyhedra are chiral – they do not coincide with their mirror images. One of the versions of each is shown in the illustrations below (Figures 87, 89, and 91), but in each case there is a mirror image as well.

*Figure 87: Johnson solid 44: the gyroelongated triangular bicupola.*

The gyroelongated bicupolae are, as stated, chiral; they have symmetry $D_n$ ($n$ = 3 for $J_{44}$; 4 for $J_{45}$; 5 for $J_{46}$). The triangular faces fall into *four* different transitivity classes for $J_{44}$ (*three* for $J_{45}$ and $J_{46}$), not the *three* for $J_{44}$ (*two* for $J_{45}$ and $J_{46}$) that one might expect: the triangles of the lateral faces of the cupolae form one, not surprisingly, and the triangular bases of $J_{44}$ form another in that specific case, but what might be unexpected is that the triangles of the antiprismatic belt fall into two *different* transitivity classes: those which are adjacent to the

*Gyroelongated bicupolae, cupolarotundae, and birotundae: Johnson solids 44-48.*

square faces and those which are adjacent only to other triangular faces. We could designate these as $3c$ (in the cupola), $3b_4$ (in the belt, adjacent to a square face), and $3b_3$ (in the belt, but adjacent only to triangular faces). As a result, the edges between the antiprismatic belt triangles, which might all appear alike to a casual observer, fall into *three* separate transitivity classes: $3b_3\text{-}3b_3$, $3b_3\text{-}3b_4$, and $3b_4\text{-}3b_4$. (Since they all come from regular antiprisms, the dihedral angles would be equal, and one can simply talk of a $3b\text{-}3b$ dihedral angle, but for purposes of discussing *symmetry*, these are three different types of edge.)

In fact, from the nature of Johnson solids, all the triangular faces, regardless of the number of transitivity classes into which they belong, are strictly congruent; since they must be equilateral and the only remaining faces must be squares, there can only be *one* edge length in the *entire* polyhedron, which requires *any* two of the triangles to be congruent.

The duals of any polyhedron will have the same symmetry as that polyhedron if the center of polar reciprocation is chosen properly. Therefore, the duals of $J_{44}$ (Figure 88), $J_{45}$ (Figure 90), and $J_{46}$ (Figure 92) have the same chiral $D_n$ symmetry as the original solids ($n$ = 3 for $J_{44}$; 4 for $J_{45}$; 5 for $J_{46}$).

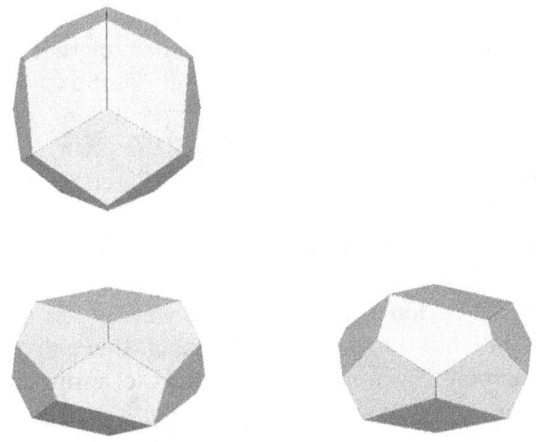

*Figure 88: The dual of Johnson solid 44 (the gyroelongated triangular bicupola).*

*The Johnson Solids and Their Duals*

All three of the duals of the gyroelongated bicupolae are diapical polyhedra, with kite-quadrilaterals meeting at both apices, and a double belt of irregular pentagons between the two sets of kite-quadrilaterals. If the order of the principal axis of rotation is $n$, the number of kite-quadrilaterals at each apex is $n$, and the number of irregular pentagons in the belt is $4n$, for a total of $6n$ faces. This, of course, corresponds to the $6n$ vertices of the gyroelongated bicupola itself, and this count in turn derives from the fact that each cupola contributes $3n$ vertices, and none of the vertices of the antiprism remain separate when the two cupolae are fused to the antiprism.

When one looks at Figure 88, it is clear that the pentagonal faces that form the lateral belt are not all alike but consist of two different, mirror-image polygons. (The polygons that meet at the ends of the threefold axis are kite-quadrilaterals, which *are* mirror-symmetric, but that is irrelevant.) The same is true of Figures 90 and 92. However, the chirality of these polyhedra is not obvious at first glance, any more than the chirality of the original Johnson solids. In the Johnson solids, there are $n$ planes which are mirror planes for that part of the solid consisting of *one* cupola and the antiprismatic belt, and another $n$ planes which are mirror planes for that part of the solid consisting of *the other* cupola and the antiprismatic belt; these are not the same planes, so no mirror plane exists for the entire polyhedron. In the duals, the same is true, allowing for the duality: there are $n$ planes which are mirror planes for that part of the solid consisting of *one* set of kite-quadrilaterals and the belt containing the two sets of asymmetric pentagons, and another $n$ planes which are mirror planes for that part of the solid consisting of *the other* set of kite-quadrilaterals and the belt containing the two sets of asymmetric pentagons; again, these are not the same planes, so no mirror plane exists for the entire polyhedron. Close examination of Figures 87 and 88 brings this out. In Figure 87, looking at the front view, a left-right reflection preserves all the top two thirds of the polyhedron unchanged, namely the upper cupola and the antiprismatic belt, but the lower cupola is rotated 60°, with squares replacing triangles and vice versa. If one looks at the side view, a similar reflection (which is in fact a front-back reflection of the entire polyhedron) does just the opposite: the *lower* cupola and the antiprismatic belt are unchanged, but the *upper* cupola is

*Gyroelongated bicupolae, cupolarotundae, and birotundae: Johnson solids 44-48.*

rotated 60°, with squares replacing triangles and vice versa. In the case of Figure 88, a similar phenomenon occurs, but it is the upper set of kite-quadrilaterals and the upper half of the belt of asymmetric pentagons that is unchanged in the front view, and the lower ones in the side view.

Despite this, and though the pentagonal faces of the dual belong in two different transitivity classes (though this is very hard to see), since the pentagons are congruent to each other, both make the same dihedral angles with the kite-quadrilaterals, and all dihedral angles between two pentagons are equal regardless of transitivity class membership, Table 20 does not need to distinguish between the two transitivity classes of pentagon.

As irregular as the pentagonal faces of the dual are, with no two sides equal, they still have three of their five angles equal. However, as the other two are far from equal, they are not isosceles pentagons, as defined in my earlier book, *Polyhedra: a New Approach*.

It will sometimes be useful to refer to an "upper half" and a "lower half" of a solid like the dual of Johnson solid 44. While the elongated bicupolae like $J_{35}$ have planes that can be considered planes of fusion between two identical halves, there is not a plane that divides the halves of $J_{44}$ into two (unless you allow it to cut through faces and edges). Nevertheless, the three kite-quadrilaterals at one apex, plus the six irregular pentagons adjacent to them, form a set of nine that can be considered a half of $J_{44}$. Two of these will fit together to form the full solid. The pentagons of either half form two different transitivity classes, but each one of the pentagons in either half can be paired with one adjacent to it in the other half that belongs to the same transitivity class. (Each pentagon in either half is also adjacent to a pentagon in the other half that belongs to the *other* transitivity class, as well.) So the boundary between the two halves is a zig-zag sequence of edges; if the two transitivity classes of pentagon were denoted 5a and 5b, the set of edges dividing one half from the other would cycle 5a-5a, 5a-5b, 5b-5b, 5b-5a, and repeat. (Here 5a-5b is an edge between a 5a pentagon in the upper half and a 5b pentagon in the lower; 5b-5a is the reverse.) Exactly the same will be found in the duals of $J_{45}$ and $J_{46}$, except that sets of three become sets of four or five, sets of six become sets of eight or ten, and the halves consist of twelve or fifteen faces instead of nine.

The relationship between the duals of elongated and gyroelongated solids is derived from the fact that the central belt of an elongated solid consists of the lateral portion of a prism, while the central belt of a gyroelongated solid consists of the lateral portion of an antiprism. The dual of a prism is a bipyramid, which consists of two (pyramidal) halves fused at a plane, while the dual of an antiprism is an antibipyramid, which consists of two halves, each of which looks like a pyramid at the end remote from the central region of the polyhedron, joined by the kind of zig-zag boundary which has just been described above.

Johnson solid 45, the gyroelongated square bicupola, illustrated in Figure 89 below, resembles $J_{44}$ except that the symmetry is fourfold rather than threefold.

*Figure 89: Johnson solid 45: the gyroelongated square bicupola.*

The reader is therefore referred back to the earlier discussion, rather than there being a separate discussion of the properties of this solid here.

The dual of the gyroelongated square bicupola, illustrated below in Figure 90, is identical to the dual of the gyroelongated triangular bicupola, except for the symmetry being fourfold rather than threefold. The comments above still apply with that single change made, so the reader is referred to that discussion. One interesting point, however, is that the dual of $J_{45}$ has dihedral angles that are almost equal between all its faces. There is no reason why this is obviously required, and in fact if it were a consequence of symmetry one would expect exact equality rather than the approximate equality, so it must be chalked up to coincidence.

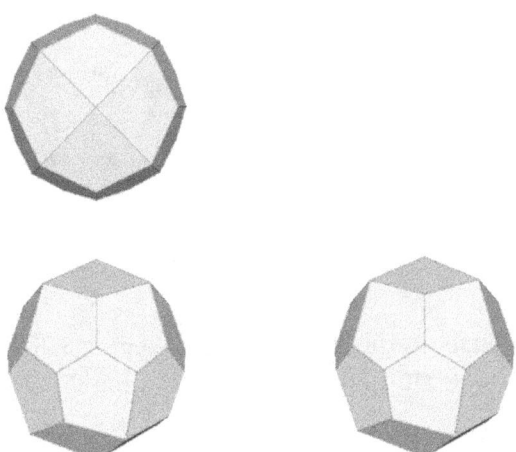

Figure 90: The dual of Johnson solid 45 (the gyroelongated square bicupola).

Johnson solid 46, the gyroelongated pentagonal bicupola, illustrated in Figure 91 below, again resembles $J_{44}$ except that the symmetry is fivefold rather than threefold. The reader is therefore referred back to the earlier discussion, as was done for the gyroelongated square bicupola, rather than there being a separate discussion of the properties of this solid here. Examination of Figures 87, 89, and 91 will provide a visual comparison of the three gyroelongated bicupolae, or if the details are desired, they can be found in Table 20 below, which, as in other such cases, illustrates the 3-4-5 trend rule (introduced on p. 12 in Chapter 2). (The clearest illustration of the 3-4-5 trend rule in this case is to look at the dihedral angles between the triangles forming the gyroelongation, on the one hand, and the triangles and squares from the cupolae, on the other. They are shown as 3c-3L and 3L-4 in Table 20, and both decrease substantially as the order of the cupola increases from 3 to 5.)

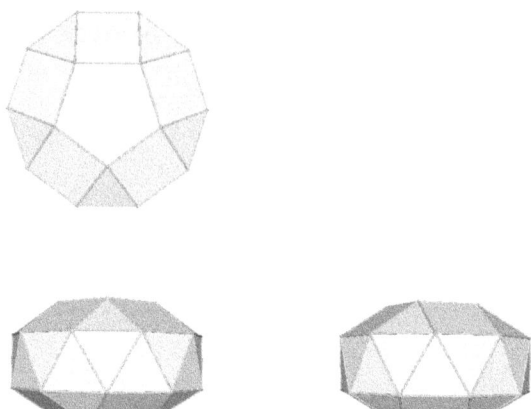

Figure 91: Johnson solid 46: the gyroelongated pentagonal bicupola.

The dual of the gyroelongated pentagonal bicupola is illustrated below in Figure 92.

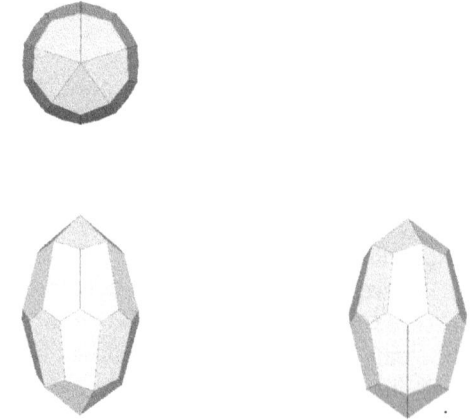

Figure 92: The dual of Johnson solid 46 (the gyroelongated pentagonal bicupola).

As was noted for the pyramids, the 3-4-5 trend rule applies

to the gyroelongated bicupolae, a fact which will be apparent on looking at the figures. It is, however, interesting to note that the figures for the surface areas and volumes of the duals do not follow a continuous trend. The area of the dual of the gyroelongated square bicupola is not between the other two, but is larger than either, as seen in Table 20, and likewise for the volumes. What appears to be the case is that going from the dual of $J_{44}$ to the dual of $J_{45}$, the height of the faces increases more than the width narrows, but the narrowing of the widths is more important as one goes to the dual of $J_{46}$. However, this applies only to the duals, not the original Johnson solids.

| Johnson solid number | 44 | 45 | 46 |
|---|---|---|---|
| Name | Gyroelongated triangular bicupola | Gyroelongated square bicupola | Gyroelongated pentagonal bicupola |
| Symmetry* | $D_3$ | $D_4$ | $D_5$ |
| Faces | [2+6+6+6](3) + [6](4) | [8+8+8](3) + [2+8](4) | [10+10+10](3) + [10](4) + [2](5) |
| Edges* | 42 | 56 | 70 |
| Vertices | 18 | 24 | 30 |
| Dihedral angles | 3c-3L: 169.4282° (169° 25′ 42″), 3L-3L: 145.2219° (145° 13′ 19″), 3c-4: 125.2644° (125° 15′ 52″), 3L-4: 153.6350° (153° 38′ 6″)♣ | 3c-3L: 151.3301° (151° 19′ 48″), 3L-3L: 153.9624° (153° 57′ 45″), 3c-4: 144.7356° (144° 44′ 8″), 3L-4: 141.5945° (141° 35′ 40″), 4-4: 135°¤ | 3c-3L: 132.6240° (132° 37′ 26″), 3L-3L: 159.1865° (159° 11′ 11″), 3c-4: 159.0948° (159° 5′ 41″), 3L-4: 126.9641° (126° 57′ 51″), 4-5: 148.2825° (148° 16′ 57″)△ |
| Area | 14.66025404 = 6 + 5√3 | 20.39230485 = 10 + 6√3 | 26.43133586 |
| Volume | 4.69456439 | 8.15357483 | 11.39737851 |
| Dual | | | |
| Name | Unnamed | Unnamed | Unnamed |
| Faces | [6](4) + [6+6](5) | [8](4) + [8+8](5) | [10](4) + [10+10](5) |
| Vertices | 26 | 34 | 42 |
| Face angles† | 4: [2]72.0234° (72° 1′ 24″), 100.9045° (100° 54′ 16″), 115.0487° (115° 2′ 55″); 5: [3]111.4707° (111° 28′ 15″), 85.8470° (85° 50′ 49″), 119.7409° (119° 44′ 27″) | 4: [2]81.7315° (81° 43′ 53″), 80.6778° (80° 40′ 40″), 115.8592° (115° 51′ 33″); 5: [3]115.0798° (115° 4′ 47″), 80.2131° (80° 12′ 47″), 114.5476° (114° 32′ 51″) | 4: [2]91.5661° (91° 33′ 58″), 57.9949° (57° 0′ 18″), 119.8628° (119° 51′ 46″); 5: [3]116.8115° (116° 48′ 41″), 74.6316° (74° 37′ 54″), 114.9340° (114° 56′ 2″) |

*Gyroelongated bicupolae, cupolarotundae, and birotundae: Johnson solids 44-48.*

| Johnson solid number | 44 | 45 | 46 |
|---|---|---|---|
| Dihedral angles | 4-4: 137.2455° (137° 14′ 44″), 4-5: 133.7817° (133° 46′ 54″), 5-5: 125.2644° (125° 15′ 52″) | 4-4: 136.1530° (136° 9′ 11″), 4-5: 136.7965° (136° 47′ 47″), 5-5: 137.3705° (137° 22′ 14″) | 4-4: 134.0269° (134° 1′ 37″), 4-5: 143.4192° (143° 25′ 9″), 5-5: 145.2534° (145° 15′ 12″) |
| Area | 9.77502247 | 18.52313868 | 12.45487821 |
| Volume | 2.54666256 | 6.99274183 | 3.71930477 |

Table 20: Properties of Johnson solids 44, 45, and 46 and their duals.

*The symmetry and number of edges are always the same for each solid and its dual, so are not repeated in the "dual" section.

†Face angles are only given for the duals, as the Johnson solids have regular faces, whose angles can be found in Table 1 on p. 3.

♦Although the triangular faces divide into *four* transitivity classes, it is unnecessary for this tabulation to make any distinction except between the triangular faces derived from the cupolae (designated 3c) and those derived from the antiprismatic elongation (designated 3L). In particular, all the triangular faces of the elongation make the *same* dihedral angle with the cupola squares, as they also do with each other.

□Although the triangular faces divide into *three* transitivity classes and the squares into *two*, it is unnecessary for this tabulation to make any distinction except between the triangular faces derived from the cupolae (designated 3c) and those derived from the antiprismatic elongation (designated 3L). In particular, all the triangular faces of the elongation make the *same* dihedral angle with the cupola squares, as they also do with each other, while the only edges that the cupola roof squares share are with the other cupola squares, which can be unambiguously designated 4-4.

△Although the triangular faces divide into *three* transitivity classes, it is unnecessary for this tabulation to make any distinction except between the triangular faces derived from the cupolae (designated 3c) and those derived from the antiprismatic elongation (designated 3L). In particular, all the triangular faces of the elongation make the *same* dihedral angle with the cupola squares, as they also do with each other.

*Gyroelongated bicupolae, cupolarotundae, and birotundae: Johnson solids 44-48.*

The next two Johnson solids are similarly constructed, except that one or both of the cupolae are replaced by a rotunda. Since the only rotunda that can be constructed with regular polygons as faces is the pentagonal ($J_6$), only a pentagonal version of the gyroelongated cupolarotunda and gyroelongated birotunda can be constructed within the limitations of the Johnson solids; the former is $J_{47}$ (see Figure 93) and the latter is $J_{48}$ (see Figure 95).

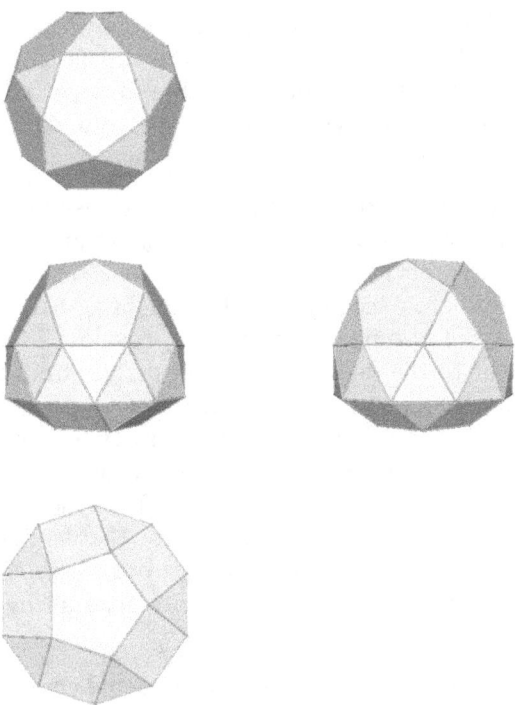

*Figure 93: Johnson solid 47: the gyroelongated pentagonal cupolarotunda.*

Like the gyroelongated bicupolae, the individual pieces fused together are mirror-symmetric, but the combinations are chiral. The fact that the top is different from the bottom means that $J_{47}$ only has $C_5$ symmetry, but the two congruent rotundae make for the higher $D_5$ symmetry of $J_{48}$, identical to $J_{46}$.

In order to determine the face/vertex/edge count, one can follow the same process which has been followed elsewhere in this book. A pentagonal cupola contributes 12 faces, 15 vertices, and 25 edges. A pentagonal rotunda contributes 17 faces, 20 vertices, and 35 edges. And a decagonal antiprism contributes 22 faces, 20 vertices, and 40 edges. Combined, there are 51 faces, 55 vertices, and 100 edges. However, the fusion process eliminates four faces, one from each side of each of the fusion planes, so the total number of faces is 47. The fusion also merges ten vertices of the antiprism with ten of the cupola, as well as the other ten vertices of the antiprism with ten of the rotunda, reducing the total number of vertices to 35. Similarly the ten edges bounding one of the decagonal bases of the antiprism are merged with the ten edges bounding the cupola base, while the ten edges bounding the other of the decagonal bases are merged with the ten edges bounding the rotunda base, reducing the total number of edges to 80.

The division of triangular faces into transitivity classes in this polyhedron produces *seven* different transitivity classes (each consisting of five triangular faces): one set is adjacent to the rotunda roof, a second can be thought of as having been adjacent to the rotunda base (which was lost in the fusion process) prior to the fusion, a third set derives from the lateral faces of the cupola, and then the faces derived from the antiprismatic elongation divide into four classes: one set adjacent to the triangles from the cupola, one set adjacent to the square faces, one set adjacent to the triangles from the rotunda, and one set adjacent to the pentagonal faces. The pentagonal faces divide into just two transitivity classes: one, deriving from the roof of the rotunda, forms a transitivity class all by itself; the remaining five constitute a second.

The dual of the gyroelongated pentagonal cupolarotunda is illustrated in Figure 94 below. One end of the polyhedron (shown at the bottom in Figure 94) is very similar to the dual of $J_{46}$, with five identical kite-quadrilaterals at an (anti)apex, surrounded by a set of five irregular pentagons. This array of ten polygons is attached to a piece which resembles the rotunda duals which have been covered before, with a set of five apical kites and, nestled in the spaces between the outer edges of those apical kites, five lateral kites. In turn, there are ten irregular pentagons, resembling the ring of alternating-chirality irregular pentagons found in the dual of $J_{25}$, in the spaces between the outer edges of the lateral kites. It is not at all surprising to find that the two pieces joined look much like the duals of $J_{25}$ (at the top) and $J_{24}$ (at the bottom), less their antiapices.

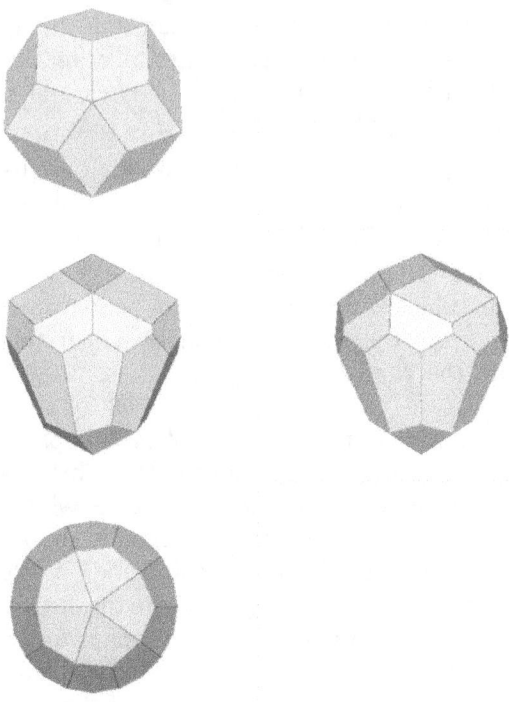

Figure 94: The dual of Johnson solid 47 (the gyroelongated pentagonal cupolarotunda).

On page 146 above, a rule was cited describing the duals of polyhedra that can be thought of as elongated versions of polyhedra obtained by the fusion of two component polyhedra, and it was there named the *fusion-elongation-dual rule*. While $J_{47}$ is not an elongated version of a polyhedron obtained by the fusion of two omponent polyhedra, it is a *gyroelongated* version of a polyhedron obtained by the fusion of two component polyhedra, and a similar rule can be seen to apply. It will not be formulated as explicitly here as the rule on p. 146 was there, but it should be relatively easy to formulate after a comparison of the dual of $J_{47}$ (Figure 94) with the duals of $J_{24}$ (Figure 48) and $J_{25}$ (Figure 50), for example. It could be termed the *fusion-gyroelon-*

*gation-dual rule*. (Since both the cupola dual part of Figure 48 and the rotunda dual part of Figure 50 are placed at the top of the respective polyhedra, but for Figure 94 it is only possible to choose one or the other orientation, it will be necessary to invert the vertical axis in Figure 50 to see it in a corresponding position to Figure 94.) And it can be seen that the apical/lateral kite rule, discussed on p. 30, also applies, relating the acute and obtuse angles of the apical and lateral kites and noting that both are close to rhombi in shape.

It was earlier noted that unlike the duals of the bicupolae or the birotundae, which have central sections with all parallel edges, which are also parallel to the principal axis of rotation (of course this is required by the $S_{2nv}$ or $D_{nh}$ symmetry!), the duals of the cupolarotundae are rather pear-shaped, with the part coming from the rotunda distinctly wider than the part coming from the cupola. (See Figures 52, 54, 56, 58, 60, and 62 for the duals of the bicupolae and Figure 68 for the dual of the only birotunda among the Johnson solids. Compare Figures 64 and 66 for the duals of the cupolarotundae.) This was explained as an artifact of the larger number of vertices which the rotunda contributes as compared with the cupola, so when one determines the centroid of the cupolarotunda for use as the center of polar reciprocation, it is not in the fusion plane but slightly below. A very similar comment applies to the gyroelongated cupolarotundae, although the central plane that one might consider for an alternative location of the center of polar reciprocation is not the single fusion plane that was referred to in the case of the duals of the cupolarotundae themselves; each of the *two* fusion planes, originally bases of the antiprism, where the cupola and rotunda are fused to it produces a different result, causing one set of ten edges to be parallel to the fivefold axis. If the polar reciprocation is done using a center that is actually in the center of one of these two planes, as opposed to the convention followed in this book where the centroid of the vertices is used as the center of the polar reciprocation process, the dual becomes more nearly cylindrical than pear-shaped. This is most notable if the rotunda base plane is chosen. In such a case, the dual becomes much more like a combination of the dual of a pentagonal bicupola and the dual of a pentagonal birotunda (choosing, as appropriate, the "ortho" or "gyro" orientation), with the appropriate pieces removed before combining them.

The last of the Johnson solids to be discussed in this chapter, $J_{48}$, the gyroelongated pentagonal birotunda, is illustrated in Figure 95 below. It hardly needs a long description, as the name clearly describes it: a fusion of two pentagonal rotundae to the bases of a pentagonal antiprism. The antiprism forces the two rotundae to be rotated 36° compared to each other, leading to an alternation in the positions of elements of the two rotundae, but like the other four polyhedra discussed in this chapter, the gyroelongated pentagonal birotunda is chiral, with $D_5$, not $S_{10v}$, symmetry. Although, since in all Johnson solids any triangular faces must all be congruent equilateral triangles (since all faces are regular, and there is no way that more than one edge length can occur in a Johnson solid), there are clearly three kinds of triangular face, when they are grouped into transitivity classes: ten (in two sets of five) surrounding what would be considered the rotunda roofs, ten (also in two sets of five) derived from the triangular faces adjacent to the rotunda bases (the bases themselves, of course, are lost in the fusion process), and the remaining ten from the antiprismatic elongation.

Figure 95: Johnson solid 48: the gyroelongated pentagonal birotunda.

The pentagonal faces also can be divided into transitivity classes, but only two: the two faces which were roofs of the rotundae, and the other ten (in two sets of five) which were lateral faces of the rotundae.

In order to determine the face/vertex/edge count, one can follow the same process which has been followed elsewhere in this book, including the previous discussion in this chapter. A pentagonal rotunda contributes 17 faces, 20 vertices, and 35 edges, and there are two of them. And a decagonal antiprism contributes 22 faces, 20 vertices, and 40 edges. Combined, there are 56 faces, 60 vertices, and 110 edges. However, the fusion process eliminates four faces, one from each side of each of the fusion planes, so the total number of faces is 52. The fusion also merges ten vertices of the antiprism with ten of each rotunda, that is twenty altogether, reducing the total number of vertices to 40. Similarly the ten edges bounding one of the decagonal bases of the antiprism are merged with the ten edges bounding the base of each rotunda base, again deleting twenty edges and reducing the total number of edges to 90.

The dual of the gyroelongated pentagonal birotunda is illustrated in Figure 96 below. Since, like the gyroelongated bicupolae but unlike the gyroelongated pentagonal cupolarotunda, the gyroelongated pentagonal birotunda has two identical polyhedra fused to the antiprism, the symmetry is *D*-type rather than *C*-type, specifically $D_5$, as was noted above.

In discussing the dual of the gyroelongated pentagonal birotunda, it shows that the requirement that we discuss the Johnson solids and duals in order makes for some awkwardness in this discussion. For it would have made the discussion simpler to treat this polyhedron *before* the section dealing with the dual of $J_{47}$, so that the dual of $J_{47}$ could simply be described as half of the dual of $J_{46}$ united with half of the dual of $J_{48}$. However, the most concise way to treat the dual of $J_{48}$ now is to refer to the bottom part of the dual of $J_{47}$ with two of the same figures combined together. One can also appeal to the fusion-gyroelongation-dual rule, earlier introduced in the discussion of the dual of the gyroelongated pentagonal cupolarotunda, together with the dual already given for the pentagonal rotunda (Johnson solid 6), to help in the characterization of this polyhedron. The appearance, in any case, is as one could guess from all these considerations, easy to describe: a diapical polyhedron with a set of five kite-quadrilaterals (the apical kites) at each apex, another set of five kite-quadrilaterals (differing in shape from the first in accordance with the apical/lateral kite rule, discussed on p. 30) adjacent to the five apical kites, and a double ring of irregular pentagons between the two opposing sets of kite-quadrilaterals.

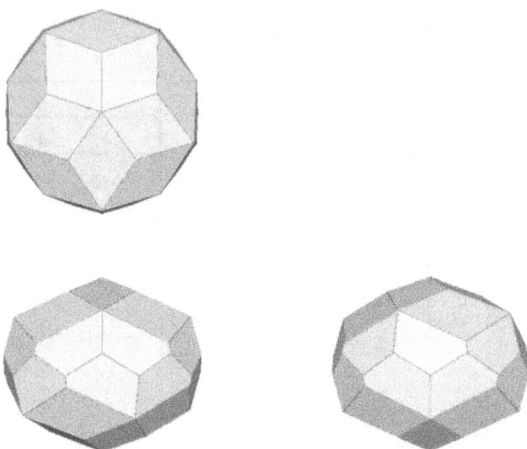

*Figure 96: The dual of Johnson solid 48 (the gyroelongated pentagonal birotunda).*

As in the case of the dual of $J_{47}$, applying the apical/lateral kite rule, one can relate the acute and obtuse angles of the apical and lateral kites, and it can be seen that both are close to rhombi in shape. The pentagons are very irregular, but it is interesting to note that the dihedral angles between any two of them (whether two of the same side, either at their short edges or at their longer edges, or two on opposite sides of the polyhedron) are all equal.

The five Johnson solids described in this chapter are, in fact, the *only* chiral Johnson solids, which is perhaps surprising. Interestingly, the five have four different symmetries: $C_5$, $D_3$, $D_4$, and $D_5$. So except for $D_5$ (which is the symmetry of both $J_{46}$ and $J_{48}$) these symmetries *each* apply to *only one* of the Johnson solids. Earlier, it was noted that only one Johnson solid, $J_{36}$, has $S_{6v}$ symmetry; one can list $S_{6v}$, $C_5$, $D_3$, and $D_4$ as the only symmetry types that apply to one single Johnson solid each.

The properties of Johnson solids 47 and 48, and their duals, are given in Table 21 below.

| Johnson solid number | 47 | 48 |
|---|---|---|
| Name | Gyroelongated pentagonal cupolarotunda | Gyroelongated pentagonal birotunda |
| Symmetry* | $C_5$ | $D_5$ |
| Faces | [5+5+5+5+5+5+5](3) + [5](4) + [1+1+5](5) | [10+10+10+10](3) + [2+10](5) |
| Edges* | 80 | 90 |
| Vertices | 35 | 40 |
| Dihedral angles | 3c-3L: 132.6240° (132° 37' 26"), 3L-3L: 159.1865° (159° 11' 11"), 3L-3r: 174.4343° (174° 26' 4"), 3c-4: 159.0948° (159° 5' 41"), 3L-4: 126.9641° (126° 57' 51"), 3L-5r: 158.6816° (158° 40' 54"), 3r-5r: 142.6226° (142° 37' 21"), 4-5c: 148.2825° (148° 16' 57")♦ | 3L-3L: 159.1865° (159° 11' 11"), 3L-3r: 174.4343° (174° 26' 4"), 3L-5: 158.6816° (158° 40' 54"), 3r-5: 142.6226° (142° 37' 21")△ |
| Area | 32.19878637 | 37.96623688 |
| Volume | 15.99109616 | 20.58481381 |
| **Dual** | | |
| Name | Unnamed | Unnamed |
| Faces | [5+5+5](4) + [5+5+5+5](5) | [10+10](4) + [10+10](5) |
| Vertices | 47 | 52 |

*Gyroelongated bicupolae, cupolarotundae, and birotundae: Johnson solids 44-48.*

| Johnson solid number | 47 | 48 |
|---|---|---|
| Face angles† | 4a: [2]116.6111° (116° 36' 40"), 62.1681° (62° 10' 5"), 64.6096° (64° 36' 35"); 4b: [2]63.5615° (63° 33' 41"), 115.1861° (115° 11' 10"), 117.6909° (117° 41' 27"); 4c: [2]88.9788° (88° 58' 44"), 62.1291° (62° 7' 45"), 119.9132° (119° 54' 48"); 5c: [2]115.9502° (115° 57' 1"), 79.7533° (79° 45' 12"), 112.7779° (112° 46' 40"), 115.5684° (115° 34' 6"); 5r: [2]118.9086° (118° 54' 31"), 65.4542° (65° 27' 15"), 117.4485° (117° 26' 55"), 119.2801° (119° 16' 48")¤ | 4a: [2]116.9663° (116° 57' 59"), 59.4266° (59° 25' 36"), 66.6407° (66° 38' 27"); 4L: [2]64.6465° (64° 38' 48"), 111.4180° (111° 25' 5"), 119.2890° (119° 17' 20"); 5: [3]116.8115° (116° 48' 41"), 69.5259° (69° 31' 33"), 120.0397° (120° 2' 23")* |
| Dihedral angles | 4a-4a: 146.3410° (146° 20' 28"), 4a-4b: 145.9697° (145° 58' 11"), 4c-4c: 141.6314° (141° 37' 53"), 4b-5r: 144.9113° (144° 54' 41"), 4c-5c: 145.8160° (145° 48' 58"), 5c-5c: 147.3251° (147° 19' 30"), 5c-5r: 146.1829° (146° 10' 58"), 5r-5r: 144.1134° (144° 6' 48")¤ | 4a-4a: 151.0115° (151° 0' 41"), 4a-4L: 150.3135° (150° 18' 49"), 4L-5: 147.9951° (147° 59' 42"), 5-5: 145.2534° (145° 15' 12")* |

| Johnson solid number | 47 | 48 |
|---|---|---|
| Area | 21.17254145 | 27.67591813 |
| Volume | 8.59055684 | 12.90244292 |

Table 21: Properties of Johnson solids 47 and 48 and their duals.

*The symmetry and number of edges are always the same for each solid and its dual, so are not repeated in the "dual" section.

†Face angles are only given for the duals, as the Johnson solids have regular faces, whose angles can be found in Table 1 on p. 3.

♦Although the triangular faces divide into *seven* transitivity classes and the pentagons into *three*, it is unnecessary for this tabulation to make any distinction among the triangular faces, except between those derived from the cupola (designated 3c), those derived from the rotunda (designated 3r), and those derived from the antiprismatic elongation (designated 3L); and it is unnecessary for this tabulation to make any distinction among the pentagonal faces, except between the one derived from the cupola (designated 5c) and those derived from the rotunda (designated 5r).

△Although the triangular faces divide into *four* transitivity classes and the pentagons into *two*, it is unnecessary for this tabulation to make any distinction among the triangular faces, except between those derived from the rotunda (designated 3r) and those derived from the antiprismatic elongation (designated 3L); and it is unnecessary for this tabulation to make any distinction among the pentagonal faces.

▫The designation 4a refers to the five apical kites deriving from the rotunda; 4b to the five lateral kites deriving from the rotunda; 4c to the five kites deriving from the cupola. Although the pentagonal faces divide into *four* transitivity classes, it is unnecessary for this tabulation to make any distinction among them, except between those derived from the cupola (designated 5c) and those derived from the rotunda (designated 5r).

°The designation 4a refers to the apical kites; 4L to the lateral kites. Although the pentagonal faces divide into *two* transitivity classes, it is unnecessary for this tabulation to make

*Gyroelongated bicupolae, cupolarotundae, and birotundae: Johnson solids 44-48.*

any distinction among them.

## Chapter 11: (Laterally) augmented prisms: Johnson solids 49-57.

Norman Johnson designated these polyhedra "augmented prisms." He apparently did not notice that the term could also apply, as was stated in Chapter 4, to the polyhedra he called "elongated pyramids" (and, as in Chapter 6, to the elongated bipyramids) as well. In this chapter, Johnson's names are kept, but it should be noted that the term "augmented" should really be qualified by the descriptive adverb "laterally" to distinguish them from the elongated pyramids that are "basally monoaugmented" prisms, and the word "laterally" will be inserted in parentheses in many places.

It is appropriate at this point to discuss the process of fusion, as it is used to build many of the Johnson solids. This discussion could, of course, have been inserted at a number of points in this book, particularly Chapters 4 and 18, and in fact much of what will be said in this discussion repeats statements made earlier in the book, but I have chosen to put it at this point, which I believe is the best location to make some of the points that need to be made when discussing the use of fusion to construct Johnson solids. (It will be useful to keep these points in mind when the reader gets to Chapter 18, of course.)

First of all, let us revisit the definition of a Johnson solid. One can start with a more general concept, that of a regular-polygon-faced polyhedron, by which we mean a polyhedron *all* of whose faces are regular polygons. Regular polygons can be fitted together in an infinite number of ways, of course, and as I did in my earlier book, *Polyhedra: a New Approach*, I tend to avoid discussing those polyhedra which are not convex. For this purpose, the most important criterion to introduce in order to insure convexity is that the dihedral angles between any pair of faces that share an edge is strictly less than 180°. (If it is *precisely* 180°, the two faces are in one plane, so they are not really two distinct faces, and one really has a polyhedron with a face that is not a regular polygon, but a figure composed of two regular polygons fused together on a side.) If one starts with two convex regular-polygon-faced polyhedra and fuses them, the resulting polyhedron is still, necessarily, a regular-polygon-faced polyhedron, since all faces of the new polyhedron were present in the original polyhedra. But the dihedral angles at the edges where fusion took place will be equal to the *sum* of the dihedral angles at the corresponding edges of the original polyhedra. So in any case where two convex regular-polygon-faced polyhedra are fused to produce a new one, one can be certain that *if* all of the dihedral angles at the faces being fused add to less than 180°, the result will be another convex regular-polygon-faced polyhedron.

Now it should be noted that I have been referring to "convex regular-polygon-faced polyhedra" rather than "Johnson solids." The Johnson solids are all the convex regular-polygon-faced polyhedra *with the exception* of the Platonic and Archimedean solids, prisms, and antiprisms. But when it is stated above that "where two convex regular-polygon-faced polyhedra are fused to produce a new one, ... if all of the dihedral angles at the faces being fused add to less than 180°, the result will be another convex regular-polygon-faced polyhedron," the polyhedra being fused may include the Platonic and Archimedean solids, prisms, and antiprisms, and not merely the Johnson solids. As was seen in Chapters 4 and 6, in particular, the fusion of a pyramid (or two pyramids) and a prism can yield a Johnson solid (and the pyramid does not need to be a Johnson solid such as $J_1$ or $J_2$; it can be a regular tetrahedron – a Platonic solid – so *neither* polyhedron need be a Johnson solid.) In this chapter, as has already been stated, the emphasis is on polyhedra obtained by fusing a prism and one or more pyramids, just as was done in Chapters 4 and 6, but where the lateral faces, rather than the bases, of the prisms are fused with the bases of pyramids.

In Chapters 4 and 6, the limitation on the numbers of sides that the original prism could have came from the fact that regular pyramids could only be constructed with numbers of sides from 3 to 5, if their lateral faces were to be equilateral triangles. In this chapter, the faces of the original prism to which the pyramid(s) is/are fused may only be squares, so the pyramids are square pyramids, and the restrictions on the prisms come from a different source. If one examines Table 2 on p. 13, it can be seen that the dihedral angle at the base of a square pyramid is 54° 44' 8". If the pyramid is fused onto a lateral face of a prism, the resulting polyhedron will not be convex unless the dihedral angle between the lateral face in question and every adjacent face of the prism is less than 180° − 54° 44′ 8″ = 125° 15′ 52″. And looking at Table 1 on p. 3, it can be seen that this requires $n \leq 6$. (If a pyramid is fused onto each of two adjacent lateral faces, the resulting polyhedron will not be convex unless the dihedral angle between the lateral faces in question is less than 180° − 2[54° 44′ 8″] = 70° 31′ 44″. This requires $n = 3$.) Thus laterally monoaugmented $n$-gonal prisms could occur for $n = 3, 4, 5$, or 6; however if $n = 4$, there is no clear distinction between lateral faces and bases, so the resulting augmented prism is the already-seen elongated square pyramid, $J_8$ (Figure 15, p. 43). The other cases, for $n = 3, 5$, and 6, give the (laterally) augmented triangular, pentagonal, and hexagonal prisms: $J_{49}$ (Figure 97), $J_{52}$ (Figure 103), and $J_{54}$ (Figure 107). (Johnson's terminology does not include "laterally," but my own preference is to include it.) Once more, it should be emphasized that the limitation to $n \leq 6$ is purely due to the requirement that all the triangular faces of the pyramid fused to the prism be equilateral. It is possible to fuse a square (or rectangular) pyramid to a lateral face of a prism *no matter how many sides the prism bases have*, if the pyramid is made flat enough.

Figure 97 below shows the (laterally) augmented triangular prism ($J_{49}$), the first of the augmented prisms with which this chapter is concerned. One should note that the symmetry is $C_{2v}$, as opposed to the $D_{3h}$ of the prism prior to the augmentation. And since in this book it is usually preferred to orient polyhedra so that the top view looks down on the principal axis of rotation, and the front view puts the principal axis vertical, the view in Figure 97 has the pyramidal augmentation at the top, and one of the triangular faces showing in the front view.

In the front view, the two triangular faces that are visible look rather different. In fact they are both congruent equilateral triangles (as they must, given the requirements of a Johnson solid), and the difference only comes from the fact that the lower triangle is being viewed approximately head on, while the upper triangle is being viewed very obliquely. In the same way, the square face in the side view does not look like a precise square, but again this is an artifact of the fact that this face is being viewed from an oblique angle.

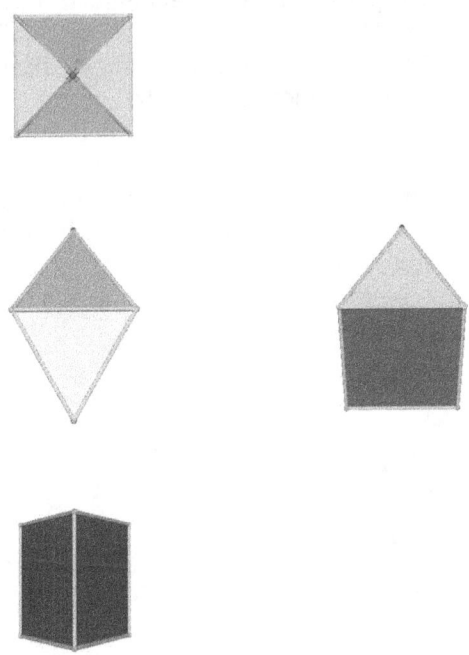

Figure 97: Johnson solid 49: the (laterally) augmented triangular prism.

It should be noted that for each of the polyhedra discussed in this chapter, the dual is very easily described. In Chapter 5, it was noted that the bipyramids and prisms are duals; in Chapter 4 (see p. 40) it was noted that the dual of an augmented polyhedron is obtained by truncating the dual of the unaugmented polyhedron. So in the case of $J_{49}$, the original triangular prism has, as its dual, a triangular bipyramid (much like $J_{12}$, as illustrated in Figure 23, but with different proportions), and the dual of $J_{49}$ would thus be a triangular bipyramid with one of its lateral vertices truncated, as illustrated in Figure 98 below. (A triangular bipyramid with one of its *apical* vertices truncated would be the dual of $J_7$, shown in Figure 14, as the vertices which are duals of the prism bases are the apices of the bipyramid.)

It may be hard to see Figure 98 as a triangular bipyramid

with one lateral vertex truncated, because it is shown in an orientation corresponding to Figure 97 above. The truncated lateral vertex is at the top, and is square; the $C_{2v}$ symmetry of the polyhedron would permit it to be rectangular, but in fact it is precisely a square. The threefold symmetry originally possessed by the bipyramid is gone, so one can only speak of a trivalent vertex rather than a threefold axis, and it can be seen that although the angles at the apex (visible in the front view) are all equal, the two faces that have been cut by the truncation (and the other two, not visible from the front but symmetrically arranged) are obviously parts of much larger triangles than the unaltered triangles (the one pointing down from the trivalent vertex of the front view, and a corresponding one behind).

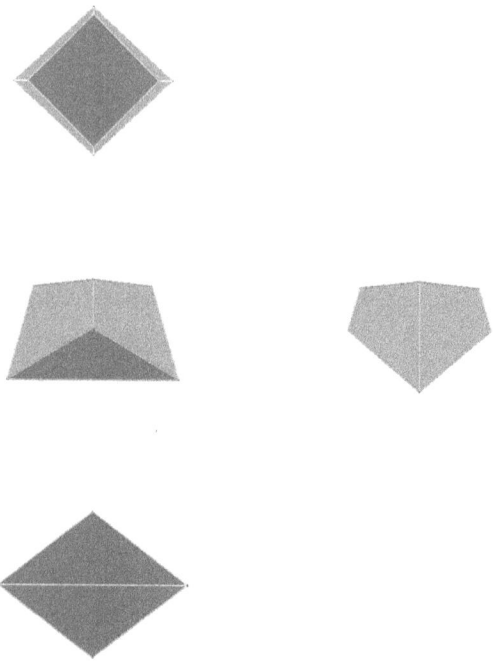

Figure 98: The dual of Johnson solid 49 (the laterally augmented triangular prism): a triangular bipyramid with one lateral face truncated.

## (Laterally) augmented prisms: Johnson solids 49-57.

Actually, the dual of Johnson solid 49 can be seen in yet another way. If one starts with a square pyramid, whose height in the particular case we are considering here is 2.23103644, and whose base is a square of side 0.92396130, let the apex be *A* and the base vertices be *B, C, D,* and *E*. Cut off the apex to produce a frustum of a square pyramid with the vertices *F, G, H,* and *J*. The plane *FGHJ* must be 0.69522587 away from the plane *BCDE*, which places it 1.53581057 from *A*. From this frustum of a pyramid, cut off triangular pieces truncating away vertices *C* and *E*, putting new vertices *K* on *CG* and *L* on *EJ*. The positions of *K* and *L* are such that they are in a plane 0.31802025 from the plane *FGHJ*, and 0.37720562 from plane *BCDE*. The resulting solid has a square base *FGHJ*, four lateral irregular quadrilaterals *BFGK, BFJL, DHGK,* and *DHJL*, and two isosceles triangular faces *BDK* and *BDL*. (However, using the term "base" for *FGHJ* implies the top and bottom of Figure 98 should be interchanged.)

If one starts with a square pyramid as in the construction just described, it becomes clear why some of the parameters in Table 22 show more equalities than the $C_{2v}$ symmetry of this polyhedron would require. Because *BCDEFGHJ* is a frustum of a square pyramid with $C_{4v}$ symmetry, *FGHJ* must be a perfect square. And the dihedral angles along edges *BF, CG, DH,* and *EJ* must be equal, so even if vertices *C* and *E* are truncated in the last step, the dihedral angles along edges *GK* and *JL* remain equal to the dihedral angles along edges *BF* and *DH*. While the truncation reduces the symmetry from $C_{4v}$ to $C_{2v}$, it does not alter the fact that those dihedral angles are equal or that *FGHJ* is a perfect square.

There are in theory three ways that *two* pyramids could be fused to a triangular prism, forming a figure that could be called a *biaugmented triangular prism*. If the pyramids must have regular polygons for faces (a requirement for a Johnson solid), however, it is easy to see that one of those arrangements is not possible. A triangular pyramid, which is what would need to be fused to one of the bases of the prism, would be in fact a regular tetrahedron, which has a dihedral angle of 70° 31' 44" between any two faces. A square pyramid, which is needed to fuse to one of the lateral faces, has a dihedral angle between any triangular face and its base, as stated earlier, of 54° 44' 8". If both these fusions would take place, the dihedral angle at the edge between the two pyramids would be the sum of these plus the 90° already present between each base of the prism and any lateral face, and 54° 44′ 8″ + 70° 31′ 44″ + 90° = 231° 3′ 27″, far in excess of 180°, so such a polyhedron would not be convex. So the only two possibilities are that the fusion occurs to the two opposite bases (forming the already discussed $J_{14}$, illustrated in Figure 27 on p. 66, in Chapter 6) or to two of the lateral faces. And this latter case is the only one that Johnson called a biaugmented triangular prism, which is illustrated in Figure 98 below and designated as $J_{50}$.

Figure 99: Johnson solid 50: the (laterally) biaugmented triangular prism.

For (laterally) biaugmented $n$-gonal prisms, it would be noted again that the only possible arrangement for $n = 4$, where the augmented faces are opposite, again was already dealt with as the elongated square bipyramid, $J_{15}$ (illustrated in Figure 29). Again, there are possible cases for $n = 3$, 5, or 6: the (laterally)

*(Laterally) augmented prisms: Johnson solids 49-57.*

biaugmented triangular, pentagonal, and hexagonal prisms: $J_{50}$ (Figure 99), $J_{53}$ (Figure 105), $J_{55}$ (Figure 109), and $J_{56}$ (Figure 111). They will be considered in their proper sequence, which means that they will not be discussed sequentially, but $J_{50}$ is the first to be discussed in this chapter.

The description of $J_{50}$ does not need a lot of detail; it can simply be summarized in the same way that other Johnson solids formed by fusion have been.

The dual of Johnson solid 50 is illustrated in Figure 100 below.

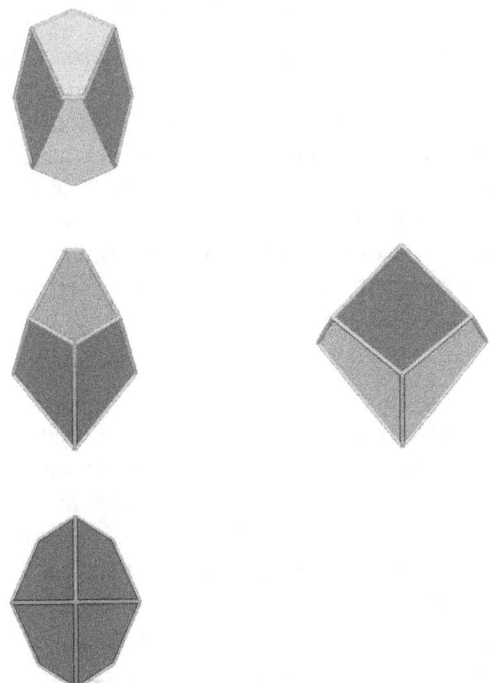

Figure 100: The dual of Johnson solid 50 (the laterally biaugmented triangular prism): a triangular bipyramid with two lateral faces truncated.

It was noted in the discussion of the dual of $J_{49}$ that the face that is dual to the pyramid apex was a perfect square. In the case of the dual of $J_{50}$, this is not true; the two faces dual to the apices of the pyramids constituting the augmentations are *nearly* squares, but in fact are kite-quadrilaterals, with two sides of 0.41087703 and two sides of 0.48607184, and having angles of 79.4652° (79° 27' 55"), 98.2584° (98° 15' 30"), and 91.1382° (91° 8' 18") (twice). In the dual of $J_{51}$, they are perfect squares again. It would appear that this distinction results from symmetry considerations, but the precise reasoning that makes the duals of the pyramid apices of $J_{49}$ and $J_{51}$ perfect squares, while those of $J_{50}$ are kite-quadrilaterals, is hard to come by. It is certainly true that, of the three duals, that of $J_{50}$ is the only one that does not have a $C_2$ axis passing through the center of these quadrilateral faces. But a $C_2$ axis is not a $C_4$ axis, and one might wonder why these quadrilaterals are squares and not rhombi.

As was shown in the case of biaugmented triangular prisms above, it is impossible, if one insists on equilateral triangles for the faces of the pyramids to be fused to a triangular prism, to have the augmentations on both a base and a lateral face of the prism, so the only way one can fuse three pyramids to the prism is for them to be fused to *all three lateral faces*. So it is not necessary, in the context of Johnson solids, to insist on adding the word "laterally," as I have in discussing $J_{49}$ and $J_{50}$ above.

Unlike the cases of $J_{49}$ and $J_{50}$ above, the "triaugmented triangular prism" (Johnson solid 51, illustrated in Figure 101 below) in Johnson's terminology retains the $D_{3h}$ symmetry of the original prism, and because the convention in this book is to show polyhedra with the principal axis of rotation in a vertical orientation, the top view in Figure 101 is actually comparable to the front views in Figures 97 and 99. Please keep that in mind when comparing them.

The only two faces of the original triangular prism that remain in the triaugmented triangular prism are the two triangular bases. And all the square faces have been replaced by sets of four triangular faces, so $J_{51}$ is another member of the set of eight polyhedra termed *deltahedra*: the polyhedra all of whose faces are equilateral triangles. Since it is a deltahedron with fourteen faces, it has been called (*e. g.*, in Anthony Pugh's book, *Polyhedra: a Visual Approach*) a *tetracaidecadeltahedron*. It is one of the five deltahedra among the Johnson solids. (See also p. 61

in Chapter 5.)

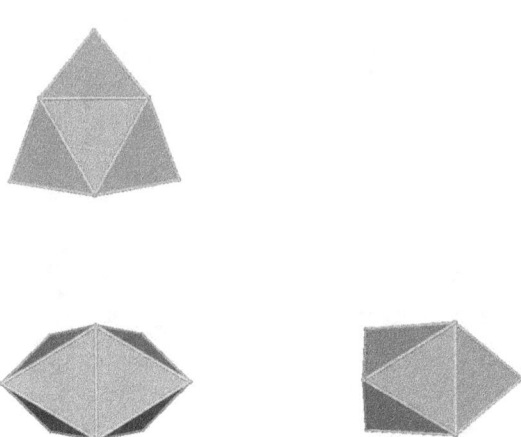

Figure 101: Johnson solid 51: the (laterally) triaugmented triangular prism.

The dual of $J_{51}$ (illustrated in Figure 102 below) has not been given a name, to my knowledge, before I wrote my earlier book, *Polyhedra: a New Approach*. But when the name was given to it, there was no particular thought being given to the subject of Johnson solids. It was simply one case ($n = 3$) of the polyhedra obtained by truncating a bipyramid at all vertices in the plane where the two pyramids were fused (i. e., at all vertices other than the two apices). In the terminology of the book, therefore, the dual of $J_{51}$ is a peritruncated triangular bipyramid. As was noted above, truncation is a dual operation to augmentation and bipyramids are the duals of prisms, so that *all* the augmented prisms have duals that are bipyramids truncated at some of the lateral vertices. However, because of the restrictions arising from the Johnson limitation to regular polygonal faces, the only peritruncated bipyramid that is a Johnson dual is this one.

Since duals share the symmetry of the polyhedra of which they are duals, the same comments about recognizing the difference in orientation of the views when comparing Figure 101 with Figures 97 and 99 applies to comparisons of Figure 102 with Figures 98 and 100.

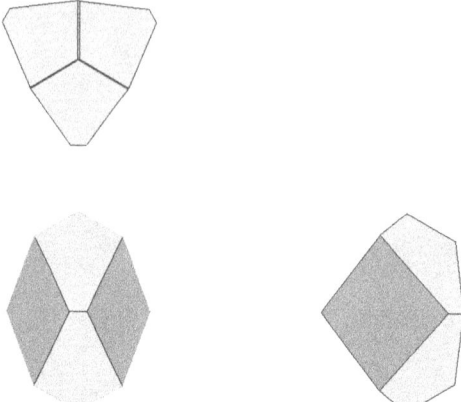

*Figure 102: The dual of Johnson solid 51 (the triaugmented triangular prism): the peritruncated triangular bipyramid.*

Table 22 provides a detailed comparison of the properties of Johnson solids 49 through 51 and their duals.

*(Laterally) augmented prisms: Johnson solids 49-57.*

| Johnson solid number | 49 | 50 | 51 |
|---|---|---|---|
| Name | Augmented triangular prism | Biaugmented triangular prism | Triaugmented triangular prism |
| Symmetry* | $C_{2v}$ | $C_{2v}$ | $D_{3h}$ |
| Faces | [2+2+2](3) + [2](4) | [2+2+2+4](3) + [1](4) | [2+6+6](3) |
| Edges* | 13 | 17 | 21 |
| Vertices | 7 | 8 | 9 |
| Dihedral angles | 3a-3b: 109.4712° (109° 28' 16"), 3b-3c: 144.7356° (144° 44' 8"), 3a-4: 114.7356° (114° 44' 8"), 3c-4: 90°, 4-4: 60°♣ | 3a-3a: 169.4712° (169° 28' 16"), 3a-3b: 109.4712° (109° 28' 16"), 3a-3c: 144.7356° (144° 44' 8"), 3b-4: 114.7356° (114° 44' 8"), 3c-4: 90°△ | 3a-3b: 109.4712° (109° 28' 16"), 3a-3c: 144.7356° (144° 44' 8"), 3b-3b: 169.4712° (169° 28' 16")* |
| Area | 4.59807621 = 2 + (3/2)√3 | 5.33012702 = 1 + (5/2)√3 | 6.06217783 = (7√3)/2 |
| Volume | 0.66871496 | 0.90441722 | 1.14011948 |
| Dual | | | |
| Name | Laterally monotruncated triangular bipyramid | Laterally bitruncated triangular bipyramid | Peritruncated triangular bipyramid |
| Faces | [2](3) + [4+1](4) | [4+2](4) + [2](5) | [3](4) + [6](5) |
| Vertices | 8 | 11 | 14 |

*The Johnson Solids and Their Duals*

| Johnson solid number | 49 | 50 | 51 |
|---|---|---|---|
| Face angles[†] | 3: 89.3269° (89° 19′ 37″), [2]45.3366° (45° 20′ 12″), 4a: 54.0528° (54° 3′ 10″), [2]101.4624° (101° 27′ 45″), 103.0225° (103° 1′ 21″), 4s: [4]90°[⊛] | 4a: 53.5401° (53° 32′ 24″), 95.0677° (95° 4′ 4″), 104.0982° (104° 5′ 54″), 107.2940° (107° 17′ 38″); 4k: [2]91.1382° (91° 8′ 18″), 79.4652° (79° 27′ 55″), 98.2584° (98° 15′ 30″); 5: [2]107.7933° (107° 47′ 36″), [2]110.1243° (110° 7′ 27″), 104.1648° (104° 9′ 53″)[¤] | 4: [4]90°, 5: [4]110.7048° (110° 42′ 17″), 97.1808° (97° 10′ 51″)[⊛] |
| Dihedral angles | 3-3: 110.4149° (110° 24′ 53″), 3-4a: 103.2135° (103° 12′ 49″), 4a-4a: 92.3564° (92° 21′ 23″), 4a-4s: 101.6989° (101° 41′ 56″)[⊛] | 4a-4a: 104.7440° (104° 44′ 38″), 4a-4k: 111.0881° (111° 5′ 17″), 4a-5: 96.5368° (96° 32′ 12″), 4k-5: 105.4712° (105° 28′ 16″), 5-5: 84.3377° (84° 20′ 16″)[¤] | 4-5: 112.2077° (112° 12′ 28″), 5-5: 98.2132° (98° 12′ 48″)[⊛] |
| Area | 2.77162939 | 1.34585638 | 3.07853797 |
| Volume | 0.32043845 | 0.11290929 | 0.40228191 |

Table 22: Properties of Johnson solids 49, 50, and 51 and their duals.

[*]The symmetry and number of edges are always the same for each solid and its dual, so are not repeated in the "dual" section.

[†]Face angles are only given for the duals, as the Johnson

solids have regular faces, whose angles can be found in Table 1 on p. 3.

♦The symbol 3a designates the triangular faces from the pyramidal augmentation that are adjacent to the two square faces remaining from the prism; 3b designates the triangular faces from the pyramidal augmentation that are adjacent to the triangles derived from the prism; 3c designates the triangular faces that originate as the bases of the prism.

⊛The symbol 4a designates the four irregular quadrilaterals that are among the faces of this polyhedron, 4s is the one that is square. Note that the 4a quadrilaterals are *not* kite-quadrilaterals, even though they have two equal angles, because the equal angles are not opposite each other. All the 4a-4a dihedral angles are equal, even though there are two different 4a-4a edge lengths: 0.33137587 and 0.72442267 (assuming 1 for the edge of the original Johnson solid).

△Although the triangular faces fall into *four* transitivity classes, only three sets need be distinguished: 3a represents the four triangles that share exactly one vertex with the square face; 3b represents the four triangles that are bisected by the mirror plane halfway between the original triangular bases of the prism; and 3c represents those two triangles that derive from the bases of the prism.

▫The symbol 4k refers to the two kite-quadrilaterals dual to the apices of the pyramids that were considered the augmentations of the prism; 4a refers to the remaining four quadrilaterals, which are irregular.

✢The symbol 3a refers to the six triangles deriving from the pyramidal augmentations that are adjacent to the triangles that derive from the bases of the prism; the symbol 3b to the remaining six triangles deriving from the pyramidal augmentations; the symbol 3c to the two triangles that derive from the bases of the prism.

❋The pentagonal faces have four identical angles, with the side opposite the one angle that is different much shorter than the others. The two sides that include the small angle are intermediate in length between the shortest and the remaining two. These faces form two types of 5-5 edges, but the dihedral angles at both are identical.

*(Laterally) augmented prisms: Johnson solids 49-57.*

Johnson solid 52, the (laterally) augmented pentagonal prism (illustrated in Figure 103 below), is like $J_{49}$ in that its rotational symmetry around the prism axis is lost as a result of the augmentation, but a twofold axis through the apex of the augmenting pyramid, and two mirror planes, one bisecting the bases and one halfway between them, become the only symmetry elements remaining. $J_{52}$ has therefore only $C_{2v}$ symmetry.

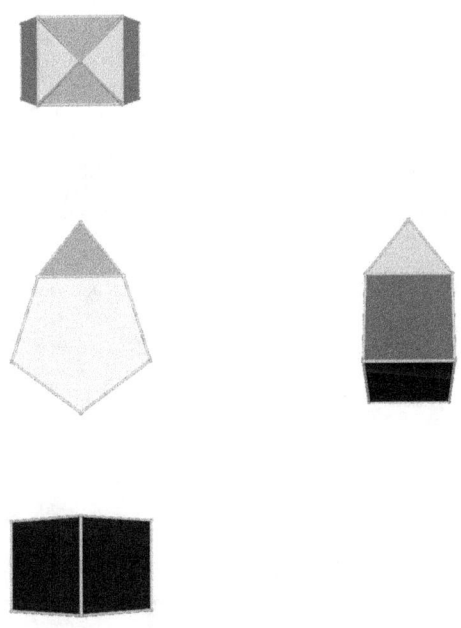

*Figure 103: Johnson solid 52: the (laterally) augmented pentagonal prism.*

The dual of $J_{52}$ (illustrated in Figure 104 below) is like the dual of $J_{49}$ in that the face dual to the pyramid apex, which by the $C_{2v}$ symmetry of the polyhedron need only be a rhombus, is in fact a perfect square. And the four irregular quadrilateral faces adjacent to the square face, though forming two kinds of edge with each other (at their shortest and their long sides), have identical dihedral angles at those edges, again like the dual of $J_{49}$. However, the polyhedron can be considered as a laterally monotruncated *pentagonal* bipyramid, rather than a laterally monotruncated *triangular* bipyramid — thus there are a number of additional triangular faces.

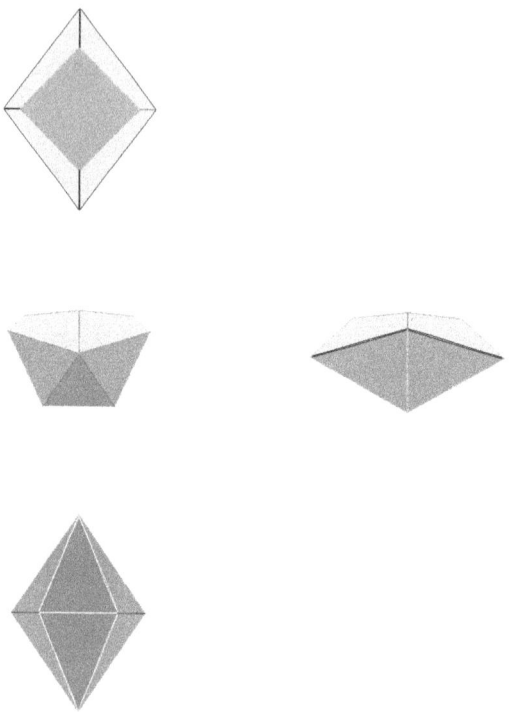

*Figure 104: The dual of Johnson solid 52 (the laterally augmented pentagonal prism): a pentagonal bipyramid with a lateral vertex truncated.*

Johnson solid 53, the (laterally) biaugmented pentagonal prism, (illustrated in Figure 105 below) has a second pyramid fused to a lateral face of a pentagonal prism. There is only one way to do this, as augmenting two adjacent lateral faces produces a non-convex polyhedron, and when there are five lateral faces, there is no other classification of pairs than adjacent or nonadjacent. (There is nothing like the "meta"/"para" distinction found in the hexagonal case.) The symmetry, as in the "triangular" case, is reduced to $C_{2v}$, and the orientation in Figure 105 reflects this. There are now three different types of triangular face, because of the four faces comprising each of the two pyramidal augmentations, two are adjacent to the pentagonal faces (bases of the original prism), one to the single square (in the bottom view of Figure 105) and one to the paired square faces (in the top view of the figure).

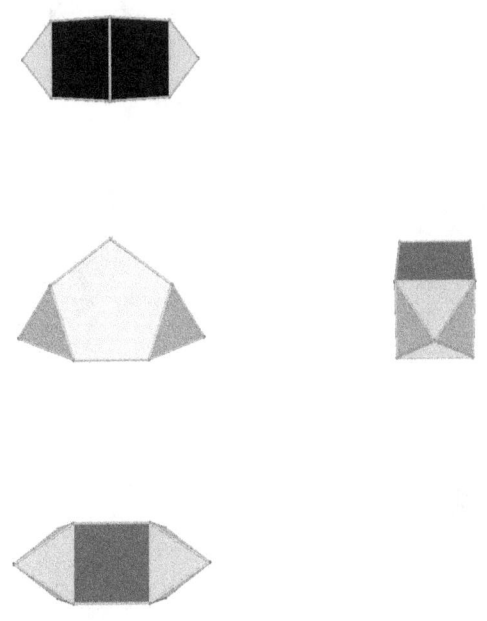

Figure 105: Johnson solid 53: the (laterally) biaugmented pentagonal prism.

The dual of $J_{53}$ (illustrated in Figure 106 below) is, as expected, a pentagonal bipyramid with two nonadjacent lateral vertices truncated. Because the centroid is not along what was originally the fivefold axis of the prism, the two quadrilaterals produced by the truncation are not exactly squares, but kite-quadrilaterals very close to square in their measurements. The triangular faces, which straddle the mirror planes, are naturally isosceles, but the remaining faces are irregular quadrilaterals belonging to two different transitivity classes. (They would be identical if the polar reciprocation were done with a center at the centroid of the original prism, but since the two augmentations pull the centroid nearer to one set and further from the other set of four irregular quadrilaterals, they end up somewhat different in measurements.)

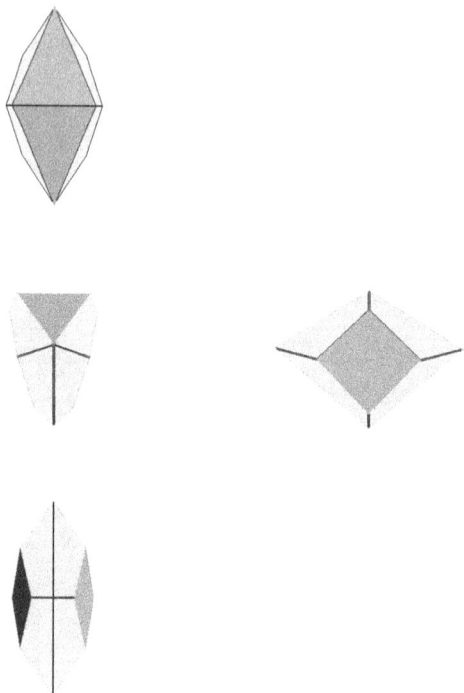

Figure 106: The dual of Johnson solid 53 (the laterally biaugmented pentagonal prism): a pentagonal bipyramid with two nonadjacent lateral vertices truncated.

*(Laterally) augmented prisms: Johnson solids 49-57.*

| Johnson solid number | 52 | 53 |
|---|---|---|
| Name | Augmented pentagonal prism | Biaugmented pentagonal prism |
| Symmetry* | $C_{2v}$ | $C_{2v}$ |
| Faces | [2+2](3) + [2+2](4) + [2](5) | [2+2+4](3) + [1+2](4) + [2](5) |
| Edges* | 19 | 23 |
| Vertices | 11 | 12 |
| Dihedral angles | 3-3: 109.4712° (109° 28′ 16″), 3-4: 162.7356° (162° 44′ 8″), 3-5: 144.7356° (144° 44′ 8″), 4-4: 108°, 4-5: 90°* | |
| Area | 9.17300561 | 9.90505642 |
| Volume | 1.95617966 | 2.19188192 |
| **Dual** | | |
| Name | Laterally monotruncated pentagonal bipyramid | Laterally bitruncated pentagonal bipyramid (two non-adjacent vertices truncated) |
| Faces | [2+4](3) + [1+4](4) | [2](3) + [2+4+4](4) |
| Vertices | 10 | 13 |
| Face angles† | 3a: [2]71.6926° (71° 41′ 33″), 36.6149° (36° 36′ 54″); 3b: 39.6149° (39° 36′ 54″), 65.0513° (65° 3′ 5″), 75.3358° (75° 20′ 2″); 4a: [2]121.8717° (121° 52′ 18″), 44.1947° (44° 11′ 41″), 72.0619° (72° 3′ 43″); 4s: [4]90°□ | 3: [2]70.9040° (70° 54′ 14″), 38.1921° (38° 11′ 31″); 4a: 39.8565° (39° 51′ 24″), 73.0819° (73° 4′ 55″), 122.7538° (122° 45′ 14″), 124.3078° (124° 18′ 28″); 4b: 42.4448° (42° 26′ 41″), 66.6147° (66° 36′ 53″), 124.4716° (124° 28′ 18″), 126.4689° (126° 28′ 8″); 4k: [2]90.2127° (90° 12′ 46″), 85.7509° (85° 45′ 3″), 93.8238° (93° 49′ 26″)⊛ |

| Johnson solid number | 52 | 53 |
|---|---|---|
| Dihedral angles | 3a-3a: 125.8195° (125° 49′ 10″), 3a-3b: 123.9940° (123° 59′ 39″), 3b-3b: 121.7819° (121° 46′ 55″), 3b-4a: 118.0074° (118° 0′ 27″), 4a-4a: 112.7422° (112° 44′ 32″), 4a-4s: 128.4448° (128° 26′ 41″)¤ | 3-3: 123.0848° (123° 5′ 5″), 3-4a: 121.8828° (121° 52′ 58″), 4a-4a: 120.5401° (120° 32′ 24″), 4a-4b: 118.2891° (118° 17′ 21″), 4a-4k: 133.4520° (133° 27′ 7″), 4b-4b: 115.5773° (115° 34′ 38″), 4b-4k: 133.3334° (133° 20′ 0″)⊗ |
| Area | 6.02709983 | 7.16012625 |
| Volume | 1.10850220 | 1.41927230 |

Table 23: *Properties of Johnson solids 52 and 53 and their duals.*

*The symmetry and number of edges are always the same for each solid and its dual, so are not repeated in the "dual" section.

†Face angles are only given for the duals, as the Johnson solids have regular faces, whose angles can be found in Table 1 on p. 3.

‡Although the square faces of both $J_{52}$ and $J_{53}$ (and the triangular faces of $J_{52}$) divide into *two* transitivity classes and the triangular faces of $J_{53}$ into *three*, all 4-5 edges have the same 90° dihedral angle, and all 4-4 edges have the same 108° dihedral angle, while the designations 3-3, 3-4, and 3-5 incur no ambiguity in the case of $J_{52}$, nor does 3-5 in the case of $J_{53}$. While there are more than one kind of 3-3 and 3-4 edge in $J_{53}$, again, the values of the dihedral angles are the same, 109.4712° (109° 28′ 16″), for all 3-3 edges, and the same, 162.7356° (162° 44′ 8″), for all 3-4 edges.

¤The designation 3a refers to the two isosceles triangles which share no edges with any quadrilaterals; 3b to the four scalene triangles which separate the 3a faces from the quadrilaterals; 4a to the four irregular quadrilaterals surrounding the

*(Laterally) augmented prisms: Johnson solids 49-57.*

square face; 4s to the single square face. Note that the 4a quadrilaterals are *not* kite-quadrilaterals, even though they have two equal angles, because the equal angles are not opposite each other.

*The designation 4a refers to the four irregular quadrilaterals which surround the two isosceles triangles; 4b to the four irregular quadrilaterals that do not share an edge with either of the two isosceles triangles; 4k refers to the two nearly-square kite-quadrilaterals which are the only faces not meeting at one of the two quinquevalent apices.

The (laterally) augmented hexagonal prism, $J_{54}$, is illustrated in Figure 107 below. The augmentation destroys the sixfold symmetry of the hexagonal prism, but the line through the apex of the augmenting pyramid and the centroid of the prism remains a twofold axis, with two planes through that axis retaining the property of being mirror planes even with the augmentation, so that the symmetry is $C_{2v}$.

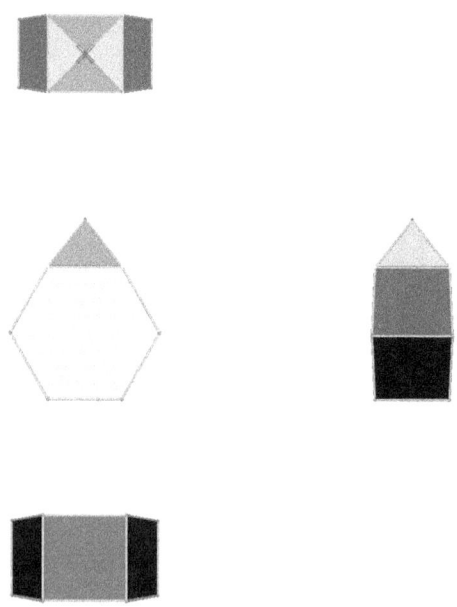

Figure 107: Johnson solid 54: the (laterally) augmented hexagonal prism.

It might be noticed that the two square faces of the augmented hexagonal prism ($J_{54}$) immediately adjacent to the triangular faces coming from the pyramid appear almost to be in the same plane as those triangular faces. In fact the dihedral angles between the faces in question are 174.7356° (174° 44′ 8″), so close to 180° that it is hard to tell the difference visually.

The dual of $J_{54}$ (illustrated in Figure 108 below) shares the

*(Laterally) augmented prisms: Johnson solids 49-57.*

$C_{2v}$ symmetry of $J_{54}$ itself, of course. The face which is dual to the apex of the augmenting pyramid would be required only to be a rhombus by the $C_{2v}$ symmetry, but is in fact a perfect square, as has been noted in some of the other cases discussed in this chapter. The four faces immediately adjacent to the square face are congruent irregular quadrilaterals, which are *not* kite-quadrilaterals, even though they have two equal angles, because the equal angles are not opposite each other; and the remaining eight faces are all triangles, which are dual to those vertices of the original hexagonal prism that are not adjacent to the augmentation.

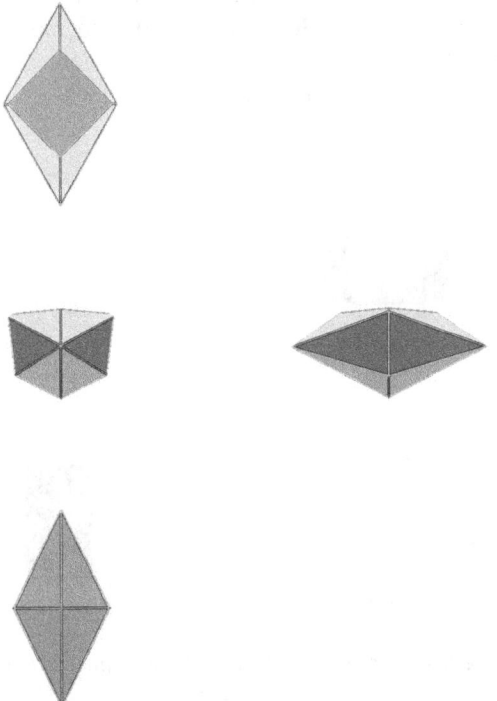

Figure 108: The dual of Johnson solid 54 (the laterally augmented hexagonal prism): a hexagonal bipyramid with a lateral vertex truncated.

Although both sets of triangles derive from vertices that are identical in the original polyhedron, and would be congruent and isosceles in the bipyramid that is the dual to the unaugmented hexagonal prism, the augmentation (or the truncation, if one concentrates on the dual itself) disturbs the symmetry enough that the two sets of triangles are not exactly isosceles nor congruent to each other.

There are two possible biaugmented hexagonal prisms, thus the two numbers 55 and 56 given here: $J_{55}$ (Figure 109) has the augmentations on opposite faces (the "para-" position, using the organic chemistry-based terminology mentioned below), and $J_{56}$ (Figure 111) has the augmentations on faces 120° apart (the "meta-" position).

Johnson solid 55, the parabiaugmented hexagonal prism, is very unusual in having $D_{2h}$ symmetry. It will be seen later that Johnson solid 91 (the bilunabirotunda) also has $D_{2h}$ symmetry, and those two are the *only* two Johnson solids with $D_{2h}$ symmetry among the entire set of ninety-two.

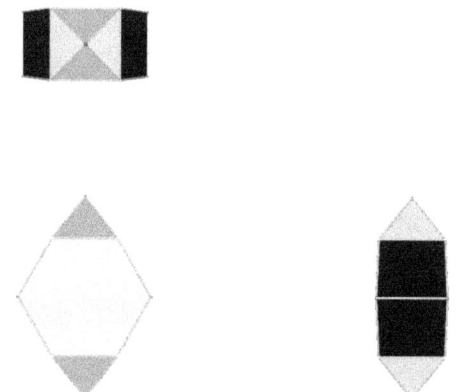

Figure 109: Johnson solid 55: the parabiaugmented hexagonal prism.

The dual of the parabiaugmented hexagonal prism is shown in Figure 110 below. Two of the faces are perfect squares, though the symmetry would only require them to be rhombi, and in addition, there are some interesting relationships among the angles of the four isosceles triangles and the angles of the eight irregular quadrilaterals (which are *not* kite-quadrilaterals, even though they have two equal angles, because the equal

*(Laterally) augmented prisms: Johnson solids 49-57.*

angles are not opposite each other), which will be discussed in detail. The isosceles triangles have their two equal angles equal to 75.5225° (75° 31′ 21″) and the remaining angle equal to 28.9550° (28° 57′ 18″), while the two angles that are not equal in the irregular quadrilaterals are also the same 28.9550° (28° 57′ 18″) and 75.5225° (75° 31′ 21″); the remaining two angles are both 127.7612° (127° 45′ 40″). Given that the sum of angles in a triangle must be 180° and the sum of angles in a quadrilateral must be 360°, all three of those numbers must be related by

$$75.5225° = 90° - 28.9550°/2,$$

$$127.7612° = 135° - 28.9550°/4,$$

and of course 135° = (3/2)90°, so these angles are all very closely related.

*Figure 110: The dual of Johnson solid 55 (the parabiaugmented hexagonal prism): a hexagonal bipyramid with two opposite lateral vertices truncated.*

There is no particular surprise in that the angles in the triangles add to 180° or that the angles in the quadrilaterals add to 360°, of course, but that the angles in both are equal certainly is unexpected. All the dihedral angles between the triangles and the irregular quadrilaterals (which means *all* dihedral angles except those involving the square faces) are equal as well, at 126.8699° (126° 52′ 12″). This is in great contrast to the dual of $J_{54}$, which has every type of dihedral angle different, and in which the triangular faces are not all alike, nor do their angles match those of the quadrilateral faces (though it can be seen that they are rather close).

Johnson solid 56, the metabiaugmented hexagonal prism, is illustrated in Figure 111 below. As is our convention in this book, it is shown in a position that makes the mirror plane through the hexagonal face, as well as the twofold axis that is its only axis of rotational symmetry, vertical.

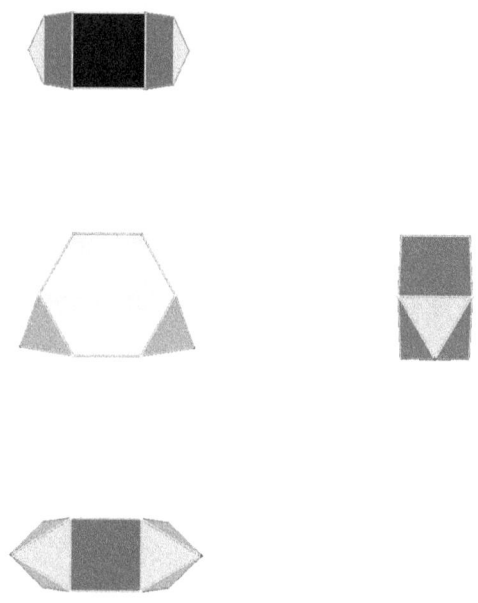

Figure 111: Johnson solid 56: the metabiaugmented hexagonal prism.

The dual of the metabiaugmented hexagonal prism is illustrated in Figure 112 below. Examining Table 24 on p. 223 below, one can see that all the faces of this polyhedron are extremely irregular. The two faces that are dual to the pyramid vertices are kite-quadrilaterals; none of the other faces even have two angles equal. It turns out that much of this irregularity is a result of using the centroid of the *entire polyhedron* as the center of the polar reciprocation process.

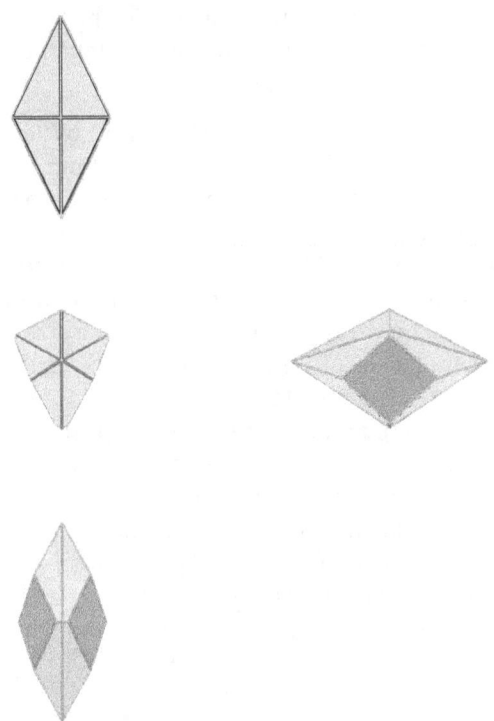

Figure 112: The dual of Johnson solid 56 (the metabiaugmented hexagonal prism): a hexagonal bipyramid with two lateral vertices, neither adjacent nor opposite, truncated.

If one uses the center of the original prism as the center of the polar reciprocation, so that all of the vertices that do not have pyramids attached to them, as well as all of the vertices that *do* have pyramids attached to them, are the same distance from the reciprocation center, one gets a different picture. In *that* dual, all the triangular faces are isosceles, with angles of 75.5225° (75° 31′ 21″) (twice) and 28.9550° (28° 57′ 18″) and the quadrilateral faces that are adjacent are still irregular, but now they have two equal angles of 127.7612° (127° 45′ 40″) each, while the remaining angles match the 75.5225° (75° 31′ 21″) and

28.9550° (28° 57′ 18″) angles of the triangles. (One might notice also that these angles are identical to the angles of the dual of $J_{55}$, which may more easily be seen to be reasonable because, in $J_{55}$, the centroid of the polyhedron *is* the centroid of the original prism). The remaining two quadrilaterals, also, instead of being kite-quadrilaterals, are perfect squares in the version of the dual where the center of the original prism is taken as the center of the polar reciprocation process.

    The prefixes "para-" and "meta-" used by Johnson in naming $J_{55}$ and $J_{56}$ seem to have been adapted from the terminology of organic chemistry. (Although another suggestion which I have seen is that "para-" comes from "parallel." But that would not explain his use of "meta." Therefore I assume that the use of these terms in organic chemistry is the real origin.) In that field, when two groups of atoms are attached to the hexagonal frame of a molecule of benzene, the prefix "ortho-" designates attachment at adjacent atoms (60° apart), "para-" at opposite atoms (180° apart), and the intermediate position (120° apart) is designated by "meta-." If pyramids were to be fused to two adjacent lateral faces of a regular hexagonal prism, the resulting polyhedron would not be convex, so that only the metabiaugmented and parabiaugmented hexagonal prisms exist in the list. (It does not appear necessary to specify "laterally" in these cases, as the "meta-" and para-" are clear enough in designating the two positions that are augmented.)

*(Laterally) augmented prisms: Johnson solids 49-57.*

| Johnson solid number | 54 | 55 | 56 |
|---|---|---|---|
| Name | Augmented hexagonal prism | Parabiaugmented hexagonal prism | Metabiaugmented hexagonal prism |
| Symmetry* | $C_{2v}$ | $D_{2h}$ | $C_{2v}$ |
| Faces | [2+2](3) + [1+2+2](4) + [2](6) | [4+4](3) + [4](4) + [2](6) | [4+4](3) + [1+1+2](4) + [2](6) |
| Edges* | 22 | 26 | 26 |
| Vertices | 13 | 14 | 14 |
| Dihedral angles | \multicolumn{3}{c}{3-3: 109.4712° (109° 28' 16"), 3-4: 174.7356° (174° 44' 8"), 3-6: 144.7356° (144° 44' 8"), 4-4: 120°, 4-6: 90°*} | | |
| Area | 11.92820323 = 5 + 4√3 | 12.66025404 = 4 + 5√3 | 12.66025404 = 4 + 5√3 |
| Volume | 2.83377847 | 3.06948073 | 3.06948073 |
| Dual | | | |
| Name | Laterally monotruncated hexagonal bipyramid | Laterally bitruncated hexagonal bipyramid (opposite vertices truncated) | Laterally bitruncated hexagonal bipyramid (neither adjacent nor opposite vertices truncated) |
| Faces | [4+4](3) + [1+4](4) | [4](3) + [2+8](4) | [4](3) + [2+4+4](4) |
| Vertices | 11 | 14 | 14 |

## The Johnson Solids and Their Duals

| Johnson solid number | 54 | 55 | 56 |
|---|---|---|---|
| Face angles[†] | 3a: 26.8289° (26° 49′ 44″), 74.9543° (74° 57′ 15″), 78.2168° (78° 13′ 1″); 3b: 29.1988° (29° 11′ 56″), 71.5397° (71° 32′ 23″), 79.2616° (79° 15′ 42″); 4a: [2]125.9541° (125° 57′ 15″), 31.5238° (31° 31′ 26″), 76.5680° (76° 34′ 5″); 4s: [4]90°[△] | 3: [2]75.5225° (75° 31′ 21″), 28.9550° (28° 57′ 18″); 4a: [2]127.7612° (127° 45′ 40″), 28.9550° (28° 57′ 18″), 75.5225° (75° 31′ 21″); 4s: [4]90°[♦] | 4a: 29.1652° (29° 9′ 55″), 79.0045° (79° 0′ 16″), 124.6845° (124° 41′ 4″), 127.1459° (127° 8′ 45″); 3: 26.9687° (26° 58′ 7″), 74.9828° (74° 58′ 58″), 78.0486° (78° 2′ 55″); 4/r/: 31.3226° (31° 19′ 21″), 72.2044° (72° 12′ 16″), 126.6338° (126° 38′ 2″), 129.8393° (129° 50′ 22″); 4k: [2]90.3529° (90° 21′ 11″), 84.4420° (84° 26′ 31″), 94.8521° (94° 51′ 8″)[¤] |

— 222 —

*(Laterally) augmented prisms: Johnson solids 49-57.*

| Johnson solid number | 54 | 55 | 56 |
|---|---|---|---|
| Dihedral angles | 3a-3a: 131.3641° (131° 21′ 51″), 3a-3b: 129.5503° (129° 33′ 1″), 3b-3b: 127.2023° (127° 12′ 8″), 3b-4a: 124.8371° (124° 50′ 13″), 4a-4a: 121.7415° (121° 44′ 29″), 4a-4s: 136.4954° (136° 29′ 43″)△ | 3-3, 3-4a, 4a-4a: 126.8699° (126° 52′ 12″); 4a-4s: 140.7685° (140° 46′ 7″)♦ | 3-3: 131.0630° (131° 3′ 47″), 3-4a: 129.3430° (129° 20′ 35″), 4a-4a: 127.1569° (127° 9′ 25″), 4a-4b: 124.9550° (124° 57′ 18″), 4a-4k: 138.8741° (138° 52′ 27″), 4b-4b: 122.1296° (122° 7′ 47″), 4b-4k: 139.2077° (139° 12′ 28″)¤ |
| Area | 9.74063508 | 14.55124830 | 9.75173040 |
| Volume | 2.25428568 | 4.05952452 | 2.21412071 |

Table 24: Properties of Johnson solids 54, 55, and 56 and their duals.

˙The symmetry and number of edges are always the same for each solid and its dual, so are not repeated in the "dual" section.

†Face angles are only given for the duals, as the Johnson solids have regular faces, whose angles can be found in Table 1 on p. 3.

*Although the square faces of both $J_{54}$ and $J_{56}$ divide into *three* transitivity classes and the triangular faces of all three of the polyhedra in this table into *two*, all 4-4 edges have the same 120° dihedral angle, and all 4-6 edges have the same 90° dihedral angle, while the designations 3-3, 3-4, and 3-6 incur no ambiguity.

△The designation 3a refers to the four triangular faces opposite to the square face; 3b to the remaining four triangular faces. The designation 4a refers to the four irregular quadrilaterals; 4s to the square.

♦The designation 4a refers to the eight irregular quadrilaterals; 4s to the two squares.

"The designation 4a refers to the four irregular quadrilaterals that share edges with the triangular faces; 4b to the four remaining irregular quadrilaterals; and 4k to the two kite-quadrilaterals, which are the only polygons which do not meet at the hexavalent vertices.

*(Laterally) augmented prisms: Johnson solids 49-57.*

Because there is no way of placing more than two augmentations on a prism whose bases have 5 or fewer sides without making at least two adjacent, and adjacent augmentations can only occur if $n = 3$, only a triangular or hexagonal prism can be triaugmented on lateral faces, and a hexagonal prism must have the augmentations on alternate lateral faces. This leads to $J_{57}$, illustrated below in Figure 113, which can be unambiguously labeled as a triaugmented hexagonal prism.

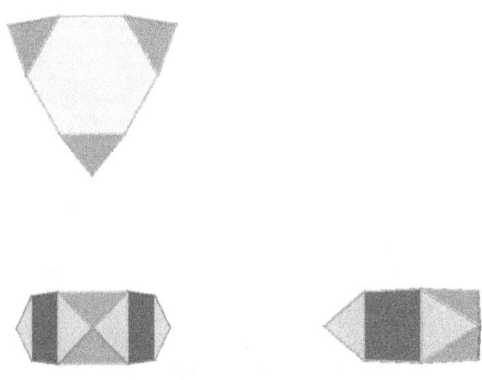

Figure 113: Johnson solid 57: the triaugmented hexagonal prism.

While the symmetry of $J_{54}$ through $J_{56}$ is only twofold, $J_{57}$ has a threefold axis which is actually the sixfold axis of the original prism, so Figure 113 has a completely different orientation from the earlier figures. The twofold axes through the pyramid apices and the mirror plane that runs through all of those apices still remain, so the overall symmetry is $D_{3h}$.

The dual of the triaugmented hexagonal prism is illustrated in Figure 114 below. It shares the $D_{3h}$ symmetry of $J_{57}$ itself. Although the symmetry is $D_{3h}$, with the only axes of rotational symmetry being the threefold axis through the two apices and three twofold axes perpendicular to them, the three quadrilaterals around the middle of the polygon are perfect squares, though the symmetry would only require them to be rhombi. By contrast, the remaining twelve quadrilaterals are among the most irregular polygons met with in this book. If the length of all the edges of the original Johnson solid is taken as 1, the shortest edges of those twelve irregular quadrilaterals are ~0.1166, and the longest are ~2.3094, almost twenty times as long!

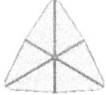

*Figure 114: The dual of Johnson solid 57 (the triaugmented hexagonal prism).*

This completes the discussion of the laterally augmented prisms.

*(Laterally) augmented prisms: Johnson solids 49-57.*

| Johnson solid number | 57 |
|---|---|
| Name | Triaugmented hexagonal prism |
| Symmetry* | $D_{3h}$ |
| Faces | [6+6](3) + [3](4) + [2](6) |
| Edges* | 30 |
| Vertices | 15 |
| Dihedral angles | 3-3: 109.4712° (109° 28′ 16″), <br> 3-4: 174.7356° (174° 44′ 8″), <br> 3-6: 144.7356° (144° 44′ 8″), <br> 4-6: 90°♣ |
| Area | 13.39230485 = 3 + 6√3 |
| Volume | 3.30518299 |
| **Dual** | |
| Name | Unnamed |
| Faces | [3+12](4) |
| Vertices | 17 |
| Face angles† | 4a: [2]127.7612° (127° 45′ 40″), <br> 28.9550° (28° 57′ 18″), <br> 75.5225° (75° 31′ 21″); <br> 4s: [4]90°* |
| Dihedral angles | 4a-4a: 126.8699° (126° 52′ 12″), <br> 4a-4s: 140.7685° (140° 46′ 7″), |
| Area | 14.08090576 |
| Volume | 3.77988570 |

Table 25: *Properties of Johnson solid 57 and its dual.*

*The symmetry and number of edges are always the same for each solid and its dual, so are not repeated in the "dual" section.

†Face angles are only given for the duals, as the Johnson solids have regular faces, whose angles can be found in Table 1 on p. 3.

♣Although the triangular faces divide into *two* transitivity classes, there is no ambiguity in the designation of edges as 3-3, 3-4, or 3-6. A 3-3 edge is only between one triangle in each class and one in the other, while only one class participates in 3-4 edges and only the other participates in 3-6 edges.

*The symbol 4a designates the twelve irregular quadrilaterals that are among the faces of this polyhedron, 4s designates the three that are square. Note that the 4a quadrilaterals are *not* kite-quadrilaterals, even though they have two equal angles, because the equal angles are not opposite each other.

# Chapter 12: Augmented dodecahedra: Johnson solids 58-61.

Although (as will be noted in Chapter 14) Norman Johnson used the term "augmented" in a broader sense in naming some of the solids in his list, *augmentation* normally means, and will mean in this chapter, the fusion of a pyramid to a face of a polyhedron. For this to produce a Johnson solid, all faces of the original polyhedron (except, possibly the one to which the pyramid is fused) must be regular polygons, and the pyramid itself must have lateral faces that are also regular polygons. (In theory, the two faces that disappear in the fusion process need not be regular; in fact, there is no example of a polyhedron that can be obtained by fusing a pyramid whose base is not regular to a polyhedron with all other faces regular but having one irregular face to match the base of the pyramid.) For the starting polyhedron, let us consider the five Platonic solids.

If it were not necessary to make all the faces regular (the defining characteristic of the Johnson solids) any one of the five Platonic solids could be augmented on any face, or on any combination of faces. This was briefly mentioned on p. 43 where the first of these augmented Platonics, $J_8$, was described.) However, the requirement of regularity leads to some stringent restrictions.

The dihedral angle at the base of a regular tetrahedron (regular triangular pyramid) is 70.5288° (70° 31′ 44″), the dihedral angle at the base of a regular square pyramid ($J_1$) is 54.7356° (54° 44′ 8″), and the dihedral angle at the base of a regular pentagonal pyramid ($J_2$) is 37.3774° (37° 22′ 39″). This means that if a triangular face of a polyhedron has the dihedral angle with any of its neighboring faces exceeding 109.4712° (109° 28′ 16″), the resulting polyhedron will not be convex, and if it is exactly equal to that figure, the faces of the pyramid and starting polyhedron will lie in the same plane, not form two separate faces in the augmented polyhedron; the corresponding angle for a square face is 125.2644° (125° 15′ 52″), and for a pentagonal face it is 142.6226° (142° 37′ 21″). (All these numbers are simply obtained by subtracting the pyramid base dihedral angles from 180°.)

The dihedral angles for the three Platonic solids with triangular faces are: tetrahedron, 70.5288° (70° 31′ 44″) as was just noted; octahedron 109.4712° (109° 28′ 16″); icosahedron 138.1897° (138° 11′ 23″). So only the tetrahedron can be augmented; but augmenting the tetrahedron produces a triangular bipyramid ($J_{12}$), which has already been described in Chapter 5. And since the addition of that pyramid increases the dihedral angles between the faces coming from the pyramid and each of the remaining faces of the original tetrahedron to 141.0576° (141° 3′ 27″), no additional faces of the tetrahedron can be augmented.

The dihedral angles for the only Platonic solid with square faces, the cube, are 90°, which is small enough to allow augmentation; but augmenting the cube produces an elongated square pyramid ($J_8$), which has already been described in Chapter 4. Since this increases the dihedral angles adjacent to the elongation to 144.7356° (144° 44′ 8″), none of the remaining original faces of the cube except the one opposite to the augmented one can take a second augmentation, and augmenting that face produces an elongated square bipyramid ($J_{15}$), which has already been described in Chapter 6.

Thus, the only Platonic solid remaining to consider is the dodecahedron. The same considerations as in the cases above show that any face (all of which are pentagonal) can be augmented, and while no adjacent faces can be, any two or more faces can be augmented if no two are adjacent. This leads to the four Johnson solids covered in this chapter.

Before actually proceeding to the enumeration of the properties of the four Johnson solids and their duals covered in this chapter, a discussion of why there are exactly four is in order. The dodecahedron has twelve faces and can be considered the simplest of the polyhedra with "icosahedral symmetry," symbolized $I_h$ in Schönflies notation. Any object with $I_h$ symmetry must have twelve (or a multiple of twelve) elements (in the case of the dodecahedron, these are the faces; however, since we are talking in generalities, the term "element" will continue to be used), and each of these elements is the location of a fivefold axis (although there are in fact only *six* fivefold axes, because each fivefold axis passes through two opposite elements). Adjacent to any one of the twelve elements is a ring of five more, which can be termed the *nearer ring*, and adjacent to this ring is another ring of five, which can be termed the *further ring*. (Each member of this further ring is opposite to one of the members of the nearer ring.) This leaves the final, twelfth element, which is directly opposite the one we started with. If we want to perform a modification to one element, in this case augmenting a face, we have *one* member of a set.

Now suppose we have a restriction such as the one mentioned earlier, that no two adjacent faces of the dodecahedron can be augmented (or in the more general terminology we have been using, that no two adjacent elements of the object can be modified). Then, the first modification just described excludes five elements (the nearer ring adjacent to the modified element) from being the second to be modified, leaving six possibilities: the five in the further ring, and the element opposite the one first modified. If the opposite element is the second one modified (which has been termed "para" based on the organic chemistry terminology copied by Johnson), *every* remaining element, all ten, are excluded from further modification. The five in the nearer ring are adjacent to the first one modified, and the five in the further ring are adjacent to the second one modified. We can consider this case (for the dodecahedron as our object and augmentation as our modification, the *parabiaugmented dodecahedron*) the second member of the set we are constructing.

Alternatively, if the second modification is done to one of the elements in the further ring (it does not matter which, because the fivefold axis, which remained after the first modification, means that all the members of either ring are equivalent), this produces the "meta" case (for the dodecahedron as our object and augmentation as our modification, the *metabiaugmented dodecahedron*), the third member of the set we are constructing.

Modifying one of the elements in the further ring in this manner excludes from modification three more of the twelve elements: the element opposite the originally modified element and two of the elements in the further ring adjacent to the one modified second. With eight elements thus excluded (five by the first modification and three by the second), and two already modified, only two elements remain as possible subjects for modification. In fact, the object at this point has a mirror plane remaining, in a position that makes these two equivalent, so either of these elements can be modified producing the same result, the fourth member of the set we are constructing (for the dodecahedron as our object and augmentation as our modification, the *triaugmented dodecahedron*).

In subsequent chapters we will see other such sets of four Johnson solids obtained by modifying an icosahedrally-symmetric polyhedron in this way, but the rest of this chapter will be devoted specifically to the augmented dodecahedra.

The first augmentation produces the augmented dodecahedron, Johnson solid 58, illustrated in Figure 115 below.

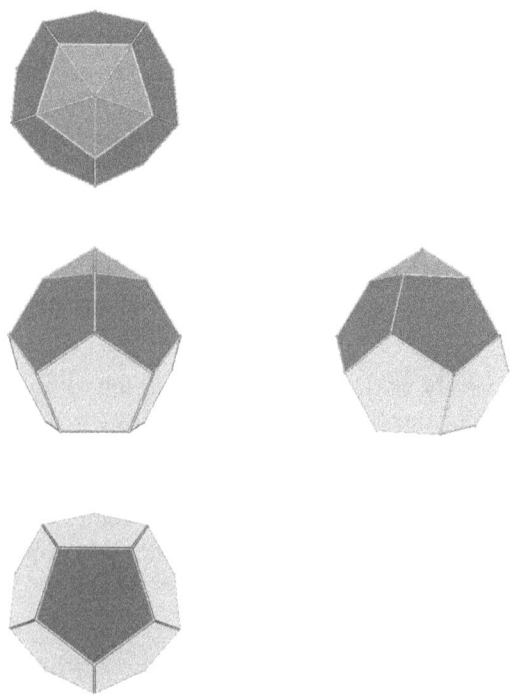

*Figure 115: Johnson solid 58: the augmented dodecahedron.*

*Augmented dodecahedra: Johnson solids 58-61.*

The augmentation destroys all the symmetries of the dodecahedron except for the fivefold axis through the augmented face and the five mirror planes through that fivefold axis (perpendicular to the face opposite the augmented one). Thus the symmetry of the augmented dodecahedron is $C_{5v}$.

For consistency with the rest of this book, Tables 26 and 27 have been provided with all the same information as is found in the other similar tables in this book; however, much of the information can be easily generated by simply considering the way these polyhedra are constructed. The dihedral angles, in particular, do not depend on the transitivity classes to which the faces belong; it can be seen that in all the polyhedra discussed in this chapter, the triangular faces derive from the triangular faces of the pentagonal pyramid ($J_2$), which make a dihedral angle of 138.1897° (138° 11′ 23″) with each other, so that not only in the case of $J_{58}$, but of all the others in this chapter, the 3-3 dihedral angle is also 138.1897° (138° 11′ 23″). Similarly, the pentagonal faces of all the polyhedra discussed in this chapter derive from those of a regular dodecahedron, so that, again, not only in the case of $J_{58}$, but of all the others in this chapter, the 5-5 dihedral angle is the same 116.5651° (116° 33′ 54″) as is found in the dodecahedron. The 3-5 dihedral angle has to be built up from two pieces. It arises from the fusion of the 3-5 dihedral angle (at the base of $J_2$) of 37.3774° (37° 22′ 39″) with the same dodecahedral dihedral angle of 116.5651° (116° 33′ 54″) just mentioned, giving a dihedral angle of 153.9424° (153° 56′ 33″) for all the Johnson solids discussed in this chapter (not merely $J_{58}$). (If you do the arithmetic, you will get 153.9425°; the difference only arises from rounding.) Because this is the case, also, it is not necessary to put in the footnotes found in most of the tables in this book explaining that there is no need to distinguish triangular (or pentagonal) faces in different transitivity classes.

Similar comments apply to the areas and volumes given in Tables 26 and 27. For the areas, it merely needs to be known that the area of each pentagonal face of a regular dodecahedron is $\frac{1}{4}\sqrt{(25+10\sqrt{5})} = 1.72047740$ (which makes the surface area of the whole dodecahedron equal to $3\sqrt{[25+10\sqrt{5}]} = 20.64572881$), while the area of the five triangular faces that replace them in each augmentation is simply $(5\sqrt{3})/4 = 2.16506351$. So if the number of augmentations is equal to $n$, the total area is simply

$$[(12 - n)\sqrt{(25 + 10\sqrt{5})} + (5n\sqrt{3})]/4,$$

or, expressed in terms of numbers,

$$1.72047740(12 - n) + 2.16506351n.$$

(For $n = 1$, as in $J_{58}$, this gives 21.09031492, as shown in Table 26.) For the volumes, the task is easier. One simply adds the volume of the original dodecahedron, $(15 + 7\sqrt{5})/4 = 7.66311896$, to the volumes of the added pyramids, each of which is shown in Table 2 to be $(5 + \sqrt{5})/24 = 0.30150283$. This gives a total of

$$[(90 + 5n) + (42 + n)\sqrt{5}]/24,$$

or, expressed numerically,

$$7.66311896 + 0.30150283n.$$

Once more, for $n = 1$, the result is 7.96462179, in agreement with Table 26.

To construct the dual of the augmented dodecahedron, two facts need to be considered: the dual of a dodecahedron is an icosahedron, and augmentation and truncation are dual processes. Therefore, the dual of the augmented dodecahedron must be a (mono)truncated icosahedron. (I add the prefix because "truncated icosahedron" is usually understood to imply truncation at all twelve of the vertices of an icosahedron.) This solid is illustrated in Figure 116 below. It should be compared with Figure 21 on p. 53 (Chapter 4). Note that the top of Figure 116 corresponds to the bottom of Figure 21; but allowing for this, both can be seen to be derived from an icosahedron by cutting off one vertex. In Figure 21, the icosahedron has been "diminished" in Johnson's terminology, with the whole pentagonal pyramid at one vertex cut off (the next chapter will be concerned with diminished icosahedra in general); in Figure 116, the cut leaves part of the lateral faces of that pyramid, in the form of isosceles trapezoids where triangles were to be found in the original icosahedron.

As we have noticed before in the case of other duals of augmented polyhedra whose symmetry is broken by the augmentation, the triangular faces, which would all be congruent equilateral triangles if one simply truncated one vertex of a regular icosahedron, are only isosceles in this case because the centroid of $J_{58}$, used as the center of the polar reciprocation process, does not exactly coincide with the centroid of the original regular dodecahedron, but they are very close to equilateral triangles, though falling into three different transitivity classes. Similarly, the isosceles trapezoids, which would be 120°-120°-60°-60° angled from slicing off a piece of an equilateral triangle, do not exactly have those angles, for the same reason. The pentagonal face is, however, regular because the centroid does lie on one of the fivefold axes of the original regular dodecahedron, and so this fivefold axis remains.

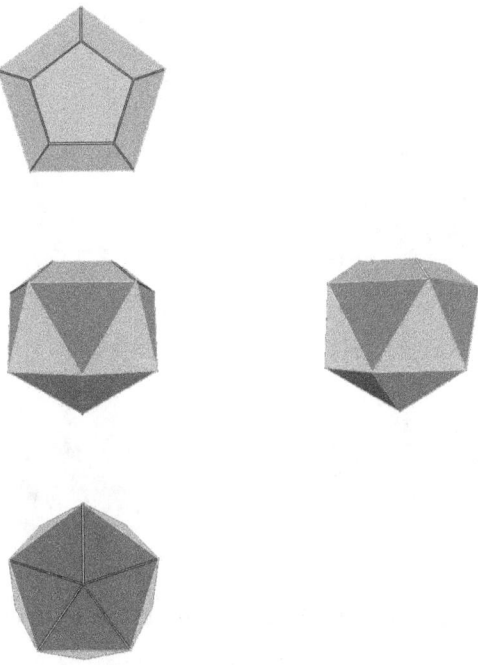

Figure 116: The dual of Johnson solid 58: a monotruncated icosahedron.

Similarly, it is found that all the dihedral angles are close to the uniform dihedral angle of 138.1897° (138° 11′ 23″) = $\cos^{-1}$ (−

$\tfrac{1}{3}\sqrt{5}$) of the regular icosahedron.

As was the case for $J_{55}$ and $J_{56}$ (see p. 220), in the naming of $J_{59}$ and $J_{60}$ (the para- and metabiaugmented dodecahedron) Johnson borrowed the prefixes from organic chemistry. The explanation on p. 220 defines them, so it will not be repeated here. But here it should be noted that while a fusion of a second pyramid to the augmented dodecahedron in the "para" position preserves the fivefold axis (in fact, it restores it to the tenfold alternating axis it was in the original dodecahedron), fusion in the "meta" position eliminates the fivefold axis, but creates a twofold axis (or more correctly, restores one of the twofold axes present in the original dodecahedron, which was absent in the augmented dodecahedron). So the parabiaugmented dodecahedron, $J_{59}$, illustrated in Figure 117, has $S_{10v}$ symmetry, while the metabiaugmented dodecahedron, $J_{60}$, illustrated in Figure 119, has only $C_{2v}$ symmetry.

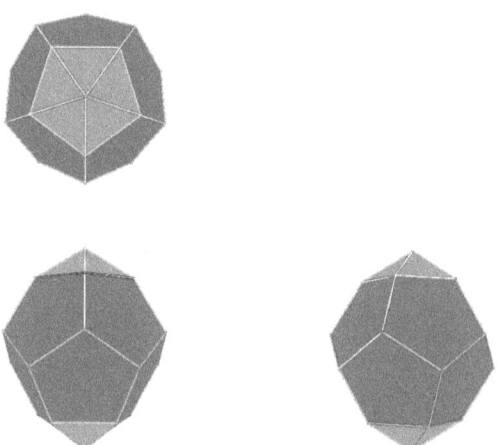

Figure 117: Johnson solid 59: the parabiaugmented dodecahedron.

Of course, since dual polyhedra have the same symmetry as the polyhedra whose duals they are, the dual of the parabiaugmented dodecahedron, illustrated in Figure 118, has $S_{10v}$ symmetry, while the dual of the metabiaugmented dodecahedron, illustrated in Figure 120, has only $C_{2v}$ symmetry.

Whether one wishes to consider the dual of the parabiaug-

*Augmented dodecahedra: Johnson solids 58-61.*

mented dodecahedron as a "parabitruncated icosahedron," *i. e.* an icosahedron with two opposite vertices truncated, or a "gyroelongated bifrustum of a pentagonal bipyramid" thus relating it to the elongated and gyroelongated bipyramids in Chapter 6, is a matter of taste; both descriptions are perfectly accurate. (And, of course, this is not the first time that more than one name has been given for the same polyhedron!)

When the two augmentations of the dodecahedron are opposite, the displacement of the centroid found in the polar reciprocation that produced the dual of $J_{58}$ does not occur, and so the triangular faces of the dual of $J_{59}$ are equilateral, the isosceles trapezoidal faces have the 60° and 120° angles that one expects by cutting off one vertex of an equilateral triangle by a line parallel to the opposite side, and, of course, the pentagonal faces are regular.

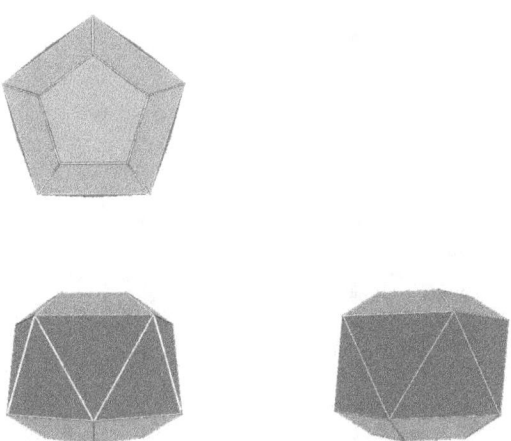

*Figure 118: The dual of Johnson solid 59 (the parabiaugmented dodecahedron): an icosahedron truncated at two opposite vertices.*

Johnson solid 60, the metabiaugmented dodecahedron, has only the $C_{2v}$ symmetry referred to above. The view in Figure 119 does not correspond to the view in Figure 117, because the practice in this book is to consider the view looking down at the principal axis of rotation as the top view, and the principal axis of the parabiaugmented dodecahedron is the tenfold alternating axis, while for the metabiaugmented dodecahedron, it must be taken to be one of the twofold axes. (It must be conceded that the choice of "top" and "bottom" is arbitrary, and since the $C_{2v}$ symmetry actually provides two twofold axes, the choice of which view is considered "front" and which is "side" is also an arbitrary decision, but some choice must be made in each case.)

While the face/vertex/edge counts of the metabiaugmented dodecahedron must be identical to that of the parabiaugmented dodecahedron, the difference in symmetry means that the division of faces into transitivity classes is different. Of course, this has been encountered before.

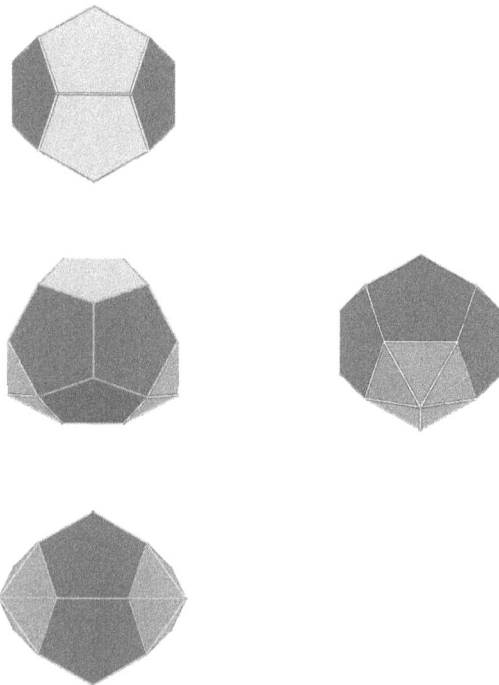

Figure 119: Johnson solid 60: the metabiaugmented dodecahedron.

The dual of the metabiaugmented dodecahedron has the same $C_{2v}$ symmetry as $J_{60}$, of course. It is illustrated in Figure 120 below. As one might expect from what has elsewhere been said about duals, it is a doubly truncated icosahedron, with the truncated vertices being in the same relative positions as the augmented faces of the original metabiaugmented dodecahedron.

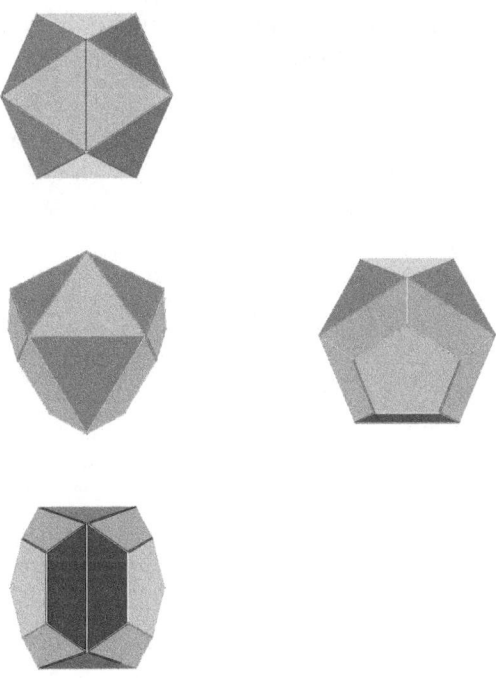

*Figure 120: The dual of Johnson solid 60 (the metabiaugmented dodecahedron).*

A comparison of the duals of $J_{59}$ and $J_{60}$ shows the effects of having the centroid (used as the center of polar reciprocation) in a different place from the centroid of the original regular dodecahedron. The two augmentations in $J_{59}$ cause no movement of the centroid because they are balanced, equally far from the original centroid and in diametrically opposite directions. The two augmentations in $J_{60}$, however, displace the centroid significantly. So while in the dual of $J_{59}$, triangles remain equilateral, and trapezoids are isosceles 120°-120°-60°-60° trapezoids, in the dual of $J_{60}$ we find four different types of triangular face, three of which are isosceles but none of which is equilateral, yet all are very close to equilateral; and three different kinds of quadrilateral face, only *one* of which is a true isosceles trapezoid, and even that with angles somewhat changed from 120°-120°-60°-60°; yet all three are very close to 120°-120°-60°-60° isosceles trapezoids. The pentagonal faces are similarly close to, but not exactly, regular. In fact, the faces are so close to the regular shapes that are seen in $J_{59}$ that looking at Figure 120 does not spot the irregularities, and only consulting the data in Table 26 shows them.

*Augmented dodecahedra: Johnson solids 58-61.*

| Johnson solid number | 58 | 59 | 60 |
|---|---|---|---|
| Name | Augmented dodecahedron | Parabiaugmented dodecahedron | Metabiaugmented dodecahedron |
| Symmetry* | $C_{5v}$ | $S_{10v}$ | $C_{2v}$ |
| Faces | [5](3) + [1+5+5](5) | [10](3) + [10](5) | [2+4+4](3) + [2+2+2+4](5) |
| Edges* | 35 | 40 | 40 |
| Vertices | 21 | 22 | 22 |
| Dihedral angles | 3-3: 138.1897° (138° 11' 23''), 3-5: 153.9424° (153° 56' 33''), 5-5: 116.5651° (116° 33' 54'') | | |
| Area | 21.09031492 = [11√(25+10√5) + (5√3)]/4 | 21.53490102 = 5[√(25+10√5) + √3]/2 | 21.53490102 = 5[√(25+10√5) + √3]/2 |
| Volume | 7.96462179 = (95 + 43√5])/24 | 8.26612463 = (100 + 44√5)/24 | 8.26612463 = (100 + 44√5)/24 |
| Dual | | | |
| Name | Monotruncated icosahedron | Bitruncated icosahedron (opposite vertices truncated); bifrustum of a gyroelongated pentagonal bipyramid | Bitruncated icosahedron (Vertices truncated that are separated by one pair of triangular faces) |
| Faces | [5+5+5](3) + [5](4) + [1](5) | [10](3) + [10](4) + [2](5) | [2+2+2+4](3) + [2+4+4](4) + [2](5) |
| Vertices | 16 | 20 | 20 |

— 241 —

## The Johnson Solids and Their Duals

| Johnson solid number | 58 | 59 | 60 |
|---|---|---|---|
| Face angles[†] | 3a: [2]59.4032° (59° 24′ 11″), 61.1936° (61° 11′ 37″); 3b: [2]61.0725° (61° 4′ 21″), 57.8550° (57° 51′ 18″); 3c: [2]59.0017° (59° 0′ 6″), 61.9967° (61° 59′ 48″); 4: [2]60.6905° (60° 41′ 26″), [2]119.3095° (119° 18′ 34″); 5: [5]108°♣ | 3: [3]60°; 4: [2]60°, [2]120°; 5: [5]108° | See list below[□△] |

— 242 —

*Augmented dodecahedra: Johnson solids 58-61.*

| Johnson solid number | 58 | 59 | 60 |
|---|---|---|---|
| Dihedral angles | 3a-3a: 140.0613° (140° 3' 41"), 3a-3b: 139.4345° (139° 26' 4"), 3b-3c: 138.2714° (138° 16' 17"), 3c-4: 137.0512° (137° 3' 4"), 4-4: 136.1842° (136° 11' 3"), 4-5: 140.5961° (140° 35' 46")✦ | 3-3, 3-4, 4-4: 138.1897° (138° 11' 23"), 4-5: 142.6226° (142° 37' 21") | 3a-3b: 138.3088° (138° 18' 32"), 3b-3d: 139.3554° (139° 21' 19"), 3c-3c: 140.3621° (140° 21' 43"), 3c-3d: 139.9963° (139° 59' 47"), 3a-4b: 137.0768° (137° 4' 37"), 3d-4a: 138.9756° (138° 58' 32"), 4a-4a: 138.2579° (138° 15' 29"), 4a-4b: 137.5713° (137° 34' 17"), 4b-4i: 136.2999° (136° 18' 0"), 4i-4i: 135.7779° (135° 46' 41"), 4a-5: 141.7960° (141° 47' 46"), 4b-5: 141.5545° (141° 33' 16"), 4i-5: 141.4405° (141° 26' 26")¤ |
| Area | 18.14426431 | 21.88925685 | 17.11516896 |
| Volume | 6.60519800 | 8.74295858 | 6.04115569 |

Table 26: *Properties of Johnson solids 58, 59, and 60 and their duals.*

˙The symmetry and number of edges are always the same for each solid and its dual, so are not repeated in the "dual" section.

†Face angles are only given for the duals, as the Johnson solids have regular faces, whose angles can be found in Table 1 on p. 3.

✦The symbol 3a designates the five isosceles triangles meeting at the apex opposite the pentagonal face; 3b designates the five isosceles triangles immediately adjacent to them; 3c designates the five isosceles triangles adjacent to the isosceles trapezoids that surround the pentagonal face.

"For this purpose, the "first mirror plane" means the one that cuts through the two pentagonal faces, the "second mirror plane" means the one halfway between the pentagonal faces. The designation 3a refers to the two isosceles triangles, bisected by the second mirror plane, that share common edges each with two different quadrilaterals; 3b refers to the two isosceles triangles, bisected by the second mirror plane, which have no common edges with any quadrilaterals; 3c refers to the two isosceles triangles with a common edge in the second mirror plane; 3d refers to the four scalene triangles which share edges with the 3c triangles; 4a refers to the four quadrilaterals with an edge in the first mirror plane; 4b refers to the four quadrilaterals which neither have an edge in the first mirror plane nor are bisected by it; 4i refers to the only two precise isosceles trapezoids among the quadrilateral faces, which share a common edge in the second mirror plane.

$^{\triangle}$The list of face angles of the dual of $J_{60}$ is too large to fit in one cell. It is given here instead:

- 3a: [2]61.0548° (61° 3′ 17″), 57.8904° (57° 53′ 26″);
- 3b: [2]59.0720° (59° 4′ 19″), 61.8560° (61° 51′ 21″);
- 3c: [2]60.3651° (60° 21′ 54″), 59.2698° (59° 16′ 11″);
- 3d: 58.4044° (58° 24′ 16″), 60.2861° (60° 17′ 12″), 61.3089° (61° 18′ 32″);
- 4a: 60.1001° (60° 6′ 0″), 61.7839° (61° 47′ 2″), 118.9768° (118° 58′ 37″), 119.1391° (119° 8′ 21″);
- 4b: 58.5693° (58° 34′ 9″), 61.6034° (61° 36′ 12″), 119.6812° (119° 40′ 52″), 120.1462° (120° 8′ 46″);
- 4i: [2]59.6082° (59° 36′ 29″), [2]120.3918° (120° 23′ 31″);
- 5: [2]106.5709° (106° 34′ 15″), [2]108.6023° (108° 36′ 8″), 109.6536° (109° 39′ 13″).

*Augmented dodecahedra: Johnson solids 58-61.*

While no third augmentation of the parabiaugmented dodecahedron is possible without losing convexity, there is one remaining pentagonal face of the metabiaugmented dodecahedron that can be augmented by fusing a regular pentagonal pyramid to it. This fusion produces Johnson solid 61, the triaugmented dodecahedron, which is illustrated in Figure 121 below. With the three augmentations, it has lost all but a single threefold axis (and associated mirror planes) of the original dodecahedron. So the symmetry is $C_{3v}$, and it is oriented in the figure, in accordance with that symmetry, placing the threefold axis vertically. (Note that the front view, which shows one of the pyramidal augmentations, does not have it at the precise center. This is a function of the angle of view, which places the eye precisely at right angles to the threefold axis. A similar view of Johnson solid 71 is to be found in Figure 141 below.)

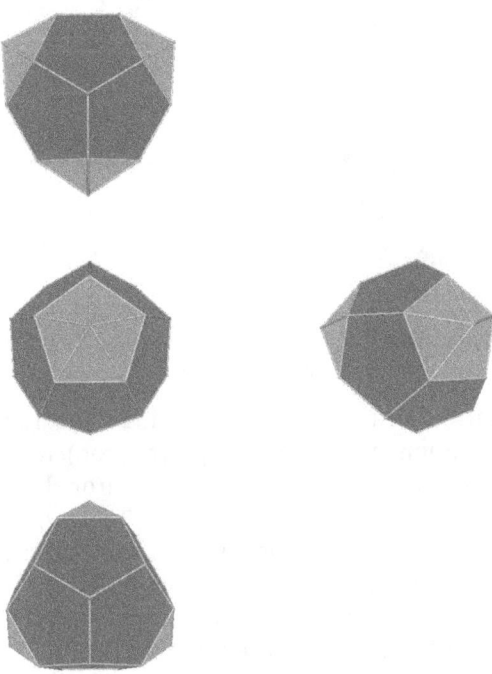

*Figure 121: Johnson solid 61: the triaugmented dodecahedron.*

The dual of the triaugmented dodecahedron can be seen to

be, as expected, an icosahedron with three of its vertices, in locations equivalent to the faces which were augmented in $J_{61}$, as expected, truncated. It is illustrated in Figure 122 below.

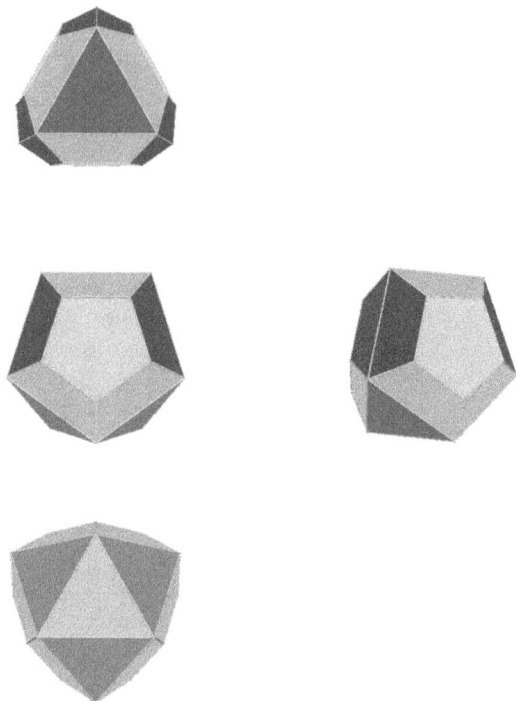

Figure 122: The dual of Johnson solid 61 (the triaugmented dodecahedron).

The centroid of the triaugmented dodecahedron (by which polar reciprocation produced this polyhedron) is not very far from the centroid of the original dodecahedron, but it does not coincide exactly with it, and consulting Table 27 you can see that the three isosceles triangles which are not centered on the threefold axis are very near to equilateral, so close that only extremely precise calculation or measurement can distinguish them. (The two triangles that *are* centered on the threefold axis are forced to be equilateral by the threefold symmetry of the polyhedron.) Similarly, the two sets of quadrilaterals that are not bisected with mirror planes of the polyhedron are not quite 120°-120°-60°-60° isosceles trapezoids, but are so close that they

*Augmented dodecahedra: Johnson solids 58-61.*

are almost indistinguishable from them, or from each other. And the three pentagonal faces are again so close to regular pentagons (they are technically only isosceles pentagons, using the definition in my earlier book, *Polyhedra: a New Approach*) that only extremely precise calculation or measurement can distinguish them from regular pentagons.

| | |
|---|---|
| Johnson solid number | 61 |
| Name | Triaugmented dodecahedron |
| Symmetry* | $C_{3v}$ |
| Faces | [3+6+6](3) + [3+3+3](5) |
| Edges* | 45 |
| Vertices | 23 |
| Dihedral angles | 3-3: 138.1897° (138° 11′ 23″), <br> 3-5: 153.9424° (153° 56′ 33″), <br> 5-5: 116.5651° (116° 33′ 54″) |
| Area | 21.97948713 = <br> [9√(25+10√5) + (15√3)]/4 |
| Volume | 8.56762746 = (105 + 45√5)/24 |
| **Dual** | |
| Name | Symmetrically tritruncated icosahedron |
| Faces | [1+1+3](3) + [3+6+6](4) + [3](5) |
| Vertices | 24 |
| Face angles† | 3a, 3b: [3]60°; <br> 3i: [2]60.3597° (60° 21′ 35″), <br> 59.2806° (59° 16′ 50″); <br> 4a: 59.8852° (59° 53′ 7″), <br> 60.9214° (60° 55′ 17″), <br> 119.5271° (119° 31′ 38″), <br> 119.6662° (119° 39′ 58″); <br> 4b: 59.0272° (59° 1′ 38″), <br> 60.8104° (60° 48′ 38″), <br> 119.9342° (119° 56′ 3″), <br> 120.2282° (120° 13′ 41″); <br> 4i: [2]59.6366° (59° 38′ 12″), <br> [2]120.3634° (120° 21′ 48″); <br> 5: [2]107.1784° (107° 10′ 42″), <br> [2]108.3329° (108° 19′ 59″), <br> 108.9773° (108° 58′ 38″)♣ |

*Augmented dodecahedra: Johnson solids 58-61.*

| Johnson solid number | 61 |
|---|---|
| Dihedral angles | 3a-3i: 139.2601° (139° 15′ 36″),<br>3b-4i: 137.0726° (137° 4′ 21″),<br>3i-4a: 138.8708° (138° 52′ 15″),<br>4a-4a: 138.6185° (138° 37′ 7″),<br>4a-4b: 138.2193° (138° 13′ 9″),<br>4b-4b: 137.7842° (137° 47′ 3″),<br>4b-4i: 137.5205° (137° 31′ 14″),<br>4a-5: 142.5353° (142° 32′ 7″),<br>4b-5: 142.4174° (142° 25′ 3″),<br>4i-5: 142.3566° (142° 21′ 25″)♣ |
| Area | 19.14448421 |
| Volume | 7.13656782 |

Table 27: *Properties of Johnson solid 61 and its dual.*

˚The symmetry and number of edges are always the same for each solid and its dual, so are not repeated in the "dual" section.

†Face angles are only given for the duals, as the Johnson solids have regular faces, whose angles can be found in Table 1 on p. 3.

♣The designation 3a refers to the central equilateral triangle in the assemblage of four at one end of the polyhedron; 3b to the single equilateral triangle (surrounded by isosceles trapezoids) at the opposite end; 3i to the three isosceles triangles surrounding the 3a face. The designation 4a refers to the six irregular quadrilaterals (nearly, but not precisely, isosceles trapezoids) adjacent to the 3i faces; the designation 4b to the six irregular quadrilaterals (nearly, but not precisely, isosceles trapezoids) whose shortest sides are adjacent to the pentagonal face (which is true of all the quadrilateral faces) and whose longest sides are adjacent to each other; 4i refers to the only three precise isosceles trapezoids, whose longest sides are adjacent to the 3b face.

The types of calculations done in this chapter of the dihedral angles, areas, and volumes of all the polyhedra of course apply to all polyhedra obtainable by fusion, which would include a very large proportion of the Johnson solids. The reasons for having them done in this chapter are numerous.

1. All the polyhedra in the chapter are derived from *one* basic polyhedron (the regular dodecahedron) to which one or more of the *same* type of object (the pentagonal pyramid), which one can call the addition, is fused.

2. All faces of the basic polyhedron are identical, as are all the edges, so the only variable between the different fused polyhedra is the *number* of additions involved in the fusion.

3. Not only does the *basic* polyhedron have only one kind of face (the pentagon) but the only polygon remaining in the addition, after removing the faces fused to the basic polyhedron, is a single type, different from the polygon derived from the basic polyhedron. Thus, in this case, *every* pentagon derives from the dodecahedron, *every* triangle derives from the pentagonal pyramid, and no faces besides those two types exist.

Those three points are not the only ones, but they are the most important reasons for treating the subject in such detail in this chapter. It will probably interest many readers to apply the same principles to other Johnson solids obtainable by fusion, and reading Chapter 18 is recommended.

# Chapter 13: Diminished icosahedra: Johnson solids 62-64.

This chapter covers only three of the Johnson solids, and thus is one of the shortest in this book. There is, however, no reason to group them with either the four polyhedra discussed in the previous chapter or with the seven discussed in the following chapter, so they will have to have a chapter of their own.

While I am generally content with Norman Johnson's nomenclature for the solids described in his paper (and this book), I believe that the names he chose for the three polyhedra described in this chapter are not the most descriptive names he could have given them. In Chapter 16, as we will see, the term "diminished" is used as a part of the names of several Johnson solids, and it is part of a systematic nomenclature relating the solids to the rhombicosidodecahedron. In the case of the three solids treated in this chapter, however, the attempt to name them by derivation from the icosahedron in the same way is a misstep, in my belief. It leads to the description of the very simple polyhedron designated as $J_{63}$ as a derivative of a solid whose relation to it is hard to visualize: the icosahedron. Far better would be a quite different construction, to be described later in this chapter.

But the Johnson names do, in fact, point out a relationship to the polyhedra discussed in the previous chapter, and that can be said in their favor. The dual of a dodecahedron is an icosahedron, and while the dual of augmentation is not diminishing but truncation, the *diminished* icosahedra are related to the *truncated* icosahedra which are the duals of the polyhedra of the previous chapter in that the cuts are made deeper, eliminating triangular faces rather than merely cutting off a piece and leaving an isosceles trapezoidal face for each triangular face cut in the truncation.

The icosahedron has the same $I_h$ symmetry as the dodecahedron from which were derived the polyhedra discussed in Chapter 12, and if the twelve vertices of the icosahedron are considered as the elements modified in the procedure described on the discussion beginning on p. 230, it would seem that there ought to be four diminished icosahedra; however, there are only two, or at least two named as such in Johnson's list. The reason for that, however, is simple; there *are* four, but one of them has already been encountered in another guise (the gyroelongated pentagonal pyramid — $J_{11}$, see Figure 21, p. 53) and one has been deliberately excluded from the Johnson solids in the definition: the pentagonal antiprism.

If the gyroelongated pentagonal pyramid were called a diminished icosahedron and the pentagonal antiprism were called a parabidiminished icosahedron, a case could be made for Johnson's names for the polyhedra described in this chapter, analogous to the names for the polyhedra in Chapter 16. But nobody uses those names, and for that reason I think there are better choices for the names of these three.

As was the case for $J_{55}$ and $J_{56}$ (see p. 220), in the naming of $J_{62}$ (the metabidiminished icosahedron) Johnson borrowed the prefixes from organic chemistry. (Since a "parabidiminished icosahedron" would be a pentagonal antiprism, as mentioned in the previous paragraph, that does not appear in Johnson's list.) An icosahedron is, in fact, as has been noted elsewhere, a gyroelongated pentagonal bipyramid (see p. 74), or equivalently, a basally biaugmented pentagonal antiprism (see p. 65). So one can consider the reverse operation: As one envisions an icosahedron as resulting from the fusion of a pentagonal pyramid on each side of a pentagonal antiprism, one can cut a pyramid off from the icosahedron, replacing five triangles by a pentagon. If one did this at two opposite vertices, of course, one would generate the pentagonal antiprism, and if one did this only at one vertex, one would produce a gyroelongated pentagonal pyramid, or equivalently, a basally augmented pentagonal antiprism, which has already been discussed as $J_{11}$. But another option exists: one can cut off the two pyramids at two vertices which are not opposite. (Of course, they cannot be too close: that would mean attempting to remove the same triangle twice as part of the two pyramids. But there is an in-between position, in which the two vertices specifying the removed pyramids are two faces apart, so that one of the triangles at one vertex borders a triangle at the other. This is what is here termed the "meta" position.) When this is done,

*Diminished icosahedra: Johnson solids 62-64.*

the remaining piece of the icosahedron is what Johnson termed a *metabidiminished icosahedron*, numbered 62 in his list, and illustrated in Figure 123 below.

The metabidiminished icosahedron has a single twofold axis, and two mirror planes passing through that axis. Thus the symmetry is $C_{2v}$.

The construction just described corresponds to Johnson's naming of the solid, but in my opinion, it would better serve the purpose of visualizing this polyhedron if it were considered an augmentation of $J_{63}$, which will be described later in this chapter. (It is to be noted, of course, that $J_{64}$ is also described as an augmentation of $J_{63}$, but the difference is that $J_{64}$ is obtained by augmenting one specific *triangular* face, namely the isolated one, while $J_{62}$ is obtained by augmenting one of the *pentagonal* faces of $J_{63}$.)

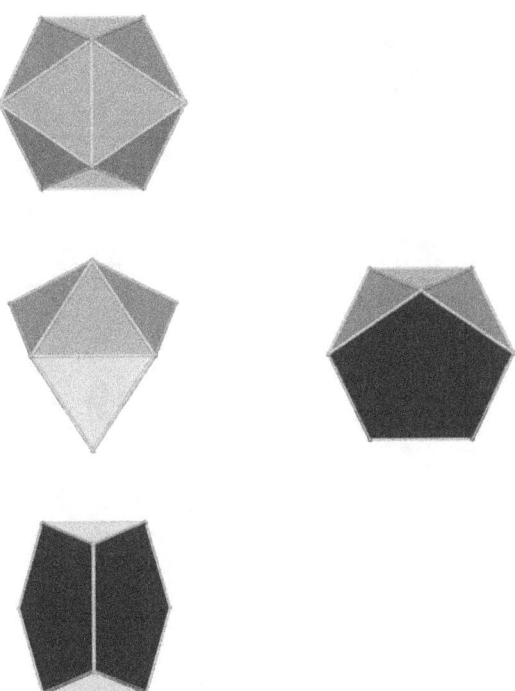

*Figure 123: Johnson solid 62: the metabidiminished icosahedron.*

The dual of the metabidiminished icosahedron is illustrated in Figure 124 below. Surprisingly, it does not resemble the metabiaugmented dodecahedron (Figure 119) very closely, although one might expect it to. Obviously, the difference between truncating (as seen in Figure 120) and diminishing makes a substantial difference in the appearance of the dual polyhedron. The dual, of course, has the same $C_{2v}$ symmetry as the original metabidiminished icosahedron.

There are two isosceles triangular, two pentagonal, and six quadrilateral faces, but the quadrilaterals divide into two transitivity classes, one consisting of the four which share an edge with one of the triangular faces, the other consisting of the remaining two quadrilaterals.

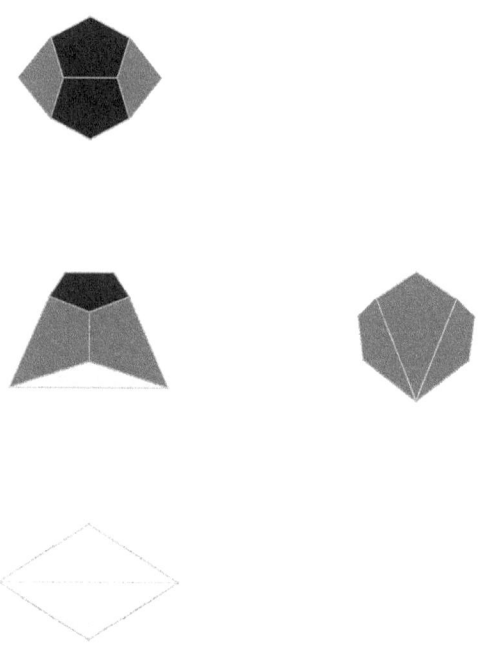

Figure 124: The dual of Johnson solid 62 (the metabidiminished icosahedron).

At first glance, the two pentagonal faces appear to be

regular, but they are in fact only isosceles (as defined in my earlier book, *Polyhedra: a New Approach*); yet they are extremely close to regular pentagons: The angles include two of 105.9046° (105° 54′ 17″), two of 108.9186° (108° 55′ 7″), and one of 110.3537° (110° 21′ 13″), which are all less than 2.5° of the regular pentagon angle of 108°. The sides include two of ~0.5065, two of ~0.5290, and one of ~0.5416, again rather close. The four quadrilaterals that share edges with the triangular faces look like kite-quadrilaterals, but like the pentagons that appear to be regular, they are only close: one pair of edges that ought to be equal are ~0.4794 and ~0.5065; the others are ~1.1580 and ~1.1222. The two remaining quadrilaterals, however, are truly kite-quadrilaterals. The four "almost-kites," in fact, come close to the two true kites in measurements; the sides of the true kites are ~0.5290 and ~1.1580.

Each of the diminishings of the icosahedron replaces five triangles by a pentagon; so that, starting with twenty triangles on the icosahedron, we first get to $J_{11}$, with fifteen triangles and a pentagon, and then $J_{62}$, with ten triangles and two pentagons. It is possible to do yet one more such replacement, producing $J_{63}$, with five triangles and three pentagons, called a *tridiminished icosahedron*, numbered 63 in Johnson's list, and illustrated in Figure 125 below. But at this point, as can be seen in the figure, one of the triangles is remote from the other four, so no further diminishing can occur. However, the tridiminished icosahedron is a very symmetrical polyhedron, with a threefold axis, and will also be mentioned in Chapter 18 of this book.

I would like, however, to consider the tridiminished icosahedron as the triangular member of a family of specially truncated pyramids, because the family relationship is much clearer than the derivation from the icosahedron (fifteen of the twenty faces of which no longer exist in $J_{63}$!)

Normally, in describing, for example, the Archimedean solids, the term "truncation" refers to a cutting point that makes the polygon regular. The term "rectification" is used to refer to a cut that proceeds to the midpoint of the edge, which, when it is combined with a similar cut from the adjacent vertex, causes that edge to disappear. I want, however, to discuss a cutting off of a pyramid from a vertex that is rectification-like with regard to one type of edge, but not all the edges that it cuts. There is no official name for such a cutting, but it is necessary to describe the construction which will now be described, which generates $J_{63}$ in a different way.

Start with a triangular pyramid (*not* a regular tetrahedron, since its lateral faces would be isosceles triangles whose face angles are 72°, 54°, and 54°!) Truncate the apex, at a distance that can be determined to make the resulting edge have an appropriate length. Do a similar cut from the base vertices, but have the cut go to the midpoints of the base edges (which, as stated before, obliterates those edges when combined with the adjacent cuts) but only far enough up the lateral edges to make the lateral faces into regular pentagons. The result is the solid illustrated in Figure 125 below.

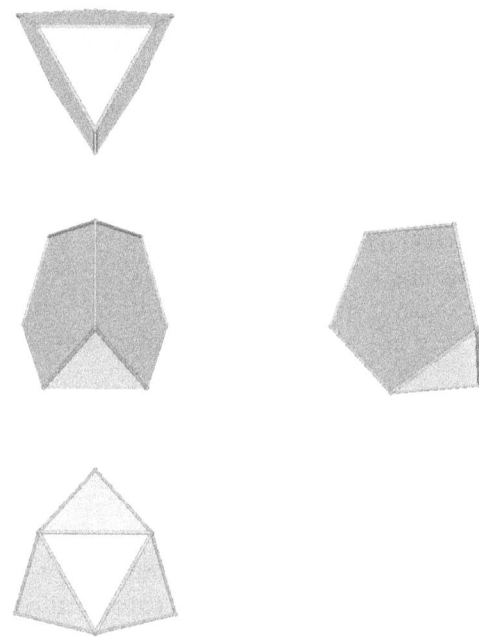

*Figure 125: Johnson solid 63: the tridiminished icosahedron.*

## Diminished icosahedra: Johnson solids 62-64.

Expressing the construction in terms of vertex coordinates: Let the starting triangular pyramid be **ABCD**, where the coordinates are:

**A**: $(0, 0, [3 + \sqrt{5}][\sqrt{3}]/4) = (0, 0, 2.26728394)$,

**B**: $([\sqrt{3}]/3, 1, -[3 + \sqrt{5}][\sqrt{3}]/12) = (0.57735027, 1, -0.75576131)$,

**C**: $([\sqrt{3}]/3, -1, -[3 + \sqrt{5}][\sqrt{3}]/12) = (0.57735027, -1, -0.75576131)$,

**D**: $(-2[\sqrt{3}]/3, 0, -[3 + \sqrt{5}][\sqrt{3}]/12) = (-1.15470054, 0, -0.75576131)$.

The reason for the z-coordinates having the strange values they do is to make the edges of the final polyhedron have the value 1. On the lines **AB**, **AC**, and **AD** mark points **E**, **F**, and **G**, at the midpoints of those lines, so their coordinates are given by:

**E**: $([\sqrt{3}]/6, 1/2, [3 + \sqrt{5}][\sqrt{3}]/12) = (0.28867513, 0.5, 0.75576131)$,

**F**: $([\sqrt{3}]/6, -1/2, [3 + \sqrt{5}][\sqrt{3}]/12) = (0.28867513, -0.5, 0.75576131)$,

**G**: $([\sqrt{3}]/3, 0, -[3 + \sqrt{5}][\sqrt{3}]/12) = (-0.57735027, 0, 0.75576131)$.

Also on the lines **AB**, **AC**, and **AD** mark points **H**, **J**, and **K**, such that **BH** = **CJ** = **DK** = 1; they will be located at:

**H**: $([\sqrt{15} + \sqrt{3}]/12, [1 + \sqrt{5}]/4, -[\sqrt{3}][\sqrt{5} - 1]/12) =$

$(0.46708618, 0.80901699, -0.17841104)$,

**J**: $([\sqrt{15} + \sqrt{3}]/12, -[1 + \sqrt{5}]/4, -[\sqrt{3}][\sqrt{5} - 1]/12) =$

$(0.46708618, -0.80901699, -0.17841104)$,

**K**: $(-[\sqrt{15} + \sqrt{3}]/6, 0, -[\sqrt{3}][\sqrt{5} - 1]/12) =$

$(-0.93417236, 0, -0.17841104)$.

Note that the factor $\sqrt{15} + \sqrt{3}$ appearing in the coordinates of **H** and **J** could have been written as $(\sqrt{3})(\sqrt{5} + 1)$, and the factor $(\sqrt{3})(\sqrt{5} - 1)$ could have been written as $\sqrt{15} - \sqrt{3}$.

In addition, mark three points **L**, **M**, and **N** with **L** at the midpoint of **CD**, **M** at the midpoint of **BD**, and **N** at the midpoint of **BC**; their coordinates will be:

**L**: $(-[\sqrt{3}]/6, 1, -[3 + \sqrt{5}][\sqrt{3}]/12) = (-0.28867513, 1, -0.75576131)$,

**M**: $(-[\sqrt{3}]/6, -1, -[3 + \sqrt{5}][\sqrt{3}]/12) = (-0.28867513, -1, -0.75576131)$,

***N***: $([\sqrt{3}]/3, 0, -[3 + \sqrt{5}][\sqrt{3}]/12)$ = (0.57735027, 0, −0.75576131).

The nine points ***E, F, G, H, J, K, L, M,*** and ***N*** will be the vertices of the polyhedron. The two triangular faces perpendicular to the $C_3$ axis, each a single-member transitivity class, are ***EFG*** and ***LMN***; the other three triangular faces, which form a transitivity class, are ***HMN, JLN,*** and ***KLM***. There are three pentagonal faces, forming a single transitivity class: ***EFJNH, FGKLJ,*** and ***GEHMK***.

Of course, in this construction, the four vertices of the original pyramid are all lost in truncations; not a single one remains in the final polyhedron, but they were necessary to provide a starting point in the construction.

Only for the triangular case can the polyhedron be made a Johnson solid (all faces regular polygons) but if one starts with an *n*-gonal pyramid in general, it is possible to produce, by this process, a whole family of solids, of $C_{nv}$ symmetry, whose faces comprise two *n*-gons, *n* pentagons (not, in general, regular, but still having a mirror plane through them), and *n* triangles (not, in general, equilateral, but isosceles). This family has no traditional name, but in my earlier book, *Polyhedra: a New Approach*, I provided terminology that is appropriate. If one ignores the truncation of the apex of the original pyramid, the solid obtained is what was in that book termed a 4-pyramoid. With the truncation of the apex included, therefore, $J_{63}$ is described as a *frustum of a triangular 4-pyramoid* — a much more descriptive name than Johnson's. (Of course, not every frustum of a triangular 4-pyramoid is a Johnson solid, just as not every pentagonal pyramid is an instance of $J_2$; the sides and angles have to be such as to make all the polygons regular.)

I have since found out that the term antihermaphrodite has been used to refer to a 4-pyramoid. It does not seem very descriptive, so will not be used in this book. However, if one wishes to refer to $J_{63}$ as a frustum of a triangular antihermaphrodite or a monotruncated triangular antihermaphrodite, either of those names is a better description than "tridiminished icosahedron," and the metabidiminished icosahedron described earlier in this chapter would then be simply described as an augmented version of the same-named solid. (Since $J_{64}$ is also an augmented version of the same solid, the name would have to

*Diminished icosahedra: Johnson solids 62-64.*

designate the position of the augmentation. As was earlier stated in the discussion of $J_{62}$, it is a question of augmenting a pentagonal or a triangular face of $J_{63}$.)

The dual of $J_{63}$ is illustrated in Figure 126 below. Its symmetry, of course, is the same $C_{3v}$ as that of $J_{63}$ itself, and it contains nine faces, six isosceles triangles and three kite-quadrilaterals. The three isosceles triangles at the apex form a pyramid-like apex configuration (see p. 20 in Chapter 3), the three kite-quadrilaterals at the antiapex form a configuration similar to that at the vertices of such Catalan (Archimedean-dual) solids as the deltoidal icositetrahedron. The remaining three isosceles triangles are lateral.

The dual of Johnson solid 63 can also be considered as constructed in another way, just as we considered two different ways of getting to Johnson solid 63 itself.

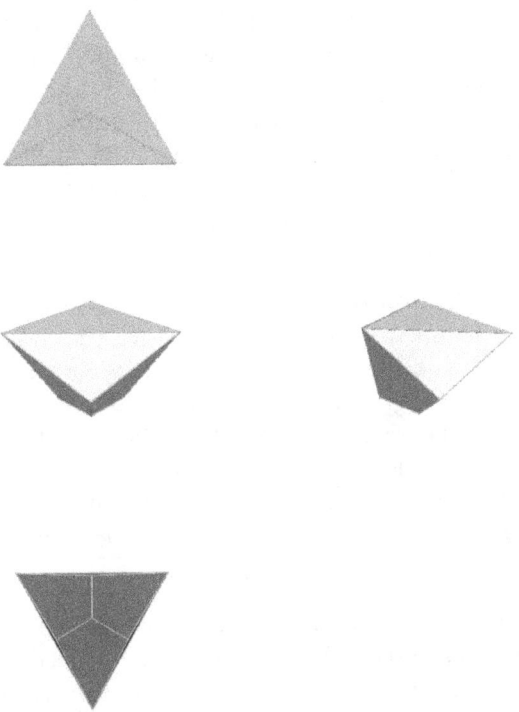

*Figure 126: The dual of Johnson solid 63 (the tridiminished icosahedron).*

Start with a triangular bipyramid (not exactly Johnson solid 12, because the faces would not be equilateral, but for the purposes of this construction, it has to be asymmetric and much squatter than $J_{12}$). Leave the top half alone, but take the triangular faces forming the lower half and pivot them outward, but preserving the $C_3$ axis, so that the single vertex at the bottom becomes three separate vertices. Through each of the sets of three points that include one equatorial vertex and two of the three vertices that the bottom was split into, pass a plane; these planes will all intersect on the threefold axis because of the $C_{3v}$ symmetry, and that point, together with the three that defined each plane, will constitute the vertices of a kite-quadrilateral. These kite-quadrilaterals, together with the six triangular faces derived from the bipyramid, are the nine faces of the $J_{63}$ dual.

Another Johnson solid, the augmented tridiminished icosahedron, $J_{64}$, is obtained by augmenting $J_{63}$ with a regular tetrahedron on the isolated triangle at the top (as oriented in Figure 125). No other triangular face can be so augmented, because the adjacent dihedral angles are such that the resulting polyhedron would not be convex. It is illustrated in Figure 127 below. Since the threefold axis of the augmenting pyramid is along the same line as the threefold axis of the tridiminished icosahedron before augmentation, the threefold symmetry is preserved and the symmetry remains $C_{3v}$.

This polyhedron is *not* the 4-pyramoid that would be considered the parent to $J_{63}$, considered as a frustum of a triangular 4-pyramoid, however; the reason is that the three faces that derive from the augmenting tetrahedron do not lie in the same planes as the pentagonal faces of $J_{63}$. As a result, there is a distinct edge between each pentagon and the adjacent triangle. The dihedral angle at each of those edges, however, is so close to 180° (171.3411° = 171° 20′ 28″) that it is very difficult to see them.

Because of the fact that the threefold axis is along the same line as the threefold axis of the tridiminished icosahedron before augmentation, the orientation of Figure 127 is the same as that of Figure 125, with which one should compare it. The bottom view of Figure 127, of course, is identical to that in Figure 125, since the augmenting pyramid is out of sight in that

*Diminished icosahedra: Johnson solids 62-64.*

view.

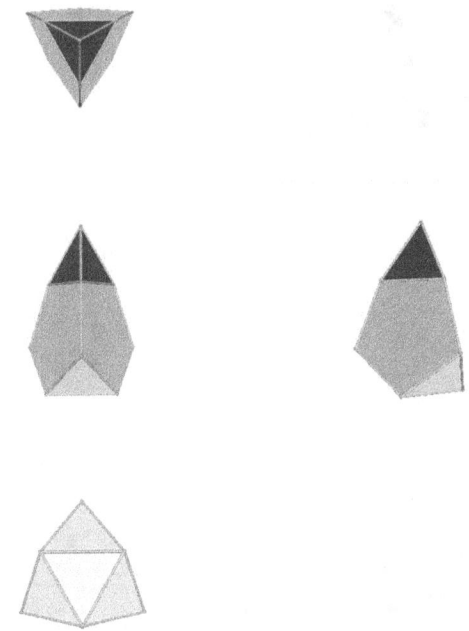

Figure 127: *Johnson solid 64: the augmented tridiminished icosahedron.*

 The dual of $J_{64}$ is most easily visualized by noting that augmentation and truncation are dual operations; the figure would, therefore, be obtained by truncating the pyramid-like apex of the polyhedron shown in Figure 126 to produce the polyhedron illustrated in Figure 128 below. (I must admit that I would not have expected the isosceles trapezoidal faces left from the apical pyramid by the truncation to be so short vertically, but the result of the polar reciprocation process is as shown in the figure.)

 The symmetry, of course, is the same $C_{3v}$ as is the symmetry of $J_{64}$ itself. And for the same reason that applied to Figures 125 and 127, Figure 128 is oriented identically to Figure 126, making comparison of the two particularly easy. (The augmentation does move the centroid, though it remains along the same threefold axis, so that there is a bigger change from Figure 126 to Figure 128 than there is from Figure 125 to Figure 127.

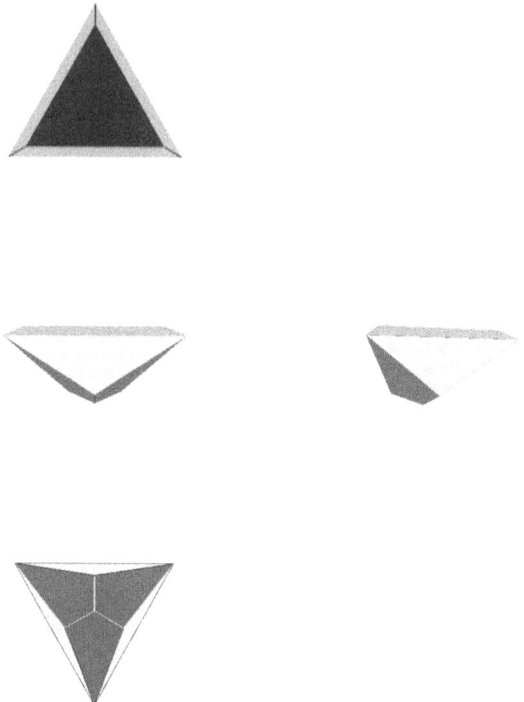

*Figure 128: The dual of Johnson solid 64 (the augmented tridiminished icosahedron).*

The properties of the three Johnson solids with which this chapter deals, and their duals, are summarized in Table 28 below.

*Diminished icosahedra: Johnson solids 62-64.*

| Johnson solid number | 62 | 63 | 64 |
|---|---|---|---|
| Name | Metabidiminished icosahedron | Tridiminished icosahedron | Augmented tridiminished icosahedron |
| Symmetry* | $C_{2v}$ | $C_{3v}$ | $C_{3v}$ |
| Faces | [2+2+2+4](3) + [2](5) | [1+1+3](3) + [3](5) | [1+3+3](3) + [3](5) |
| Edges* | 20 | 15 | 18 |
| Vertices | 10 | 9 | 10 |
| Dihedral angles | 3-3: 138.1897° (138° 11′ 23″), 3-5: 100.8123° (100° 48′ 44″), 5-5: 63.4349° (63° 26′ 6″)$^\Delta$ | | 3a-3a: 70.5288° (70° 31′ 44″), 3b-3c: 138.1897° (138° 11′ 23″), 3a-5: 171.3411° (171° 20′ 28″), 3b-5: 100.8123° (100° 48′ 44″), 5-5: 63.4349° (63° 26′ 6″)* |
| Area | 7.77108182 = $\sqrt{[25 + (5/2)\sqrt{5} + (5/2)\sqrt{(75 + 30\sqrt{5})}]}$ | 7.32649571 | 8.19252112 |
| Volume | 1.57868932 | 1.27718649 | 1.39503762 |
| **Dual** | | | |
| Name | Unnamed | Unnamed | Unnamed |
| Faces | [2](3) + [2+4](4) + [2](5) | [3+3](3) + [3](4) | [1+3](3) + [3+3](4) |
| Vertices | 12 | 8 | 10 |

Note: In row "Dihedral angles" for solid 63, the value 3-3: 138.1897°, 3-5: 100.8123°, 5-5: 63.4349° appears to span across solids 62 and 63.

## The Johnson Solids and Their Duals

| Johnson solid number | 62 | 63 | 64 |
|---|---|---|---|
| Face angles[†] | 3: [2]37.1189° (37° 7′ 8″), 105.7622° (105° 45′ 44″); 4a: 40.2677° (40° 16′ 4″), 103.4045° (103° 24′ 13″), 105.6266° (105° 37′ 36″), 110.7022° (110° 42′ 8″); 4k: [2]107.1489° (107° 8′ 56″), 42.0858° (42° 5′ 9″), 103.6165° (103° 36′ 59″); 5: [2]105.9046° (105° 54′ 17″), [2]108.9186° (108° 55′ 7″), 110.3537° (110° 21′ 13″)[♦] | 3a: [2]35.2784° (35° 16′ 42″), 109.4433° (109° 26′ 36″); 3b: [2]37.3881° (37° 23′ 17″), 105.2238° (105° 13′ 26″); 4: 38.6411° (38° 38′ 28″), 106.2596° (106° 15′ 34″), [2]107.5497° (107° 32′ 59″)[▫] | 3e: [3]60°; 3i: [2]34.0708° (34° 4′ 15″), 111.8585° (111° 51′ 30″); 4i: [2]37.6373° (37° 38′ 14″), [2]142.3627° (142° 21′ 46″); 4k: [2]108.9222° (108° 55′ 20″), 31.7946° (31° 47′ 41″), 110.3610° (110° 21′ 40″)[*] |
| Dihedral angles | 3-3: 122.3745° (122° 22′ 28″), 3-4a: 119.9568° (119° 57′ 24″), 4a-4a: 116.9529° (116° 57′ 11″), 4a-4k: 114.9826° (114° 58′ 58″), 4a-5: 113.7092° (113° 42′ 33″), 4k-5: 111.2358° (111° 14′ 9″), 5-5: 122.3745° (122° 22′ 28″)[♦] | 3a-3a: 119.9316° (119° 55′ 54″), 3a-3b: 118.0503° (118° 3′ 1″), 3b-4: 114.4449° (114° 26′ 42″), 4-4: 112.8841° (112° 53′ 3″)[▫] | 3e-4i: 138.4775° (138° 28′ 39″), 3i-4i: 114.7186° (114° 43′ 7″), 3i-4k: 120.4565° (120° 27′ 23″), 4i-4i: 109.9266° (109° 55′ 36″), 4k-4k: 122.2481° (122° 14′ 53″)[*] |

— 264 —

*Diminished icosahedra: Johnson solids 62-64.*

| Johnson solid number | 62 | 63 | 64 |
|---|---|---|---|
| Area | 5.46941991 | 7.71049847 | 6.08158408 |
| Volume | 0.97082900 | 1.53963830 | 0.98826319 |

Table 28: *Properties of Johnson solids 62, 63, and 64 and their duals.*

*The symmetry and number of edges are always the same for each solid and its dual, so are not repeated in the "dual" section.

†Face angles are only given for the duals, as the Johnson solids have regular faces, whose angles can be found in Table 1 on p. 3.

△Although the triangular faces of $J_{62}$ divide into *four* transitivity classes and those of $J_{63}$ divide into *three*, all make the same dihedral angles with the pentagonal faces as well as the same dihedral angles with each other, so no distinctions are made in this table.

‡The symbol 3a refers to the three triangles deriving from the pyramidal augmentation, found at one end of the polyhedron; the symbol 3b to those three of the the remaining four triangles (at the other end of the polyhedron) which are adjacent to the pentagonal faces; the symbol 3c to the only triangle which has no edge in common with a pentagonal face.

♦The symbol 4a refers to the four irregular quadrilaterals (almost kites, but not truly) that share an edge with the triangular faces; 4k to the two true kite-quadrilaterals.

□The symbol 3a refers to the three isosceles triangles opposite to the three kite-quadrilateral faces (*i. e.* those with no edges in common with the kites); 3b to the remaining three isosceles triangles; it is not necessary to designate the kite-quadrilaterals as 4k, since *all three* quadrilateral faces are congruent kite-quadrilaterals.

°The symbol 3e refers to the single equilateral triangle at the base; the symbol 3i to the three isosceles triangles; the symbol 4i to the three isosceles trapezoids surrounding the equilateral triangle (and separating it from the three isosceles triangles); and the symbol 4k to the three kite-quadrilaterals forming the apex.

## Chapter 14: "Augmented" Archimedean solids: Johnson solids 65-71.

For this chapter I have put the word "augmented" in quotes, because in fact it means something rather different from its usual meaning. In most cases, the term "augmentation" means the fusing of one or more *n*-gonal pyramid(s) to *n*-gonal face(s) of some polyhedron. This was seen in Chapters 11 and 12. However, in the polyhedra discussed in this chapter, one or more *n*-gonal *cupola(e)* has/have been fused to 2*n*-gonal face(s) of an Archimedean solid. It would really have been better if Johnson had used a different word for this process than "augmented": perhaps "cupolated" or such. Johnson's names are still given here, however, because they are the ones by which the polyhedra in question are generally known. As was noted in Chapter 3, a cupola can be constructed with regular-polygon faces only if the order of rotation is 3, 4, or 5; since the base (which would be fused to whatever other polyhedron is involved) is a 2*n*-gon, only polyhedra with at least one hexagonal, octagonal, or decagonal face can have a cupola fused to it to make a Johnson solid. This excludes all the Platonic solids and all but seven Archimedean solids: the truncated tetrahedron, truncated cube, truncated octahedron, great rhombicuboctahedron, truncated dodecahedron, truncated icosahedron, and great rhombicosidodecahedron. In fact, other considerations, such as the requirement that dihedral angles are less than 180° to insure convexity, limit the possibility of the construction to only three Archimedean solids: the truncated tetrahedron, truncated cube, and truncated icosahedron. The resulting fusion processes produce the seven Johnson solids described in this chapter. For ease in description, it will be advantageous to designate the set of eleven faces – five triangular, five square, and one pentagonal – derived from the cupola, which are all that remain of it after the base is lost in the fusion, as the "cupola faces" for short, and it should be understood that, whenever this phrase is used in this chapter, that will be the meaning.

Johnson solid 65, the augmented truncated tetrahedron, illustrated in Figure 129 below, is the first of the "augmented"

*"Augmented" Archimedean solids: Johnson solids 65-71.*

Archimedean solids discussed in this chapter. If one starts with a regular tetrahedron and cuts a piece off each of its four vertices, a truncated tetrahedron is produced. If the depth of the cutting is so adjusted that the original triangular faces of the tetrahedron are made into regular hexagons, an Archimedean solid, with four faces that are regular hexagons and four that are equilateral triangles, results. Any one of the four hexagonal faces of the truncated tetrahedron can be fused to the hexagonal base of a triangular cupola ($J_3$) to produce the resulting polyhedron.

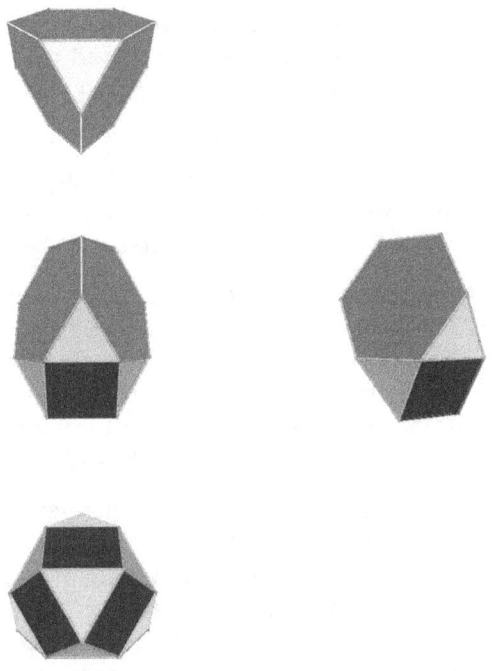

Figure 129: Johnson solid 65: the augmented truncated tetrahedron.

It would, at first glance, appear that the cupola could be fused to the truncated tetrahedron in either of two orientations. The hexagonal face of the cupola has two types of edges: the edges shared with triangles and the ones shared with squares. Any of the hexagonal faces of the truncated tetrahedron, including whichever one is chosen for the fusion, also has two types of edges: the edges shared with triangles and the ones shared with other hexagons. However, the dihedral angle in the cupola between a triangle and the hexagonal base is 70.5288° (70° 31′ 44″) and the dihedral angle in the truncated tetrahedron between a triangle and a hexagon is 109.4712° (109° 28′ 16″), and if those were combined, the total would be exactly 180°, implying that both triangles would be in the same plane. Fusing the cupola in the other orientation, so a triangular face derived from the cupola is adjacent to a hexagonal face derived from the truncated tetrahedron, combines two dihedral angles of 70.5288° (70° 31′ 44″) to produce one of 141.0576° (141° 3′ 27″), and putting a triangular face derived from the truncated tetrahedron adjacent to a square face derived from the cupola combines a dihedral angle of 109.4712° (109° 28′ 16″) with one of 54.7356° (54° 44′ 8″) to produce a dihedral angle of 164.2068° (164° 12′ 25″). Both of these are less than 180°, so the fusion retains convexity. So the latter orientation is the only way a cupola can actually be fused to a truncated tetrahedron.

The original regular tetrahedron has $T_d$ symmetry, and this symmetry is left unchanged by the truncation operation. The selection of a single one of the four hexagonal faces for the fusion, however, destroys all but one of the four threefold axes of the original, and similarly, only three of the six mirror planes of the original remain. The symmetry is thus $C_{3v}$.

It could be argued that the term "augmented truncated tetrahedron" could have no other meaning than the polyhedron just described, because a *pyramid*, rather than a cupola, could never be fused to a truncated tetrahedron. Fusing a pyramid to a triangular face of a truncated tetrahedron simply restores one of the vertices that was cut off in the truncation; the faces of the pyramid would be coplanar with the faces of the truncated tetrahedron. Fusing a pyramid to a hexagonal face is impossible because no pyramid can have more than five sides for its base, as was mentioned in Chapter 2. However, this argument is only available in the context of the Johnson restriction to regular polygonal faces. A triangular pyramid that is "flatter" than a regular tetrahedron could be fused to a triangular face, and a hexagonal pyramid can be constructed if its lateral faces are not

*"Augmented" Archimedean solids: Johnson solids 65-71.*

required to be equilateral triangles. So in a more general discussion of polyhedra, the term "augmented truncated tetrahedron" could easily mean something quite different. (This same discussion could apply to every other "augmented Archimedean polyhedron" in this chapter, so it will not be repeated, but can be assumed in each case.)

The dual of the augmented truncated tetrahedron, illustrated in Figure 130 below, has similar $C_{3v}$ symmetry to the augmented truncated tetrahedron, in keeping with the rule that dualization preserves symmetry.

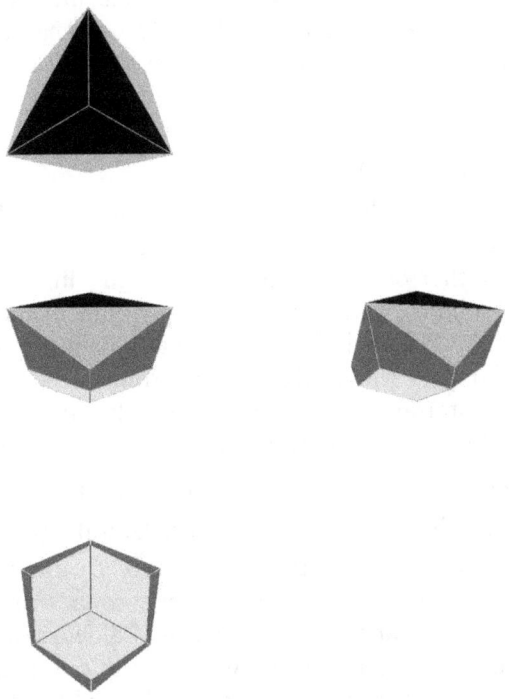

Figure 130: The dual of Johnson solid 65 (the augmented truncated tetrahedron).

Because the $C_{3v}$ symmetry involves a threefold axis and three mirror planes through that axis, all the triangular faces (which are bisected by the mirror planes) are isosceles, and the three quadrilaterals that are viewed most directly in the bottom view of Figure 130 are kite-quadrilaterals. The band of alternating-chirality quadrilaterals remaining around the middle, however, is not forced to have any symmetry at all, and they are quite irregular (see Table 29 below). The isosceles triangles are extremely long and narrow, and the two sets do not match each other.

Johnson solid 66, the augmented truncated cube, illustrated in Figure 131 below, is constructed by a similar process to the augmented truncated tetrahedron, but starting with a cube instead of a regular tetrahedron. If one starts with a cube and cuts a piece off each of its eight vertices, a truncated cube is produced. If the depth of the cutting is so adjusted that the original square faces of the cube are made into regular octagons, an Archimedean solid, with six faces that are regular octagons and eight that are equilateral triangles, results. Any one of the four octagonal faces of the truncated cube can be fused to the octagonal base of a square cupola ($J_4$) to produce the resulting polyhedron.

Again, one might have thought that the fusion of the square cupola to an octagonal face of the truncated cube could be done in either of two orientations, but again, as in the case of Johnson solid 65, only one is in fact possible. In the truncated cube, the dihedral angle between a pair of octagonal faces is exactly 90° (since these edges come from the original cube), and the dihedral angle between a triangular and an octagonal face is 125.2644° (125° 15′ 52″). In the square cupola, the dihedral angle between a triangular face and the base is 54.7356° (54° 44′ 8″), and the dihedral angle between a lateral square face and the base is exactly 45°. If the fusion were to put a triangular face derived from the cupola adjacent to a triangular face derived from the truncated cube, the total dihedral angle would then be 180°, and again the two would lie in one plane, so only the orientation where a triangular face derived from the cupola is adjacent to an octagonal face derived from the truncated tetrahedron, and a triangular face derived from the truncated tetrahedron is adjacent to a square face derived from the cupola yields a convex solid.

It is an interesting coincidence that the dihedral angle between the lateral faces (a triangle and a square) of the square cupola is 144.7356° (144° 44′ 8″), while combining the triangle-octagon dihedral angle of 54.7356° (54° 44′ 8″) at the base of the cupola with the 90° octagon-octagon dihedral angle from the truncated cube gives the same 144.7356° (144° 44′ 8″) dihedral angle at the triangle-octagon edge created by the fusion.

The original cube has $O_h$ symmetry, and this symmetry is left unchanged by the truncation operation. The selection of a single one of the six octagonal faces for the fusion, however, destroys all but one of the three fourfold axes of the original, and similarly, only four of the nine mirror planes of the original remain. The symmetry is thus $C_{4v}$.

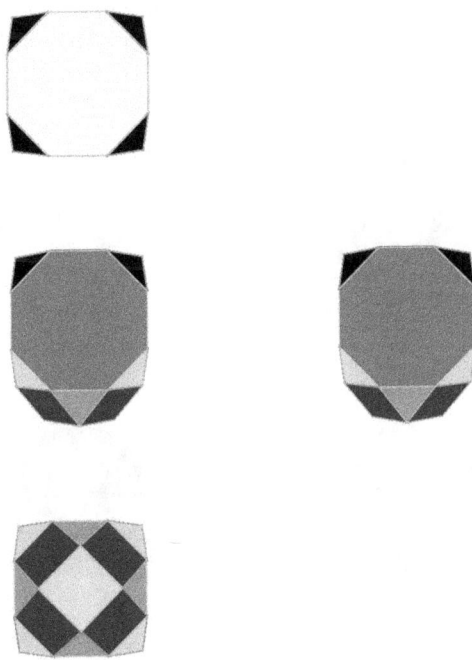

*Figure 131: Johnson solid 66: the augmented truncated cube.*

The dual of the augmented truncated cube, illustrated in Figure 132 below, has the same $C_{4v}$ symmetry as its dual. It is a figure that could be considered as the fusion of two rather different polyhedra, fused along a plane perpendicular to the fourfold axis. The face that is lost from each of the two polyhedra is a square. The top polyhedron would have, besides the square base, faces consisting of four kite-quadrilaterals around the apex, eight irregular quadrilaterals arranged in alternating chirality, and finally a set of four isosceles triangles, whose bases are the edges along the square that will be lost in the fusion, completing the upper polyhedron. Although this polyhedron, of $C_{4v}$ symmetry with four kite-quadrilateral faces, eight irregular quadrilaterals, and four isosceles triangles over a square base, has no name, it will shortly be met again, so in order to refer to it I will designate it a "$J_{66}$ dual upper half." The bottom polyhedron would have eight scalene triangles (arranged, like the irregular quadrilaterals in the top figure, in an alternating-chirality pattern) and four isosceles triangles whose bases, like the bases of the isosceles triangles of the top half of the figure, are the edges along the square that will be lost in the fusion.

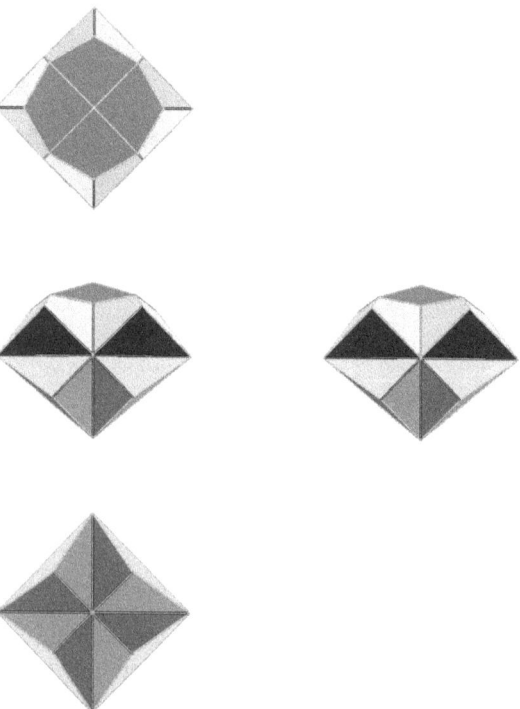

*Figure 132: The dual of Johnson solid 66 (the augmented truncated cube).*

*"Augmented" Archimedean solids: Johnson solids 65-71.*

This polyhedron illustrates a phenomenon that should be called out at this point. It can be seen that the Johnson solids described as "elongated" all have duals that can be characterized as two simpler polyhedra fused along faces that, in each of the two simpler ones, are perpendicular to the principal axis of rotation. So does $J_{66}$, even though it is not an "elongated" polyhedron. What these all have in common is that there are a *set of edges all parallel to each other and to the principal axis*. Whenever a polyhedron has such a feature, its dual will have a set of vertices that all lie in a plane, and edges that connect them, so that the dual polyhedron can be seen to be decomposable into two pieces fused along that plane.

Johnson solid 67, the biaugmented truncated cube, illustrated in Figure 133 below, is constructed in the same manner as $J_{66}$, except that two opposite octagonal faces of the truncated cube are fused to square cupolae.

Figure 133: Johnson solid 67: the biaugmented truncated cube.

It would not be possible to augment two octagonal faces that had an edge in common, because a dihedral angle in excess of 180° would be produced, leading to nonconvexity of the resulting polyhedron; therefore the term "biaugmented" is sufficient, without an indication such as "para-" as was used in such names as that of $J_{55}$. And the augmentation of two opposite faces restores some of the twofold axes and mirror planes of the original cube, though, to be sure, the symmetry is still far less than $O_h$; it is only $D_{4h}$.

Of course, the same considerations that led to the conclusion that only one orientation of the square cupola can be fused to the truncated cube apply to the second cupola, so only one possible biaugmented truncated cube can be constructed if convexity is required.

The dual of the biaugmented truncated cube, illustrated in Figure 134 below, of course has the same $D_{4h}$ symmetry. Although it has no name, it resembles a pillow or cushion.

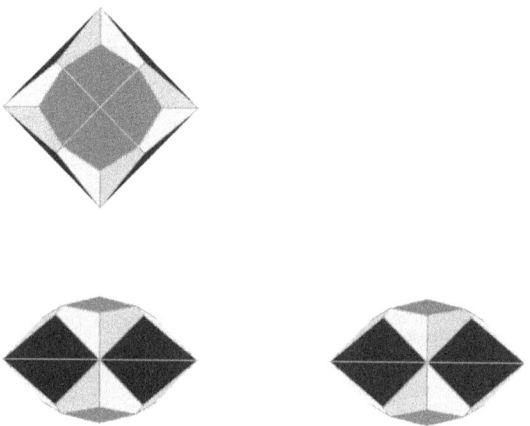

Figure 134: The dual of Johnson solid 67 (the biaugmented truncated cube).

Eight of the thirty-two faces are isosceles triangles, arranged in a central belt in Figure 134, with two angles of 31.3997° (31° 23′ 59″) and one of 117.2006° (117° 12′ 2″). Eight more are kite-quadrilaterals, four at the top and four at the bottom, whose two equal angles are 81.8321° (81° 49′ 56″); the other two angles of the kite-quadrilaterals are 109.9977° (109° 59′ 52″) and 86.3380° (86° 20′ 17″). The remaining sixteen faces are irregular quadrilaterals. Irregular they may be, but their angles match the angles of the isosceles triangles in that one is 31.3997° (31° 23′ 59″) and one is 117.2006° (117° 12′ 2″), while a third is exactly

90° more than the smaller one (121.3997°, or 121° 23' 59"). The remaining angle is a right angle. So in effect, they are figures related to the isosceles triangles at the belt of this polyhedron by cutting off one vertex by a line perpendicular to the base of the triangle.

In fact, by comparison with Figure 132, it can be seen that (not very surprisingly) the $J_{67}$ dual is simply two $J_{66}$ dual upper halves fused together, except for some of the precise values of edge lengths and angles.

Some of the measurements in the dual of $J_{67}$ have interesting values, even though most of the numbers in the measurements of the duals, when they are not compelled by symmetry, are usually not so interesting. To discuss them, let us note that the eight kite-quadrilaterals at the two apices (four at each) are the only kite-quadrilaterals among the faces, and all the other quadrilaterals are irregular, so that the term "kite" and the term "irregular quadrilateral" each uniquely specifies one type of face. Now we can state that *every* irregular quadrilateral has one 90° angle, and of the two sides that do not meet at that right-angle vertex (the two longest sides of the quadrilateral), in *every* one of these quadrilaterals one side is exactly 2 and one is $\sqrt{2}$. The side that is exactly 2 is the edge that the irregular quadrilateral shares with an isosceles triangle, so of course two of the sides of the isosceles triangles (the ones that are shared edges with the quadrilaterals) are also exactly 2, but it is also true that the remaining side of each isosceles triangular face (the one which is a shared edge with *another* isosceles triangular face) is $2 + \sqrt{2}$ exactly. So many of the edges of this polyhedron are either $\sqrt{2}$, 2, or $2 + \sqrt{2}$ exactly. This is an unusual result for a Johnson dual. It is also true that all dihedral angles that do not involve the kite-quadrilaterals are equal.)

## The Johnson Solids and Their Duals

| Johnson solid number | 65 | 66 | 67 |
|---|---|---|---|
| Name | Augmented truncated tetrahedron | Augmented truncated cube | Biaugmented truncated cube |
| Symmetry* | $C_{3v}$ | $C_{4v}$ | $D_{4h}$ |
| Faces | [1+1+3+3](3) + [3](4) + [3](6) | [4+4+4](3) + [1+4](4) + [1+4](8) | [8+8](3) + [2+8](4) + [4](8) |
| Edges* | 27 | 48 | 60 |
| Vertices | 15 | 28 | 32 |
| Dihedral angles | 3a-4: 164.2068° (164° 12′ 25″), 3c-4: 125.2644° (125° 15′ 52″), 3a-6: 164.4712° (109° 28′ 16″), 3c-6: 141.0576° (141° 3′ 27″), 6-6: 70.5288° (70° 31′ 44″)♦ | 3a-4: 170.2644° (170° 15′ 52″), 3c-4: 144.7356° (144° 44′ 8″), 3a-8: 125.2644° (125° 15′ 52″), 3c-8: 144.7356° (144° 44′ 8″), 4-4: 135°, 8-8: 90°△ | |
| Area | 14.25833025 = 3 + (13/2)√3 | 34.33828805 | 36.24191173 |
| Volume | 3.88908730 | 15.54247233 | 17.48528137 |
| Dual | | | |
| Name | Unnamed | Unnamed | Unnamed |
| Faces | [3+3](3) + [3+6](4) | [4+4+8](3) + [4+8](4) | [8](3) + [8+16](4) |
| Vertices | 14 | 22 | 30 |

*"Augmented" Archimedean solids: Johnson solids 65-71.*

| Johnson solid number | 65 | 66 | 67 |
|---|---|---|---|
| Face angles[†] | 3a: [2]32.2609° (32° 15' 39"), 115.4782° (115° 28' 42"); 3b: [2]38.0466° (38° 2' 48"), 103.9068° (103° 54' 25"); 4a: 43.1937° (43° 11' 37"), 82.9416° (82° 56' 30"), 116.4822° (116° 28' 56"), 117.3825° (117° 22' 57"); 4k: [2]71.5203° (71° 31' 13"), 102.7500° (102° 45' 0"), 114.2094° (114° 12' 34")* | 3a: [2]30.4529° (30° 27' 10"), 119.0942° (119° 5' 39"), 3b: [2]33.1089° (33° 6' 32"), 113.7821° (113° 46' 56"), 3c: 28.5031° (28° 30' 11"), 34.5069° (34° 30' 25"), 116.9900° (116° 59' 24"); 4a: 34.7369° (34° 44' 13"), 87.7224° (87° 43' 21"), 118.1824° (118° 10' 57"), 119.3582° (119° 21' 30"); 4k: [2]81.5181° (81° 31' 5"), 85.1255° (85° 7' 32"), 111.8383° (111° 50' 18")* | 3: [2]31.3997° (31° 23' 59"), 117.2006° (117° 12' 2"); 4a: 31.3997° (31° 23' 59"), 90°, 117.2006° (117° 12' 2"), 121.3997° (121° 23' 59"); 4k: 86.3380° (86° 20' 17"), 109.9977° (109° 59' 52"), [2]81.8321° (81° 49' 56")[□] |

| Johnson solid number | 65 | 66 | 67 |
|---|---|---|---|
| Dihedral angles | 3a-3a: 139.0168° (139° 1′ 0″), 3a-3b: 136.8324° (136° 49′ 57″), 3b-4a: 129.5381° (129° 32′ 17″), 4a-4a: 123.2431° (123° 14′ 35″), 4a-4k: 130.4106° (130° 24′ 38″), 4k-4k: 134.0369° (134° 2′ 13″)* | 3a-3b: 148.0663° (148° 3′ 59″), 3a-3c: 150.0266° (150° 1′ 36″), 3c-3c: 150.6647° (150° 39′ 53″), 3b-4a: 145.2490° (145° 14′ 56″), 4a-4a: 143.7178° (143° 43′ 4″), 4a-4k: 146.2453° (146° 14′ 43″), 4k-4k: 147.4969° (147° 29′ 49″)* | 3-3, 3-4a, 4a-4a: 147.3501° (147° 21′ 0″), 4a-4k: 150.6566° (150° 39′ 24″), 4k-4k: 151.6338° (151° 38′ 2″)$^\triangle$ |
| Area | 8.76076451 | 26.03353773 | 38.75583305 |
| Volume | 2.04914934 | 11.24012903 | 20.44302892 |

Table 29: Properties of Johnson solids 65, 66, and 67 and their duals.

*The symmetry and number of edges are always the same for each solid and its dual, so are not repeated in the "dual" section.

†Face angles are only given for the duals, as the Johnson solids have regular faces, whose angles can be found in Table 1 on p. 3.

*Although the triangular faces fall into *four* transitivity classes, only two sets need be distinguished: 3a represents the four triangles that originally came from the truncated tetrahedron; 3c represents the four triangles that originally came from the triangular cupola.

$^\triangle$Although the triangular faces fall into *three* transitivity classes, and the square and octagonal faces each fall into *two* in $J_{66}$, while the triangular and square faces each fall into *two* in $J_{67}$, only two sets of triangles need be distinguished: 3a represents the four triangles that originally came from the truncated cube; and 3c represents the four triangles that originally came from the square cupola. No need exists to distinguish among the square or among the octagonal faces.

*The symbol 3a refers to the three isosceles triangles at the threefold apex; 3b to the three isosceles triangles which do not surround the threefold apex; 4a to the six irregular quadrilaterals that form a band separating the three kite-quadrilaterals from the six triangular faces; 4k to the three kite-quadrilaterals meeting at a threefold (anti)apex.

⁺The symbol 3a refers to the four isosceles triangles whose equal sides are shared edges with the eight scalene triangles meeting at the antiapex of the polyhedron; 3b to the four isosceles triangles whose equal sides are shared edges with the eight irregular quadrilaterals; 3c to the eight scalene triangles meeting at the antiapex of the polyhedron; 4a to the eight irregular quadrilaterals that separate the four kite-quadrilaterals from the triangular faces of the polyhedron; 4k to the four kite-quadrilaterals meeting at the apex of the polyhedron.

□The symbol 4k refers to the eight kite-quadrilaterals at the apices of the polyhedron (four at each apex); 4a refers to the remaining sixteen quadrilaterals, which are irregular.

The construction of Johnson solid 68, the augmented truncated dodecahedron, is similar to the constructions described previously in this chapter. If one starts with a regular dodecahedron and cuts a piece off each of its twenty vertices, a truncated dodecahedron is produced. If the depth of the cutting is so adjusted that the original pentagonal faces of the dodecahedron are made into regular decagons, an Archimedean solid, with twelve faces that are regular decagons and twenty that are equilateral triangles, results.

Since the truncated dodecahedron retains all the symmetry of the original dodecahedron (symbolized $I_h$), the procedure described on the discussion beginning on p. 230, with fusing a cupola to a decagonal face being considered the modification performed, leads to the conclusion that four polyhedra are to be considered. And $J_{68}$ to $J_{71}$ are the four so derived.

Fusing a pentagonal cupola to one of the twelve decagonal faces produces the polyhedron illustrated in Figure 135 below, which Norman Johnson named the augmented truncated dodecahedron and numbered 68th in his list. Just as was the case for the previous solids discussed in this chapter, only one orientation of the cupola leads to a convex polyhedron. I must thank Adrian Rossiter for enlightening me (in this particular case, though similar arguments apply to the other polyhedra in this chapter) as to the reason why this would be the case: He informed me that the rhombicosidodecahedron is a cantellated regular dodecahedron, whose vertices are located so that their truncations have the same directions (as measured by the orientations of their normals) as those of the truncated dodecahedron. Since the pentagonal cupola is a piece of the rhombicosidodecahedron, arranging the triangles to be adjacent in the fusion would make them coplanar. Thus any pentagonal cupola has to be fused to the truncated dodecahedron in such a way that the square faces derived from the cupola are adjacent to the triangular faces of the truncated dodecahedron (the faces produced in the original truncation of the dodecahedron), and the triangular faces derived from the cupola are adjacent to the remaining decagonal faces of the truncated dodecahedron (the faces of the dodecahedron prior to the truncation). This is very clear in the top view of Figure 135, and will also be apparent in the figures, such as Figure 137, showing solids with more than one cupola fused to the truncated dodecahedron, which will complete this chapter afterward.

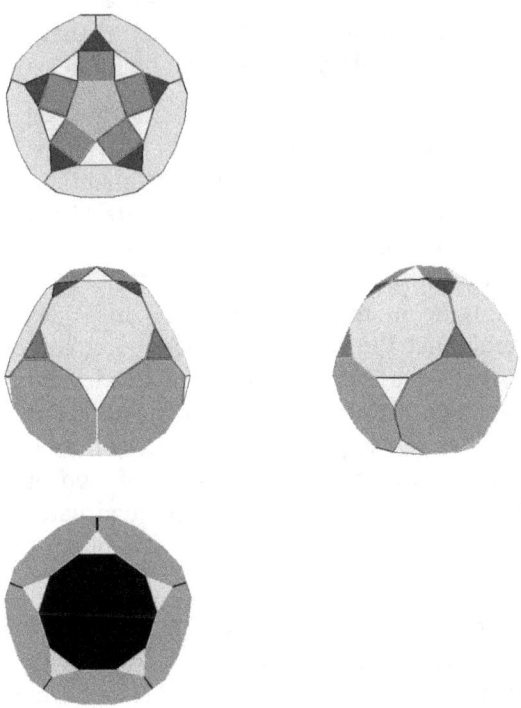

*Figure 135: Johnson solid 68, the augmented truncated dodecahedron.*

The remaining eleven decagonal faces can be classified into three sets, which have been shaded differently in Figure 135: five are immediately adjacent to the cupola, five more are neither adjacent nor opposite, and the eleventh is directly opposite the face which was fused with a cupola. This is exactly the same 5-5-1 division described in the discussion beginning on p. 230.

The dual of the augmented truncated dodecahedron is illustrated in Figure 136 below. It might be noted that the top view contains a star-like pattern consisting of fifteen of the 65 faces, which can be termed the *cupola-dual star*. The visible star is only an artifact of the fact that it is being viewed in two dimensions, because the fifteen faces do not lie in a plane, but because this configuration of fifteen faces will be seen again, it is useful to have a name for it. Because the centroid of the original augmented truncated dodecahedron is on the fivefold axis of the augmenting cupola, the cupola-dual star has exact fivefold symmetry, but we will see in the discussions of the other augmented truncated dodecahedra that very similar slightly distorted cupola-dual stars will be found.

At the center of the cupola-dual star is a set of five kite-quadrilaterals (in the distorted cupola-dual stars that will be seen later, they will not be precisely kite-quadrilaterals, but will be nearly so; in this case, they are accurate kite-quadrilaterals) and forming the points of the star are ten irregular quadrilaterals, alternating in chirality. The shortest sides of the irregular quadrilaterals lie in the same plane as the sides of the kite-quadrilaterals that form the edges meeting at the center of the star; the other edges between irregular quadrilaterals are much longer, extending from the vertices of the kites that are furthest from the center to the points of the star, and each such edge is also in a plane with the shortest sides of two other irregular quadrilaterals and the sides of two kite-quadrilaterals that form the edges meeting at the center of the star opposite the edge in question.

Because (except for the dual of $J_{69}$, where a higher degree of symmetry reduces the number of transitivity classes) the duals of the augmented truncated dodecahedra have too many different kinds of face and too many different values of their dihedral angle to fit into the cells of Table 30 below, there will be summaries in the text of the measurements rather than attempting to put them into Table 30.

In the case of the dual of the augmented truncated dodecahedron, the kite-quadrilaterals meeting at the fivefold vertex have two sides of ~1.4813 (the sides that form the edges between any two of them) and two sides of ~0.9927 (the sides that form the edges between a kite and an irregular quadrilateral); the two equal angles of the kite are 86.4374° (86° 50′ 15″), and the other two angles are 69.5850° (69° 35′ 6″) at the fivefold vertex and 116.7403° (116° 44′ 25″) at the vertex furthest from the fivefold vertex.

The irregular quadrilaterals have sides of ~0.2701, ~0.9927,

~1.3765, and ~1.9159, with the very shortest of these being the one between the apparent points of the star, the very longest being the outer edge of the star, and the side of ~0.9927 being at the same edge that was seen before as a side of a kite-quadrilateral. The face angles are 31.2059° (31° 12′ 21″) at the point of the star, 89.5393° (89° 32′ 21″) at the quadrivalent vertex where two kites and two irregular quadrilaterals come together, 119.2052° (119° 12′ 19″) at the trivalent vertex where the longest and shortest sides of the quadrilateral come together with the longest side of another irregular quadrilateral, and 120.0497° (120° 2′ 59″) at the trivalent vertex where the irregular quadrilateral, a kite, and an adjacent irregular quadrilateral come together.

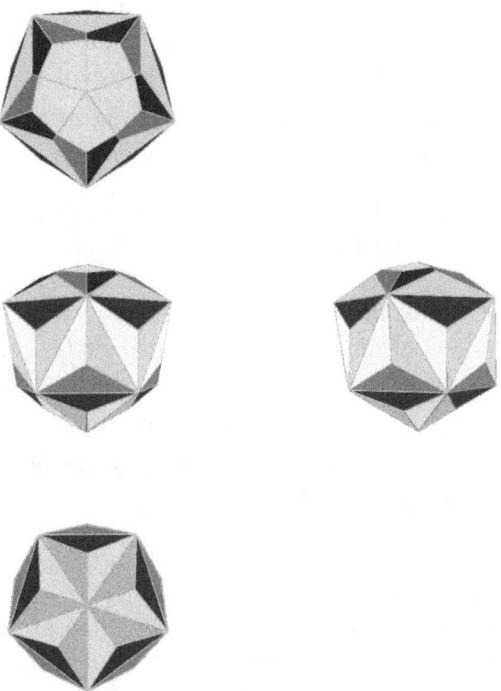

*Figure 136: The dual of Johnson solid 68 (the augmented truncated dodecahedron).*

The remaining fifty faces of this polyhedron are all triangular, but fall into a number of different transitivity classes. They are all very compressed triangles, with approximately the same shape.

- Ten of them form another star-like pattern opposite the cupola-dual star. These are scalene, but near-isosceles, triangles with sides of ~1.6811, ~1.7519, and ~2.9578, and angles of 29.8184° (29° 49' 6"), 31.2098° (31° 12' 35"), and 118.9718° (118° 58' 18"). (The 31.2098° angles are all at the decavalent antiapex.) This can be termed the *antiapical star*, but only occurs in this polyhedron and not the following three, so the term is less useful.

- Five of them, all isosceles, fit into the angles between the points of the cupola-dual star. These can be termed the *cupola-dual-star fill-in triangles*. Two of their sides are the same edges as sides of length ~1.9159 that were seen earlier in the cupola-dual star description; the third is much longer, ~3.2900. The equal angles of these triangles are 30.8379° (30° 50' 16"), the other angle, where they fit into the star, is 118.3243° (118° 19' 27").

- Five more take opposing positions, with their bases adjacent to the bases of those five. These can be termed the *cupola-dual-star fill-in-opposed triangles*. They are almost, but not exactly, the same size and shape as the cupola-dual-star fill-in triangles: their longest sides are the same edge, of length ~3.2900, but their shorter sides are slightly shorter at ~1.9030. The equal angles of these triangles are 30.1830° (30° 10' 59"); the other angle is 119.6339° (119° 38' 2").

- Five more isosceles triangles fit into the angles between the points of the antiapical star. These can be termed the *antiapical-star fill-in triangles*. They are almost the same shape as the cupola-dual-star fill-in triangles: their longest sides are of length ~3.0268 and their shorter sides are of length ~1.7519. The equal angles of these triangles are 30.2446° (30° 14' 41"); the other angle is 119.5108° (119° 30' 39").

*"Augmented" Archimedean solids: Johnson solids 65-71.*

- These five too have a set of opposing triangles, which can be termed the *antiapical-star fill-in-opposed triangles*. These are also almost the same shape as the cupola-dual-star fill-in triangles: their longest sides are of length ~3.0268, since this side is an edge shared with the longest side of the antiapical-star fill-in triangles, and their shorter sides are of length ~1.7637. The equal angles of these triangles are 30.8956° (30° 53' 44"); the other angle is 118.2089° (118° 12' 32").

- The remaining twenty form a belt around the center of the solid, but these also can be subdivided into two sets of ten: those sharing an edge with the the cupola-dual-star fill-in-opposed triangles, with sides ~1.7756, ~1.9030, and ~3.1656 and angles 29.4687° (29° 28' 7"), 31.8192° (31° 49' 9"), and 118.7121° (118° 42' 44"); and those sharing an edge with the antiapical-star fill-in-opposed triangles, with sides ~1.7637, ~1.8999, and ~3.1656 and angles 29.0039° (29° 0' 14"), 31.4880° (31° 29' 17"), and 119.5081° (119° 30' 29"). The longest sides of both are the same because they are in fact common edges between the two types of triangle.

With the triangular faces falling into seven different transitivity classes, there are nine different kinds of triangle/triangle dihedral angle:

- The dihedral angle between two adjacent antiapical-star triangles is 161.8295° (161° 49' 46").

- The dihedral angle between the antiapical-star triangles and the antiapical-star fill-in triangles is 161.7302° (161° 43' 49").

- The dihedral angle between the cupola-dual-star fill-in triangles and the cupola-dual-star fill-in-opposed triangles is 159.8812° (159° 52' 52").

- The dihedral angle between the cupola-dual-star fill-in-opposed triangles and the central belt triangles (on the side toward the cupola-dual star) is 160.3537° (160° 21'

13").

- The dihedral angle between the antiapical-star fill-in triangles and the antiapical-star fill-in-opposed triangles is 161.4485° (161° 26′ 55″).

- The dihedral angle between the antiapical-star fill-in-opposed triangles and the central belt triangles (on the side toward the antiapex) is 161.0523° (161° 3′ 8″).

- The dihedral angle between two adjacent central belt triangles (both of which are on the side toward the cupola-dual star) is 160.5365° (160° 32′ 11″).

- The dihedral angle between the central belt triangles (on the side toward the cupola-dual star) and the central belt triangles (on the side toward the antiapex) is 160.6874° (160° 41′ 15″).

- The dihedral angle between two adjacent central belt triangles (both of which are on the side toward the antiapex) is 160.8047° (160° 48′ 17″).

The only place where a triangular face shares an edge with a quadrilateral face is at the outside of the cupola-dual star, between an irregular quadrilateral and one of the triangular faces designated as a cupola-dual-star fill-in triangle. The dihedral angle at any of these edges is 159.4430° (159° 26′ 35″).

There are three types of quadrilateral/quadrilateral dihedral angles: the dihedral angle between any two kite-quadrilaterals forming the cupola-dual star is 160.2176° (160° 13′ 3″); the dihedral angle between a kite-quadrilateral and one of the irregular quadrilaterals forming the cupola-dual star is 159.9265° (159° 55′ 35″); and the irregular quadrilaterals make dihedral angles of 159.2608° (159° 15′ 39″) with each other (regardless of whether at the short or longer edges).

It might be noted that all of the dihedral angles, regardless of the kind of face between which the dihedral angles are measured, are approximately equal. Since the truncated dodecahedron is an Archimedean solid, whose dual is the triakis icosahedron (a Catalan solid, so therefore all dihedral angles are equal) it might not seem to be strange that the triangle/triangle dihedral angles are all close to the dihedral angle of the triakis icosahedron, which is

$$160.6126° = 160° \, 36' \, 45'' = \cos^{-1}[-(24 + 15\sqrt{5})/61],$$

but since the pentagonal cupola is related to the rhombicosidodecahedron, whose dual is the deltoidal hexecontahedron, also a Catalan solid, with all dihedral angles equal, but at $154.1214° = 154° \, 7' \, 17''$, one might have thought that the dihedral angles involving the cupola-dual star might be slightly smaller. However, they are not.

It might be noted that if one takes the cupola-dual stars together with the cupola-dual-star fill-in triangles, the collection of twenty faces, fifteen quadrilaterals and five triangles, has its outer edges all in one plane, so if a pentagon were added to this to form a twenty-one-faced polyhedron, and another pentagon were added to the remaining 45 faces to complete it when the first set of faces was deleted, this polygon could be considered to be a fusion of the two. And in fact the antiapical stars together with the antiapical-star fill-in triangles also form a similar configuration, so that if these fifteen triangles (ten from the antiapical star and five antiapical-star fill-in triangles) were completed by a pentagonal face and a similar pentagon were added to the remaining 31 faces (30 from the original polyhedron, plus the pentagonal face added in the first splitting) to complete it, this too can be considered a fusion. So this polyhedron can actually be considered a fusion of three polyhedra:

- One of five triangular, fifteen quadrilateral, and one pentagonal face (21 altogether) arising from the cupola-dual star,
- One of fifteen triangular and one pentagonal faces (16 altogether) arising from the antiapical star, and
- A central polyhedron, with thirty triangular and two pentagonal faces (32 altogether).

In the fusion, all the pentagonal faces are deleted, of course.

Returning to the discussion of the augmented truncated dodecahedron itself, to fuse another cupola onto any of the five decagonal faces that is adjacent to the first cupola would produce a dihedral angle exceeding 180°. This can be seen because the dihedral angle between two decagonal faces of the original truncated dodecahedron (or between *any* two adjacent faces of the regular dodecahedron before the truncation) is 116° 33′ 54″, while after the fusion, a dihedral angle is produced at the corresponding edge (now an edge between a decagon and the triangle which derives from the nearest lateral face of the cupola in the fusion) of 153° 56′ 33″ (implying that the fusion has added 37° 22′ 39″ to the dihedral angle, which in fact is the dihedral angle between a triangular and decagonal face of a pentagonal cupola).

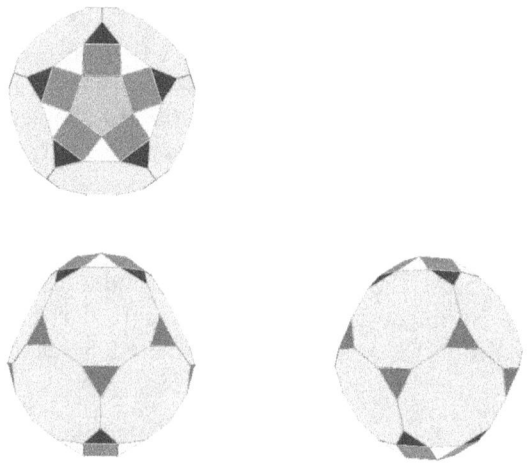

Figure 137: Johnson solid 69: the parabiaugmented truncated dodecahedron.

One might note that no attempt has been made here to fuse the cupola to the truncated dodecahedron in an orientation such that a square face of the cupola is adjacent to a decagon on the truncated dodecahedron; the reason is that the dihedral angle between a triangle from the cupola and a triangle from the truncated dodecahedron would be exactly 180°, as was discussed in the description of $J_{68}$, so the two triangles would be in a single plane. Adding another 37° 22′ 39″

to that 153° 56′ 33″ dihedral angle, by fusing a cupola to a decagon adjacent to the one first fused, would increase the dihedral angle to 191° 19′ 12″. Anything exceeding 180°, of course, implies nonconvexity of the polyhedron. So the only places where a cupola can be fused are the five decagons that are neither adjacent nor opposite to the first fusion location, producing $J_{70}$ (the metabiaugmented truncated dodecahedron, illustrated in Figure 139 below), or the decagon directly opposite the first fusion, producing $J_{69}$ (the parabiaugmented truncated dodecahedron, illustrated in Figure 137 below).

It should be noted that when the fusion takes place at two locations directly opposite each other, the symmetry of the original dodecahedron is such that it does not introduce a mirror plane, as was the case for the biaugmented truncated cube, but instead the process leads to an alternating axis (tenfold in this case), so the symmetry of the parabiaugmented truncated dodecahedron is $S_{10v}$.

Because this symmetry makes the number of transitivity classes considerably smaller than was found in $J_{68}$, this solid is actually simpler to describe than was $J_{68}$. The tenfold alternating axis means that the bottom is just like the top, except for a 36° twist. All the decagonal faces form a single transitivity class, as do all the square faces, and the triangular faces form only three, based on whether they come from the augmenting pentagonal cupola or the truncated dodecahedron; the triangles from the truncated dodecahedron need to be distinguished between those adjacent to a fusion plane and those in the central piece, approximately halfway between the fusion planes.

It might be noted that the original truncated dodecahedron has 32 faces, 60 vertices, and 90 edges, and the augmented truncated dodecahedron 42 faces, 65 vertices, and 105 edges. The fact that each augmentation increases the number of faces by ten, the number of vertices by five, and the number of edges by fifteen can be seen by following the procedures that have been used numerous times in this book to determine the face/vertex/edge counts of polyhedra obtained by fusion; thus the parabiaugmented truncated dodecahedron has 52 faces, 70 vertices, and 120 edges (as does the metabiaugmented truncated dodecahedron, which will be discussed later).

The dual of the parabiaugmented truncated dodecahedron is illustrated in Figure 138. It has the same $S_{10v}$ symmetry as the parabiaugmented truncated dodecahedron itself.

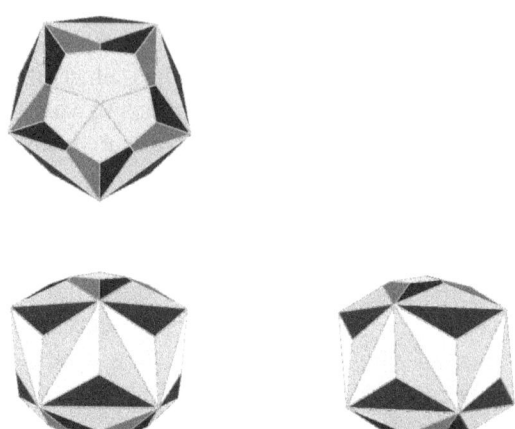

Figure 138: The dual of Johnson solid 69 (the parabiaugmented truncated dodecahedron).

Since both of the augmentations are opposite each other and the fivefold symmetry is preserved, the two cupola-dual stars found in this polyhedron are exactly fivefold-symmetric, as was the cupola-dual star in the dual of the augmented truncated dodecahedron.

In addition, since both ends are alike, there is a smaller number of different classes of face. There is no antiapical star, and, while in the case of the dual of $J_{68}$ there was a correspondence between different sets of triangular faces on opposite sides, in the case of the dual of $J_{69}$ there is a merger of classes: one can speak of cupola-dual-star fill-in triangles, but no antiapical-star fill-in triangles (and the former are now found where the latter were), and similarly one can speak of cupola-dual-star fill-in-opposed triangles, but no antiapical-star fill-in-opposed triangles, and there is only one type of central belt triangle.

As was the case in the dual of $J_{68}$, the cupola-dual stars, taken together with the cupola-dual-star fill-in triangles, form a collection whose outer vertices all lie in a plane, so that if the polyhedron were cut in those planes, one could add pentagonal faces to each set of faces to produce three polyhedra which, by fusion (which eliminates those pentagonal faces), could reconstruct this polyhedron.

It is interesting that, while the triangular faces fall into three different transitivity classes, they are all alike, with two sides of length ~2.0991 = $5(7 + \sqrt{5})/22$ and one of length ~3.6180 = $(5 + \sqrt{5})/2$. These, in fact, are also the edge lengths of the triakis icosahedron that is obtained by polar reciprocation of the unmodified truncated dodecahedron. It is also interesting to note that the face angles of the irregular quadrilaterals are related to the face angles of the triangles: the two equal angles in the triangles are 30.4803° (30° 28′ 49″) = $\cos^{-1}[¾ + (\sqrt{5})/20]$ and the remaining angle is 119.0394° (119° 2′ 22″) = $\cos^{-1}[-3(1 + \sqrt{5})/20]$; while the irregular quadrilaterals have one angle each of 30.4803° (30° 28′ 49″), one exact 90° angle, and one of 120.4803° (120° 28′ 49″) which is exactly 90° more than 30.4803° (30° 28′ 49″). Also, since the centroid of the polyhedron coincides with the centroid of the truncated dodecahedron without the augmentations, the dihedral angles involving all faces other than the cupola-dual star quadrilaterals (and even those between the irregular quadrilaterals of the cupola-dual star and the triangular faces) have the same value of 160.6126° = 160° 36′ 45″ = $\cos^{-1}[-(24 + 15\sqrt{5})/61]$ as the dihedral angle of the triakis icosahedron.

Johnson solid 70, the metabiaugmented truncated dodecahedron, is illustrated in Figure 139 below. Unlike the parabiaugmented truncated dodecahedron, its symmetry is only $C_{2v}$, which means that the faces divide into a much larger number of transitivity classes, even though both polyhedra have the same 52 faces, 70 vertices, and 120 edges. (This difference between the symmetry of "para" and "meta" solids has already, of course, been seen more than once in this book.) Other than those properties that depend on the symmetry, however, most of what was said about the parabiaugmented truncated dodecahedron applies to the metabiaugmented truncated dodecahedron. The reason for the difference between the orientations of Figure 139 and Figure 137 is the convention in this book to make the vertical axis coincide with the principal axis of rotation, which was the tenfold alternating axis in Figure 137 and is one of the twofold axes in Figure 139.

As was the case for $J_{55}$ and $J_{56}$ (see p. 220), in the naming of $J_{69}$ and $J_{70}$ (the para- and metabiaugmented truncated dodecahedron), Johnson borrowed the prefixes *meta-* and *para-* from organic chemistry.

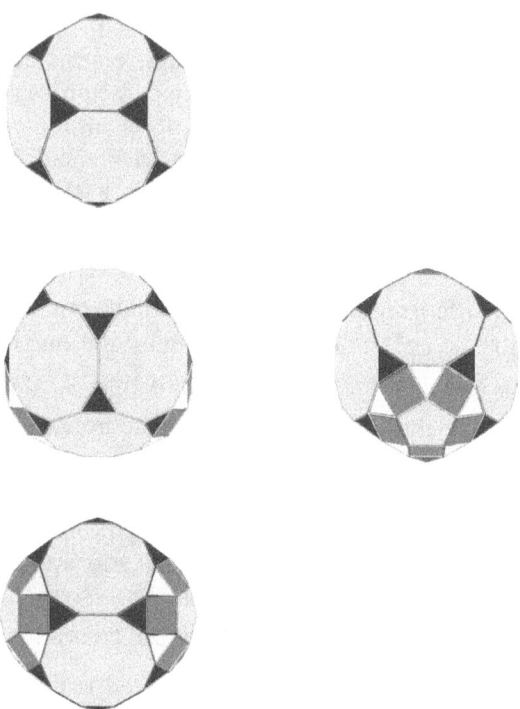

*Figure 139: Johnson solid 70, the metabiaugmented truncated dodecahedron.*

The dual of the metabiaugmented truncated dodecahedron is illustrated in Figure 140. Like the dual of the parabiaugmented truncated dodecahedron, it shows two cupola-dual stars, but because there is not a fivefold axis between the centers of the two stars, they do not retain perfect fivefold symmetry. The five kite-quadrilaterals in each star fall into three different transitivity classes, consisting of one furthest from the other star, two closest to the other star, and two intermediate. Similarly, the ten irregular quadrilaterals in each star divide into five different transitivity classes, depending on their distance from the other star.

"Augmented" Archimedean solids: Johnson solids 65-71.

The placement of the two stars means that one of the cupola-dual-star fill-in triangles from one star is opposite one from the other star (these are at the very bottom in Figure 140). So the distinction between cupola-dual-star fill-in triangles and cupola-dual-star fill-in-opposed triangles is blurred in this solid. Instead, one has one pair of cupola-dual-star fill-in triangles that are opposed to each other, two cupola-dual-star fill-in triangles around each star that are nearest to them (one point away), and two additional cupola-dual-star fill-in triangles around each star between the furthest-apart points of the stars. And all but the two cupola-dual-star fill-in triangles that oppose each other have their opposite triangles.

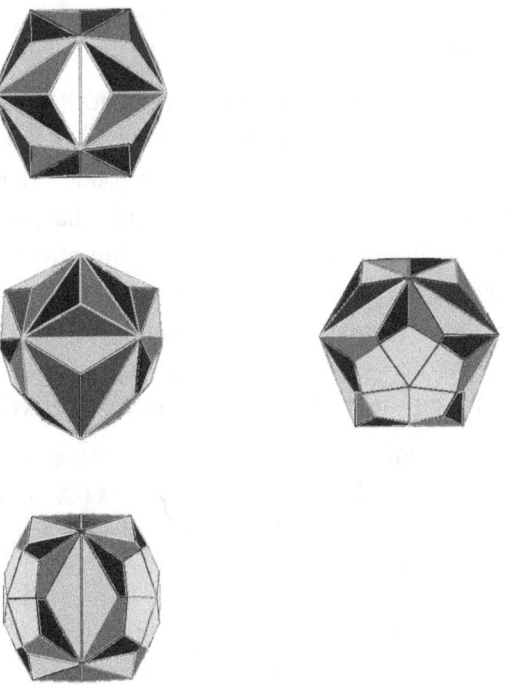

Figure 140: The dual of Johnson solid 70 (the metabiaugmented truncated dodecahedron).

Designating the mirror plane that bisects the two cupola-dual stars (shown as a horizontal line in the top and bottom views and a vertical line in the side view of Figure 140) as the *first* mirror plane and the one that comes between the two stars (a vertical line in the top, front, and bottom views in Figure 140) as the *second*, four of the quadrilaterals that would be the central kite-quadrilaterals of an undistorted cupola-dual star are slightly distorted from being kites; one from each star, however, remains a kite because it continues to be bisected by the first mirror plane.

Although the symmetry is distorted so that the cupola-dual stars are no longer fivefold symmetric, the combination of the ten quadrilaterals forming a star and the five cupola-dual-star fill-in triangles that fit them still yields a collection whose outer vertices all lie in one plane. Therefore, the same process of splitting the polyhedron into three pieces which can reconstitute it by fusion that was described in the discussions of the duals of $J_{68}$ and $J_{69}$ can be performed here, although the two planes are not parallel.

While in the dual of $J_{68}$, the kite-quadrilaterals had two sides of ~1.4813 (the sides that form the edges between any two of them) and two sides of ~0.9927 (the sides that form the edges between a kite and an irregular quadrilateral), with two equal angles of 86.4374° (86° 50′ 15″), and the other angles being 69.5850° (69° 35′ 6″) at the fivefold vertex and 116.7403° (116° 44′ 25″) at the vertex furthest from the fivefold vertex, they become three different types in the dual of $J_{70}$, all rather close in measurements to the ones in the dual of $J_{68}$, however:

- Two, straddling the first mirror plane, with sides of ~0.9616, ~0.9714, ~1.4040, and ~1.4277, and angles of 70.5231° (70° 31′ 23″), 86.1211° (86° 7′ 16″), 87.8482° (87° 50′ 54″), and 115.5075° (115° 30′ 27″),

- Two, adjacent to those, with sides of ~0.9197, ~0.9386, ~1.3735, and ~1.4040, and angles of 69.5576° (69° 33′ 27″), 85.4591° (85° 27′ 33″), 87.9844° (87° 59′ 4″), and 116.9988° (116° 59′ 56″), and

- One, in the direction furthest from where the stars touch each other, with two sides of ~0.9061 and two of

## "Augmented" Archimedean solids: Johnson solids 65-71.

~1.3735, and two angles of 86.8443° (86° 50′ 40″), with the remaining angles being 68.7094° (68° 42′ 34″), and 117.6019° (117° 36′ 7″). This quadrilateral remains a true kite.

All these quadrilaterals have their smallest angles at the quinquevalent vertex at the center of the cupola-dual star. For the purposes of the discussion of dihedral angles below, the first set will be designated "4a" faces, the second "4b" faces, and the third "4c" faces. (Note that the numbers "two" and "one" refer to a single cupola-dual star; the whole polyhedron has twice as many.)

Similarly, in the dual of $J_{68}$, the irregular quadrilaterals had sides of ~0.2701, ~0.9927, ~1.3765, and ~1.9159, and angles of 31.2059° (31° 12′ 21″), 89.5393° (89° 32′ 21″), 119.2052° (119° 12′ 19″), and 120.0497° (120° 2′ 59″). In this polyhedron, they become five different types, all rather close in measurements to the ones in $J_{68}$, however: (Again, the quantity is per star; thus where the listing says "two," there are four in the entire polyhedron)

- In the points of one star that touch the points of the other, the two quadrilaterals adjacent to each other (though the common edge is very short) have sides of ~0.2665 (the common edge between these faces in each star), ~0.9714, ~1.4054, and ~1.9354, and angles of 30.1789° (30° 10′ 44″), 90.3842° (90° 23′ 3″), 118.7445° (118° 44′ 40″), and 120.6924° (120° 41′ 33″). For the purposes of the discussion of dihedral angles below, this set will be designated "4d" faces.

- The other halves of those two points are irregular quadrilaterals with sides of ~0.2585, ~0.9616, ~1.4054 (the common edge with the 4d faces just mentioned), and ~1.9298, and angles of 29.7804° (29° 46′ 49″), 89.9724° (89° 58′ 21″), 119.4913° (119° 29′ 29″), and 120.7560° (120° 45′ 22″). For the purposes of the discussion of dihedral angles below, this set will be designated "4e" faces.

- Going away from those two points, the next-encountered irregular quadrilaterals have sides of ~0.2585 (the common edge with the 4e faces just mentioned), ~0.9386, ~1.2889, and ~1.8022, and angles of 31.4799° (31° 28' 48"), 89.9418° (89° 56' 30"), 118.4060° (118° 24' 22"), and 120.1724° (120° 10' 20"). For the purposes of the discussion of dihedral angles below, this set will be designated "4f" faces.

- Those too comprise only half of the points of the star that they help to form; the other halves of those two points are irregular quadrilaterals with sides of ~0.2475, ~0.9197, ~1.2889 (the common edge with the 4f faces just mentioned), and ~1.7868, and angles of 30.9357° (30° 56' 9"), 89.4940° (89° 29' 38"), 119.6157° (119° 36' 57"), and 119.9546° (119° 57' 16"). For the purposes of the discussion of dihedral angles below, this set will be designated "4g" faces.

- The only remaining irregular quadrilaterals form the fifth points of the star; both are the halves of the same point, and they have sides of ~0.2475 (the common edge with the 4g faces just mentioned), ~0.9061, ~1.2282 (the common edge between these two faces on each star, which lies in the first mirror plane), and ~1.7203, and angles of 31.7902° (31° 47' 25"), 89.2081° (89° 12' 29"), 119.1930° (119° 11' 35"), and 119.8087° (119° 48' 31"). For the purposes of the discussion of dihedral angles below, this set will be designated "4h" faces.

These eight classes of quadrilaterals account for thirty out of the seventy faces of the polyhedron; the remaining forty faces are all triangular, and none of them differ much in shape from the faces of the triakis icosahedron (page 291), although only those faces that are bisected by the first mirror plane are exactly isosceles. Five of the twelve transitivity classes into which the triangular faces divide, accounting for eighteen faces, have been mentioned earlier on p. 293, without going into the details of their side lengths and face angles (Again, the quantity is per star; thus where the listing says "one" or "two," there are

*"Augmented" Archimedean solids: Johnson solids 65-71.*

two or four in the entire polyhedron):

- The cupola-dual-star fill-in triangle from one star which is opposite to one from the other star forms a transitivity class (together with that one from the other star). In this case, the distinction between cupola-dual-star fill-in triangles and cupola-dual-star fill-in-opposed triangles is erased because this triangle is a cupola-dual-star fill-in triangle with regard to one star and a cupola-dual-star fill-in-opposed triangle with regard to the other. This triangle is still isosceles, having two sides of ~1.9354 (edges shared with the 4d faces) and one of ~3.3381 (the edge shared with the corresponding face on the other side of the second mirror plane), and two angles of 30.4139° (30° 24′ 50″) and one of 119.1721° (119° 10′ 20″). For the purposes of the discussion of dihedral angles below, this set will be designated "3a" faces.

- The two cupola-dual-star fill-in triangles around each star that are nearest to the 3a faces (one point away) have sides of ~1.8022 (the edge shared with the 4f face), ~1.9298 (the edge shared with the 4e face), and ~3.2157, and angles of 29.3622° (29° 21′ 44″), 31.6699° (31° 40′ 12″), and 118.9679° (118° 58′ 5″). For the purposes of the discussion of dihedral angles below, this set will be designated "3b" faces.

- The two additional cupola-dual-star fill-in triangles around each star between the furthest-apart points of the stars have sides of ~1.7203 (the edge shared with the 4h face), ~1.7868 (the edge shared with the 4g face), and ~3.0117, and angles of 30.1842° (30° 11′ 3″), 31.4814° (31° 28′ 53″), and 118.3345° (118° 20′ 4″). For the purposes of the discussion of dihedral angles below, this set will be designated "3c" faces.

- The triangles opposite to the 3b faces have sides of ~1.7953, ~1.9280, and ~3.2157 (the edge shared with the 3b face), and angles of 29.0953° (29° 5′ 43″), 31.4799° (31° 28′ 48″), and 119.4248° (119° 25′ 29″). For the purposes of the discussion of dihedral angles below, this set will be designated "3d" faces. The 3d faces nearest one star have a common edge with the 3d faces nearest the other star.

- The triangles opposite to the 3c faces have sides of ~1.7060, ~1.7794, and ~3.0117 (the edge shared with the 3c face), and angles of 29.5258° (29° 31′ 33″), 30.9336° (30° 56′ 1″), and 119.5406° (119° 32′ 26″). For the purposes of the discussion of dihedral angles below, this set will be designated "3e" faces.

This leaves twenty-two triangular faces, which divide into seven transitivity classes. These are very difficult to describe:

- Sharing an edge with the 3d faces other than the edges shared with a 3b face or an other-star 3d face are a set of two isosceles triangles, which each have two sides of ~1.7953 (the edges shared with the 3d face) and one of ~3.0798, and two angles of 30.9357° (30° 56′ 9″) and one of 118.1286° (118° 7′ 43″). For the purposes of the discussion of dihedral angles below, this set will be designated "3f" faces.

- The triangles that share their long side with the 3f faces are also isosceles, each having two sides of ~1.7814 and one of ~3.0798 (the edge shared with the 3f face), and two angles of 30.1842° (30° 11′ 3″) and one of 119.6317° (119° 37′ 54″). For the purposes of the discussion of dihedral angles below, this set will be designated "3g" faces. The 3f and 3g faces are both isosceles because they are bisected by the second mirror plane.

- There are two class of triangular faces besides the 3c faces that share an edge with the 3e faces. The triangles sharing the edge of length ~1.7794 with the 3e faces have their remaining sides of ~1.6672 and ~2.9746, and angles of 29.2647° (29° 15′ 53″), 31.4507° (31° 27′ 2″), and 119.2846° (119° 17′ 4″). For the purposes of the discussion of dihedral angles below, this set will be designated

"3h" faces.

- The triangles that share their long side with the 3h faces also share one of the other sides with the 3g faces. They have sides of ~1.6733, ~1.7814 (the edge shared with the 3g faces), and ~2.9746 (the edge shared with the 3h face), and angles of 29.5258° (29° 31' 33"), 31.6455° (31° 38' 44"), and 118.8287° (118° 49' 43"). For the purposes of the discussion of dihedral angles below, this set will be designated "3j" faces.

- The triangles, other than the 3c and 3h faces, that share an edge with the 3e faces also share one of the other sides with the 3h faces. They have sides of ~1.6672, ~1.7060, and ~2.8996, and angles of 30.3355° (30° 20' 8"), 31.1185° (31° 7' 6"), and 118.5461° (118° 32' 46"). For the purposes of the discussion of dihedral angles below, this set will be designated "3k" faces.

- The triangles that share their long side with the 3k faces have sides of ~1.6595, ~1.7007, and ~2.900 (the edge shared with the 3k face), and angles of 29.9449° (29° 56' 41"), 30.7672° (30° 46' 2"), and 119.2880° (119° 17' 17"). The side of length ~1.7007 is shared with another face of the same class. For the purposes of the discussion of dihedral angles below, this set will be designated "3m" faces.

- The two triangles that remain (sharing a long side with each other, which lies in the second mirror plane, and their short sides with two different 3m faces) are yet another set of isosceles triangles, each having two sides of ~1.6595 and one of ~2.8586, and two angles of 30.5415° (30° 32' 29") and one of 118.9170° (118° 55' 1"). For the purposes of the discussion of dihedral angles below, this set will be designated "3n" faces. The 3a and 3n faces are both isosceles because they are bisected by the first mirror plane.

Where in $J_{68}$, there were nine types of triangle/triangle dihedral angle, there are all of twelve different types in $J_{70}$.

While some of the types found in $J_{68}$ have no correlates in $J_{70}$, such as the dihedral angles among triangles in the antiapical star, others become multiplied because of the splitting of the triangles into a larger number of transitivity classes. The twelve types of triangle/triangle dihedral angle are as follows:

- 3a-3a: 159.0213° (159° 1′ 17″)
- 3b-3d: 159.9015° (159° 54′ 5″)
- 3c-3e: 161.1375° (161° 8′ 15″)
- 3d-3d: 159.9818° (159° 58′ 55″)
- 3d-3f: 160.2394° (160° 14′ 22″)
- 3e-3h: 161.3644° (161° 21′ 52″)
- 3e-3k: 161.5014° (161° 30′ 5″)
- 3f-3g: 160.7226° (160° 43′ 21″)
- 3g-3j: 161.1375° (161° 8′ 15″)
- 3h-3j: 161.3644° (161° 21′ 52″)
- 3h-3k: 161.5496° (161° 32′ 59″)
- 3j-3j: 161.2904° (161° 17′ 25″)
- 3k-3m: 161.7626° (161° 45′ 45″)
- 3m-3m: 161.8552° (161° 51′ 19″)
- 3m-3n: 161.9228° (161° 55′ 22″)
- 3n-3n: 161.9813° (161° 58′ 53″)

Where in $J_{68}$, there was only one type of triangle/quadrilateral dihedral angle, at the edges between the irregular quadrilaterals and the cupola-dual-star fill-in triangles, the splitting of the irregular quadrilaterals into five transitivity classes and of the cupola-dual-star fill-in triangles into three leads to a split as well, so that there are altogether five types of triangle/quadrilateral dihedral angle in $J_{70}$:

- 3a-4d: 159.1254° (159° 7′ 32″)

- 3b-4e: 159.6805° (159° 40' 50")
- 3b-4f: 159.9015° (159° 54' 5")
- 3c-4g: 160.7226° (160° 43' 21")
- 3c-4h: 160.8057° (160° 48' 21")

Where in $J_{68}$, there were three types of quadrilateral/quadrilateral dihedral angle, the splitting of what had been the kites into three transitivity classes and of the irregular quadrilaterals into five leads to each of the three being split as well, so that there are altogether fourteen types of quadrilateral/quadrilateral dihedral angle in $J_{70}$:

- From the kite/kite dihedral angles:
  - 4a-4a: 160.6876° (160° 41' 15")
  - 4a-4b: 160.9193° (160° 55' 9")
  - 4b-4c: 161.2056° (161° 12' 20")
- From the kite/irregular quadrilateral dihedral angles:
  - 4a-4d: 160.4124° (160° 24' 45")
  - 4a-4e: 160.3989° (160° 23' 56")
  - 4b-4f: 160.8066° (160° 48' 24")
  - 4b-4g: 160.8504° (160° 51' 1")
  - 4c-4h: 161.0883° (161° 5' 18")
- From the irregular quadrilateral/irregular quadrilateral dihedral angles:
  - 4d-4d: 159.2155° (159° 12' 56")
  - 4d-4e: 159.3847° (159° 23' 5")
  - 4e-4f: 159.7940° (159° 47' 38")
  - 4f-4g: 160.2394° (160° 14' 22")
  - 4g-4h: 160.5570° (160° 33' 25")

- 4h-4h: 160.6692° (160° 40′ 9″)

Johnson solid 71, the triaugmented truncated dodecahedron, illustrated in Figure 141 below, is the last of the Johnson solids covered in this chapter. It was mentioned earlier that there is only one way that three decagonal faces of a truncated dodecahedron can be fused to pentagonal cupolae and leave the resulting polyhedron convex, and so no qualification is needed for "triaugmented" in the name. The three fusion planes are all 120° apart around one of the threefold axes of the original truncated dodecahedron, so that threefold axis remains in the final solid.

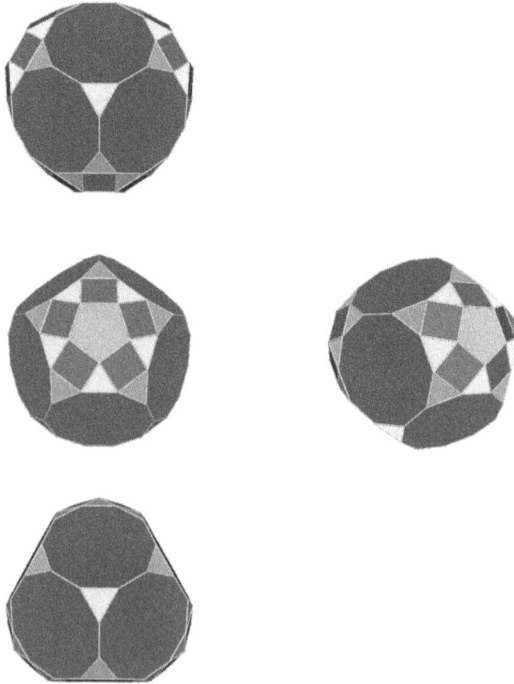

Figure 141: Johnson solid 71: the triaugmented truncated dodecahedron.

But it can be seen that the places where the cupolae are fused to the decagonal faces are not exactly halfway along the threefold axis. In Figure 141, some of those faces which derive

## "Augmented" Archimedean solids: Johnson solids 65-71.

from each of the three cupolae added in the fusion are visible in the top view, but not in the bottom. If one carefully examines the front view, which is oriented to show one of those sets of cupola faces directly in front, these faces are centered *not* at the middle of the view, but slightly *above*. Thus, although the top and bottom views are almost alike, there is no horizontal mirror plane, and the symmetry is $C_{3v}$.

The dual of the triaugmented truncated dodecahedron is illustrated in Figure 142 below. It has the same $C_{3v}$ symmetry as its dual, and triangular pyramid-like apices at top and bottom. At the three points corresponding to the centers of the places where cupolae were fused in the triaugmented truncated dodecahedron itself, one can see that there are sets of five quadrilaterals (only one of which is truly a kite-quadrilateral because the centroid of the polyhedron is not at the point where the axes of the original cupolae meet one another and the threefold axis, but slightly removed from it; if that point were actually used as the reciprocation center, they would be precise kite-quadrilaterals) and rings of alternating-chirality irregular quadrilaterals forming a distorted cupola-dual star.

As was the case in describing the duals of the other three augmented truncated dodecahedra, it is most convenient to describe faces in terms of their relationship to the cupola-dual stars. Each of the cupola-dual stars consists of fifteen quadrilaterals, five of which are either kite-quadrilaterals or very close to being so, related to the kite-quadrilaterals at the center of the cupola-dual star in $J_{68}$. The remaining ten irregular quadrilaterals of each cupola-dual star alternate in chirality and have one very short side, one much longer side, and two of intermediate length (but not near equal). Two of these irregular quadrilaterals form each point of the cupola-dual star; and while in $J_{70}$ there were two vertices where the points of each star touched the corresponding points of the other star, in $J_{71}$ there are three vertices, each of which is where a point of one star touches a point of another. From the point of view of a single star, two of its points engage one of these three special vertices, each one sharing the vertex with a different one of the other two stars. The relationship between the mirror planes and the cupola-dual stars in the dual of $J_{71}$ is very different than it was in the dual of $J_{70}$. In the dual of $J_{70}$, there were two mirror planes, one of which bisected both of the cupola-dual stars and one of which came between them, passing through the two vertices where the stars had points that touched. In the dual of $J_{71}$, there are three mirror planes, each of which bisects one of the cupola-dual stars and passes through the vertex where the the *remaining two stars* have points that touch. And unlike the two mirror planes in the dual of $J_{70}$, which do not transform into each other, so that in discussing transitivity classes it was necessary to distinguish the first from the second, the three mirror planes in the dual of $J_{71}$ are equivalent under a threefold rotation, so that there is no need to distinguish among them.

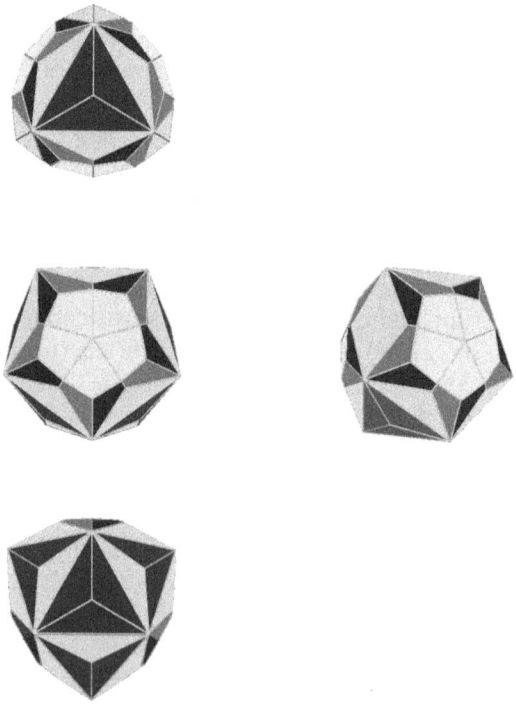

Figure 142: The dual of Johnson solid 71 (the triaugmented truncated dodecahedron).

The $C_{3v}$ symmetry of the polyhedron does not affect the classification of the quadrilaterals forming the cupola-dual star any differently than it did in the dual of $J_{70}$, although now there are three stars with each type of quadrilateral, so that instead

"Augmented" Archimedean solids: Johnson solids 65-71.

of two or four quadrilaterals of each class where each star had one or two, there are now three or six, as will be noted in Table 30. However, the number of triangular faces remaining is such that a smaller number of transitivity classes is needed to account for them. The designations 4a to 4h will have the same significance as they do in the dual of $J_{70}$:

- Two of the central near-kite-quadrilaterals of each star, straddling a mirror plane, with sides of ~1.0040, ~1.0100, ~1.4745, and ~1.4878, and angles of 70.3169° (70° 19′ 1″), 86.4811° (86° 28′ 52″), 87.4272° (87° 25′ 38″), and 115.7748° (115° 46′ 29″). For the purposes of the discussion of dihedral angles below, this set will be designated "4a" faces.

- Two more near-kite-quadrilaterals of each star, adjacent to those, with sides of ~0.9795, ~0.9901, ~1.4555, and ~1.4745, and angles of 69.7662° (69° 45′ 58″), 86.0944° (86° 5′ 40″), 87.5413° (87° 32′ 29″), and 116.5981° (116° 35′ 53″). For the purposes of the discussion of dihedral angles below, this set will be designated "4b" faces.

- One, which remains a true kite, in the direction furthest from where the stars touch each other, with two sides of ~0.9711 and two of ~1.4555, and two angles of 86.8248° (86° 49′ 29″), with the remaining angles being 69.3461° (69° 20′ 46″) and 117.0043° (117° 0′ 15″). For the purposes of the discussion of dihedral angles below, this set will be designated "4c" faces.

- In the points of one star that touch the points of another, the two quadrilaterals adjacent to each other (though the common edge is very short) have sides of ~0.2763 (the common edge between these faces in each star), ~1.0100, ~1.4606, and ~2.0115, and angles of 30.1649° (30° 9′ 53″), 90.2894° (90° 17′ 22″), 118.8608° (118° 51′ 39″), and 120.6849° (120° 41′ 6″). For the purposes of the discussion of dihedral angles below, this set will be designated "4d" faces.

- The other halves of those two points are irregular quadrilaterals with sides of ~0.2717, ~1.0040, ~1.4606 (the common edge with the 4d faces just mentioned), and ~2.0080, and angles of 29.9371° (29° 56′ 14″), 90.0484° (90° 2′ 54″), 119.2915° (119° 17′ 29″), and 120.7231° (120° 43′ 23″). For the purposes of the discussion of dihedral angles below, this set will be designated "4e" faces.

- Going away from those two points, the next-encountered irregular quadrilaterals have sides of ~0.2717 (the common edge with the 4e faces just mentioned), ~0.9901, ~1.3907, and ~1.9310, and angles of 30.9113° (30° 54′ 41″), 90.0734° (90° 4′ 24″), 118.6608° (118° 39′ 39″), and 120.3545° (120° 21′ 16″). For the purposes of the discussion of dihedral angles below, this set will be designated "4f" faces.

- Those too comprise only half of the points of the star that they help to form; the other halves of those two points are irregular quadrilaterals with sides of ~0.2649, ~0.9795, ~1.3907 (the common edge with the 4f faces just mentioned), and ~1.9232, and angles of 30.5745° (30° 34′ 28″), 89.7530° (89° 45′ 11″), 119.3583° (119° 21′ 30″), and 120.3142° (120° 18′ 51″). For the purposes of the discussion of dihedral angles below, this set will be designated "4g" faces.

- The only remaining irregular quadrilaterals form the fifth points of the star; both are the halves of the same point, and they have sides of ~0.2649 (the common edge with the 4g faces just mentioned), ~0.9711, ~1.3515 (the common edge between these two faces on each star, which lies in one of the three mirror planes), and ~1.8800, and angles of 31.1137° (31° 6′ 49″), 89.6811° (89° 40′ 52″), 119.0495° (119° 2′ 58″), and 120.1557° (120° 9′ 20″). For the purposes of the discussion of dihedral angles below, this set will be designated "4h" faces.

These faces account for 45 of the 75 faces of this polyhe-

## "Augmented" Archimedean solids: Johnson solids 65-71.

dron; the remaining thirty, all of which are triangular, divide into seven transitivity classes. Since these do not correspond well to any of the twelve transitivity classes of triangular face in the dual of $J_{70}$, the designations 3a to 3g do not correspond to similar designations in the dual of $J_{70}$, but are assigned independently:

- Filling in between the two points of each cupola-dual star that touch the other stars, and sharing an edge with both of the 4d faces, is an isosceles triangle. It has two sides of ~2.0115 (edges shared with the 4d faces) and one of ~3.4704; and two angles of 30.3861° (30° 23′ 10″) and one of 119.2278° (119° 13′ 40″). For the purposes of the discussion of dihedral angles below, this set will be designated "3a" faces.

- Another set of cupola-dual fill-in triangles share edges with the 4e and 4f faces. In this case, the distinction between cupola-dual-star fill-in triangles and cupola-dual-star fill-in-opposed triangles is erased because these triangles are cupola-dual-star fill-in triangles with regard to one star and cupola-dual-star fill-in-opposed triangles with regard to another. These triangles have sides of ~1.9310 (edge shared with the 4f face), ~2.0080 (edge shared with the 4e face), and ~3.3959 (edge shared with another of the same type), and angles of 29.7933° (29° 47′ 36″), 31.1100° (31° 6′ 36″), and 119.0967° (119° 5′ 48″). For the purposes of the discussion of dihedral angles below, this set will be designated "3b" faces.

- The last set of cupola-dual fill-in triangles share edges with the 4g and 4h faces. These triangles have sides of ~1.8800 (edge shared with the 4h face), ~1.9232 (edge shared with the 4g face), and ~3.2735, and angles of 30.2208° (30° 13′ 15″), 30.9914° (30° 59′ 29″), and 118.7878° (118° 47′ 16″). For the purposes of the discussion of dihedral angles below, this set will be designated "3c" faces.

- Opposite the 3a faces (sharing the long sides with them) are another set of isosceles triangles. These triangles each have two sides of ~2.0143 and one of ~3.4704 (edges shared with the 3a faces); and two angles of 30.5193° (30° 31′ 10″) and one of 118.9614° (118° 57′ 41″). They surround one end of the threefold axis, and their equal sides are shared with each other. For the purposes of the discussion of dihedral angles below, this set will be designated "3d" faces.

- At the other end of the threefold axis from the 3d faces are another set of isosceles triangles. These triangles each have two sides of ~1.8725 and one of ~3.2285; and two angles of 30.4455° (30° 26′ 44″) and one of 119.1089° (119° 6′ 32″). They surround one end of the threefold axis, and their equal sides are shared with each other. For the purposes of the discussion of dihedral angles below, this set will be designated "3e" faces.

- Opposite the 3e faces (sharing the long sides with them) are another set of isosceles triangles. These triangles each have two sides of ~1.8750 and one of ~3.2285 (edge shared with the 3e faces); and two angles of 30.5782° (30° 34′ 42″) and one of 118.8435° (118° 50′ 37″). For the purposes of the discussion of dihedral angles below, this set will be designated "3f" faces.

- The final set of triangular faces are opposite the 3c faces. These triangles have sides of ~1.8750 (edge shared with the 3f faces), ~1.9198 (edge shared with another face of the same type), and ~3.2735 (edge shared with the 3c faces), and angles of 29.9946° (29° 59′ 40″), 30.7881° (30° 47′ 17″), and 119.2174° (119° 13′ 3″). For the purposes of the discussion of dihedral angles below, this set will be designated "3g" faces.

The dihedral angles can be divided into sets in the same way as was done for the previous Johnson solid duals. But while the dual of the metabiaugmented truncated dodecahedron had *twelve* different types of dihedral angle, the greater symmetry of the triaugmented truncated dodecahedron means

## "Augmented" Archimedean solids: Johnson solids 65-71.

that there are only *eight*:

- 3a-3d: 159.9047° (159° 54' 17")
- 3b-3b: 160.3596° (160° 21' 34")
- 3c-3g: 161.0411° (161° 2' 28")
- 3d-3d: 159.8527° (159° 51' 10")
- 3e-3e: 161.3179° (161° 19' 4")
- 3e-3f: 161.2792° (161° 16' 45")
- 3f-3g: 161.1752° (161° 10' 31")
- 3g-3g: 161.1044° (161° 6' 16")

The fact that the 3a, 3b, and 3c faces were defined in a manner similar to the way they were defined in the dual of the metabiaugmented truncated dodecahedron, coupled with the fact that *all* the quadrilaterals divide into the same eight transitivity classes as in the dual of the metabiaugmented truncated dodecahedron, leads to the fact that the single dihedral angle of the dual of the augmented truncated dodecahedron divide into the same five that were found in the dual of the metabiaugmented truncated dodecahedron, although their numerical values differ:

- 3a-4d: 160.0371° (160° 2' 14")
- 3b-4e: 160.3146° (160° 18' 53")
- 3b-4f: 160.4391° (160° 26' 21")
- 3c-4g: 160.8680° (160° 52' 5")
- 3c-4h: 160.9278° (160° 55' 40")

Since the quadrilaterals divide into the same eight transitivity classes as in the dual of the metabiaugmented truncated dodecahedron, it should not be any surprise that the three dihedral angles of the dual of the augmented truncated dodecahedron divide into the same fourteen that were found in the dual of the metabiaugmented truncated dodecahedron, although their numerical values differ:

- From the kite/kite dihedral angles:
  - 4a-4a: 161.3771° (161° 22' 38")
  - 4a-4b: 161.4905° (161° 29' 26")
  - 4b-4c: 161.6473° (161° 38' 50")
- From the kite/irregular quadrilateral dihedral angles:
  - 4a-4d: 161.1273° (161° 7' 38")
  - 4a-4e: 161.1195° (161° 7' 10")
  - 4b-4f: 161.3327° (161° 19' 58")
  - 4b-4g: 161.3407° (161° 20' 26")
  - 4c-4h: 161.4687° (161° 28' 7")
- From the irregular quadrilateral/irregular quadrilateral dihedral angles:
  - 4d-4d: 160.1110° (160° 6' 40")
  - 4d-4e: 160.1936° (160° 11' 37")
  - 4e-4f: 160.4004° (160° 24' 1")
  - 4f-4g: 160.6384° (160° 38' 18")
  - 4g-4h: 160.8178° (160° 49' 4")
  - 4h-4h: 160.8833° (160° 53' 0")

In the discussions of the three other augmented truncated dodecahedron duals, it was noted that they could be divided each into three pieces, two of which are either a combination of a cupola-dual star and its five fill-in triangles or a combination of an antiapical star and its five fill-in triangles (completed, where it was cut from the original polyhedron, by a new pentagonal face) and the third being the remaining piece (also completed with pentagonal faces), which could be fused to reconstitute the polyhedron. A splitting into a much larger number of pieces applies to this one. First of all, there are three separate sets of cupola-dual stars combined with their fill-in triangles. But even the remaining piece has two triangular pyramidal pieces around the ends of the threefold axis, which also could be cut off, and the pieces completed with triangular faces. After all this cutting, there are five separate "little" pieces and the remaining core, which has only the 3f and 3g facces remaining from the original polyhedron: six separate pieces altogether.

| Johnson solid number | 68 | 69, 70 | 71 |
|---|---|---|---|
| Name | Augmented truncated dodecahedron | 69: Parabiaugmented truncated dodecahedron<br>70: Metabiaugmented truncated dodecahedron | Triaugmented truncated dodecahedron |
| Symmetry* | $C_{sv}$ | 69: $S_{10v}$<br>70: $C_{2v}$ | $C_{3v}$ |
| Faces | [5+5+5+5+5](3) + [5](4) + [1](5) + [1+5+5](10) | 69: [10+10+10](3) + [10](4) + [2](5) + [10](10)<br>70: [2+2+2+2+2+4+4+4+4+4](3) + [2+4+4](4) + [2](5) + [2+2+2+4](10) | [1+1+3+3+6+6+6+9](3) + [3+6+6](4) + [3](5) + [3+3+3](10) |
| Edges* | 105 | 120 | 135 |
| Vertices | 65 | 70 | 75 |
| Dihedral angles | 3a-4: 174.3401° (174° 20′ 24″),<br>3c-4: 159.0948° (159° 5′ 41″),<br>3a-10: 142.6226° (142° 37′ 21″),<br>3c-10: 153.9424° (153° 56′ 33″),<br>4-5: 148.2825° (148° 16′ 57″),<br>10-10: 116.5651° (116° 33′ 54″)♦ | | |
| Area | 102.18209222 | 103.37342429 | 104.56475635 |
| Volume | 87.36370988 | 89.68775520 | 92.01180051 |
| **Dual** | | | |
| Name | Unnamed | Unnamed | Unnamed |
| Faces | [5+5+5+5+10+10+10](3) + [5+10](4) | 69: [10+10+20](3) + [10+20](4)<br>70: [2+2+2+2+4+4+4+4+4+4](3) +[2+4+4+4+4+4+4](4) | [3+3+3+3+6+6+6](3) + [3+6+6+6+6+6+6+6](4) |
| Vertices | 42 | 52 | 62 |

"Augmented" Archimedean solids: Johnson solids 65-71.

| Johnson solid number | 68 | 69, 70 | 71 |
|---|---|---|---|
| Face angles[†] | See text beginning on page 282 | 69:<br>3: [2]30.4803° (30° 28' 49"), 119.0394° (30° 28' 49");<br>4i: 30.4803° (30° 28' 49"), 90°, 119.0394° (30° 28' 49"), 120.4803° (120° 28' 49");<br>4k: [2]86.8795° (86° 52' 46"), 69.9267° (69° 55' 36"), 116.3143° (116° 18' 52");<br>70: See text beginning on page 294 | See text beginning on page |
| Dihedral angles | | 69:<br>3-3, 3-4i, 4i-4i: 160.6126° (160° 36' 45"),<br>4i-4k: 161.3884° (161° 23' 18"),<br>4k-4k: 161.6355° (161° 38' 8");<br>70: See text beginning on page 294 | |
| Area | 87.27222616 | 69: 113.08929862<br>70: 82.36461178 | 95.73541190 |
| Volume | 73.07276650 | 69: 107.91991663<br>70: 67.08625371 | 84.18152502 |

Table 30: Properties of Johnson solids 68, 69, 70, and 71 and their duals.

*The symmetry and number of edges are always the same

for each solid and its dual, so are not repeated in the "dual" section.

†Face angles are only given for the duals, as the Johnson solids have regular faces, whose angles can be found in Table 1 on p. 3.

♦Although the triangular faces fall into *five* transitivity classes in $J_{68}$, *three* in $J_{69}$, *ten* in $J_{70}$, and *eight* in $J_{71}$, the square faces fall into *three* transitivity classes in both $J_{70}$ and $J_{71}$, and the decagonal faces fall into *three* transitivity classes in both $J_{68}$ and $J_{71}$, and *four* in $J_{70}$, only two sets of triangles need be distinguished: 3a represents the triangles that originally came from the truncated dodecahedron; and 3c represents the triangles that originally came from the pentagonal cupola. No need exists to distinguish among the decagonal faces.

## "Augmented" Archimedean solids: Johnson solids 65-71.

It might be noted that the four augmented truncated dodecahedra, Johnson solids 68 to 71, correspond to the four augmented dodecahedra, Johnson solids 58 to 61, in a general sort of way. There is one solid with a single augmentation, two with double augmentations (designated "para-" and "meta-" in the naming), and only the metabiaugmented solid provides a place where a third augmentation is possible. And in addition, the symmetries correspond as well. $J_{58}$ and $J_{68}$ are both $C_{5v}$, $J_{59}$ and $J_{69}$ are both $S_{10v}$, $J_{60}$ and $J_{70}$ are both $C_{2v}$, and $J_{61}$ and $J_{71}$ are both $C_{3v}$.

At this point there will be an interruption in the flow of this text. Starting with Johnson solid 72 there are a large number of solids derived from the rhombicosidodecahedron, and it is easier to explain the properties of these solids after the rhombicosidodecahedron has been more thoroughly investigated. The following chapter is devoted to this investigation.

## Chapter 15: Interlude: An Archimedean solid: the rhombicosidodecahedron, a basis for numerous Johnson solids.

The *rhombicosidodecahedron* is an Archimedean solid containing sixty-two faces (twenty equilateral triangular, thirty square, and twelve regular pentagonal faces) and 60 vertices. The twelve pentagons can all be considered to be, combined each with five triangles and five squares, the faces (other than the base) of a pentagonal cupola ($J_5$, see p. 17, chapter 3). A view of the rhombicosidodecahedron, with the faces of one of the cupolae indicated by a darker tone, is given in Figure 143.

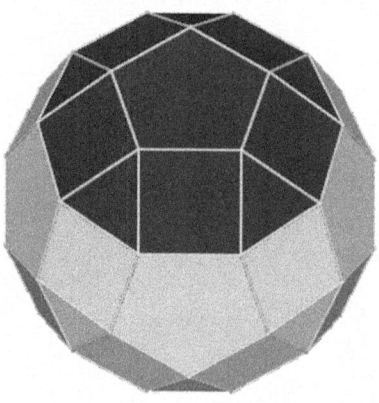

*Figure 143: A rhombicosidodecahedron, viewed from a square face.*

Another view, from a point of view directly over the pentagon that serves as the roof of the cupola, is Figure 144. This cupola can be thought of as cut off from the rhombicosidodecahedron by a plane (termed by Johnson "diminishing," see Chapter 16). Putting a decagonal face on each of the two pieces would complete the cupola and close the diminished rhombicosidodecahedron. If the cupola is then rotated by 36° and fused back where it came, the result is a gyrate rhombicosidodecahedron (Chapter 16). While there are twelve pentagonal

*Interlude: An Archimedean solid: the rhombicosidodecahedron, a basis for numerous Johnson solids.*

faces where a cupola can be seen, some of the triangles and squares forming the lateral faces are shared between nearby cupolae, as can be seen in Figure 145, where one cupola is dark, one is light, and the overlapping two triangles and square are in the darkest shade.

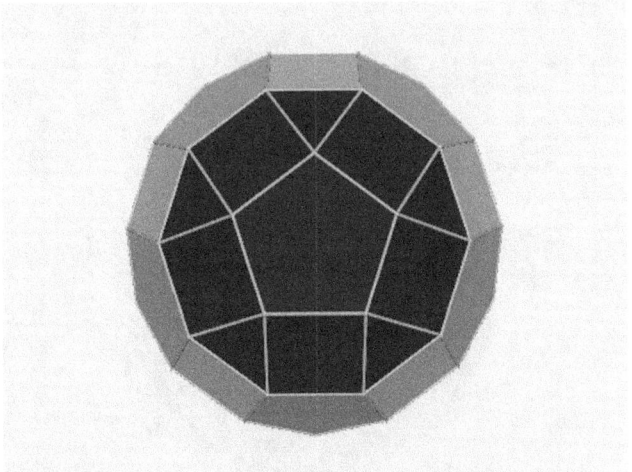

Figure 144: The same rhombicosidodecahedron, viewed from above the pentagonal face that serves as the roof of the marked cupola.

Because of this overlapping, only certain possible combinations of cupolae can be removed or rotated if more than one is to be. Therefore, only the specific cases discussed in Chapter 16 are possible..

Since the rhombicosidodecahedron has symmetry $I_h$, it has threefold axes (in fact, sixfold alternating, but for purposes of this discussion threefold) as well as fivefold, and the cupolae removed or rotated need to be either 120° apart on a threefold axis or diametrically opposite. Thus the chemical prefixes "meta-" and "para-" (see p. 220, Chapter 11) are used to indicate which positions have been modified.

Twice before we have seen a pattern where, starting with a polyhedron of icosahedral symmetry, we find that there are alterations that can be made in one or more places, but if two alterations are made, they must be either 120° apart on a three-fold axis of the original polyhedron or diametrically opposite.

These were the augmented dodecahedra, $J_{58}$ to $J_{61}$, discussed in Chapter 12, and the "augmented" (or rather cupolated) truncated dodecahedra, $J_{68}$ to $J_{71}$, discussed in Chapter 14. (The diminished icosahedra of Chapter 13 might constitute another example, but because two of the four polyhedra one would get are not included in this set, it is inappropriate to consider them here.)

Figure 145: Two overlapping cupolae on a rhombicosidodecahedron.

It is clear that when, starting with a polyhedron of icosahedral symmetry, one can make alterations only in locations that need to be either 120° apart on a threefold axis or diametrically opposite, there are *four* ways to pick the places to make such alterations. However, now we are talking of *two kinds* of alteration, and if one alteration is of one kind, the second can be the same or the other. So the four become *twelve*, as summarized in Table 31 below. When there is one alteration, there are two altered polyhedra, depending on which alteration (cupola rotation or removal) is performed. When there are two, three altered polyhedra are possible because there can be two cupola rotations, two cupola removals, or one of each. And when there are three, four altered polyhedra are possible because there can be three cupola rotations, three cupola removals, or one alteration of one kind and two of the other. So the four rows of Table 31 contain twelve polyhedra in the rightmost column.

*Interlude: An Archimedean solid: the rhombicosidodecahedron, a basis for numerous Johnson solids.*

| Number and position of alterations | Augmented dodecahedron | "Augmented" truncated dodecahedron | Altered rhombicosi-dodecahedra |
|---|---|---|---|
| 1 | $J_{58}$ | $J_{68}$ | $J_{72}, J_{76}$ |
| 2, para | $J_{59}$ | $J_{69}$ | $J_{73}, J_{77}, J_{80}$ |
| 2, meta | $J_{60}$ | $J_{70}$ | $J_{74}, J_{78}, J_{81}$ |
| 3 | $J_{61}$ | $J_{71}$ | $J_{75}, J_{79}, J_{82}, J_{83}$ |

Table 31: *Johnson solids obtained by altering icosahedrally symmetric Platonic or Archimedean solids.*

In fact, it is probably better to consider both the gyrate rhombicosidodecahedra and the diminished (and gyrate diminished) rhombicosidodecahedra (Chapter 16) from a different point of view. It was earlier mentioned that the icosahedron can be considered as a gyroelongated pentagonal bipyramid (see p. 74) or a basally biaugmented pentagonal antiprism (see p. 65). Specifically looking at the latter, one can see that the rhombicosidodecahedron is similarly a solid that can be thought of as obtained by "augmenting" (with cupolae, not pyramids, as in Chapter 14, thus the quotation marks) a core polyhedron. However, this can be done in two ways. One can take a core which is obtained by cutting off from the rhombicosidodecahedron two opposite cupolae, preserving the alternating tenfold axis of the rhombicosidodecahedron, or, by allowing the cuts to be closer together, actually take off *three* cupolae, preserving the threefold axis of the rhombicosidodecahedron. The first keeps a greater symmetry, but the second produces a simpler core (in the sense of fewer faces and vertices). It might be noted that the decagonal faces of the core polyhedron are bordered by a ring of pentagons and squares. It will actually be desirable to consider both of these two core polyhedra, as some of the polyhedra in the next two chapters can be considered as based on one core and some on the other. Each of these cores can be noted to have decagonal faces (two in the first case, three in the second) which can be left alone or "augmented," and the cupolae to "augment" them may be placed in either of two positions: when fusing a decagonal face of a cupola to a decagonal face of the core, one can put the triangular faces of the cupola adjacent to square faces of the core and square faces of the cupola adjacent to pentagonal faces of the core (the "ortho" orientation), or vice versa (the "gyrate" orientation).

The core polyhedra are themselves Johnson solids: the one preserving the alternating fivefold axis and $S_{10v}$ symmetry is the parabidiminished rhombicosidodecahedron ($J_{80}$), and the one preserving the threefold axis is the tridiminished rhombicosidodecahedron ($J_{83}$).

The full set of polyhedra is summarized in Table 32 below, where the total number of cupolae added in the "ortho" orientation, the number added in the "gyrate" orientation, and the number of faces left unfused must add to 2 if the core is $J_{80}$ and to 3 if it is $J_{83}$. (Note that the rhombicosidodecahedron itself, as well as some of the Johnson solids, appear twice in the table, being obtainable from both cores.)

*Interlude: An Archimedean solid: the rhombicosidodecahedron, a basis for numerous Johnson solids.*

| Core polyhedron | Ortho fused cupolae | Gyrate fused cupolae | Unfused decagons | Name of polyhedron | J-number |
|---|---|---|---|---|---|
| $J_{80}$ | 0 | 0 | 2 | Parabidiminished rhombicosidodecahedron | $J_{80}$ |
| $J_{80}$ | 0 | 1 | 1 | Paragyrate diminished rhombicosidodecahedron | $J_{77}$ |
| $J_{80}$ | 1 | 0 | 1 | Diminished rhombicosidodecahedron | $J_{76}$ |
| $J_{80}$ | 0 | 2 | 0 | Parabigyrate rhombicosidodecahedron | $J_{73}$ |
| $J_{80}$ | 1 | 1 | 0 | Gyrate rhombicosidodecahedron | $J_{72}$ |
| $J_{80}$ | 2 | 0 | 0 | Rhombicosidodecahedron | — |
| $J_{83}$ | 0 | 0 | 3 | Tridiminished rhombicosidodecahedron | $J_{83}$ |
| $J_{83}$ | 0 | 1 | 2 | Gyrate bidiminished rhombicosidodecahedron | $J_{82}$ |
| $J_{83}$ | 1 | 0 | 2 | Metabidiminished rhombicosidodecahedron | $J_{81}$ |
| $J_{83}$ | 0 | 2 | 1 | Bigyrate diminished rhombicosidodecahedron | $J_{79}$ |
| $J_{83}$ | 1 | 1 | 1 | Metagyrate diminished rhombicosidodecahedron | $J_{78}$ |
| $J_{83}$ | 2 | 0 | 1 | Diminished rhombicosidodecahedron | $J_{76}$ |
| $J_{83}$ | 0 | 3 | 0 | Trigyrate rhombicosidodecahedron | $J_{75}$ |
| $J_{83}$ | 1 | 2 | 0 | Metabigyrate rhombicosidodecahedron | $J_{74}$ |
| $J_{83}$ | 2 | 1 | 0 | Gyrate rhombicosidodecahedron | $J_{72}$ |
| $J_{83}$ | 3 | 0 | 0 | Rhombicosidodecahedron | — |

Table 32: Gyrate, diminished, and gyrate diminished rhombicosahedra.

While the rhombicosidodecahedron has icosahedral symmetry and both the diminishing and the rotation of cupolae to produce gyrate forms can be considered as modifications to it in the manner discussed on p. 230, there are considerably more than the four modified versions discussed there. However, there are four polyhedra that are purely gyrate rhombicosidodecahedra ($J_{72}$ to $J_{75}$) and four that are purely diminished rhombicosidodecahedra ($J_{76}$, $J_{80}$, $J_{81}$, and $J_{83}$). Four more are mixed, with both "gyrate" and "diminished" in their descriptive names.

# Chapter 16: Gyrate, diminished, and gyrate diminished rhombicosidodecahedra: Johnson solids 72-83.

When I first planned this book, there were two chapters in it, entitled "Gyrate rhombicosidodecahedra: Johnson solids 72-75" and "Diminished and gyrate diminished rhombicosidodecahedra: Johnson solids 76-83." However, after writing the bulk of the previous chapter, demonstrating that all the Johnson solids from $J_{72}$ to $J_{83}$ are in fact derived from the rhombicosidodecahedron by either diminishing it or replacing a pentagonal cupola in one orientation by one in the opposite one (with regard to the fivefold axis of the cupola), or some combination of both, and especially after completing Table 32, basing all twelve, plus the rhombicosidodecahedron, on the two core polyhedra $J_{80}$ and $J_{83}$, it just seemed to make more sense to put all twelve of these polyhedra into a single chapter.

As was the case for $J_{55}$ and $J_{56}$ (see p. 220), in the naming of $J_{73}$, $J_{74}$, $J_{77}$, $J_{78}$, $J_{80}$, and $J_{81}$ (the para- and metabigyrate rhombicosidodecahedron, the para- and metagyrate diminished rhombicosidodecahedron, and the para- and metabidiminished rhombicosidodecahedron), Johnson borrowed the prefixes from organic chemistry.

While pictures of the front, side, top, and (where appropriate) bottom views of all twelve of these solids are included in this chapter, as is done for all ninety-two of the Johnson solids, I believe that it is far less easy to grasp the differences among these polyhedra from such views than in most of the other cases, because they have large numbers of faces, many of which are not clearly in view in relation to others. (For example, the top view in Figure 154 is identical to any of twelve possible views of the unmodified rhombicosidodecahedron, while the bottom view is identical to the top view of Figure 162.) In addition, in many cases, the views that are oriented according to the symmetry elements of different members of this set are quite different, making comparisons hard to do based on the figures. (For example, Figures 150 and 152 appear quite different, but this is primarily because, based on the $C_{2v}$ and $C_{3v}$ axes of the two polyhedra, the positions of "top," "front," "side," and "bottom" in one have no relationship to those of the other.) It is my belief that one is better off, in trying to visualize these polyhedra, looking primarily at Figure 162 (the parabidiminished rhombicosidodecahedron) and Figure 168 (the tridiminished rhombicosidodecahedron), considering those two polyhedra as "core polyhedra" (in the sense of Chapter 15), and using the information in Table 32 of Chapter 15 to see how the others derive from them. (For myself, it is helpful that I have software that enables me to "turn over" images of the polyhedra on my computer; such manipulations cannot, of course, be done with a figure in a book, but if you would like to have such software for yourself, a nice package is available, at no cost, at

http://www.antiprism.com

(it can do much more, and in fact, is behind most of the illustrations in this book). I have tried to be helpful in the way I have shaded the faces of the polyhedra in the illustrations, but I will admit that even so, visualizing the relationships among these twelve Johnson solids is not easy.

While Johnson's terminology derives all the polyhedra in this chapter from the rhombicosidodecahedron, by either removing one or more cupolae and putting a decagonal face where that cupola was removed, rotating one or more of those cupolae by 36°, or both, and his numerical order comes from this process, as was just mentioned I think that the visualization of these polyhedra is facilitated by starting with one of the two core polyhedra and either doing nothing or fusing one or more cupolae to each of the two or three decagonal faces of the cores. Therefore in the description of each of the twelve Johnson solids treated in this chapter, both constructions — starting from the rhombicosidodecahedron and from the appropriate core (either $J_{80}$ or $J_{83}$) — will be described.

Johnson solids 72 through 75 involve the rotating of one, two, or three cupolae from their positions in the rhombicosidodecahedron, as described in the previous chapter. It should be noted that all the four polyhedra which are obtained by rotating one or more cupolae in a rhombicosidodecahedron are going to be identical in the face/vertex/edge counts, with 62 faces (20 triangles, 30 squares, and 12 pentagons), 60 vertices, and 120

*Gyrate, diminished, and gyrate diminished rhombicosidodecahedra: Johnson solids 72-83.*

edges; the duals must, of course, have 60 faces, 62 vertices, and 120 edges. Considered as simply adding cupolae to the maximally diminished rhombicosidodecahedra, the gyrate rhombicosidodecahedron ($J_{72}$) can be thought of as obtained by filling all but one of the decagonal sites of either one with a cupola oriented as in the basic rhombicosidodecahedron, and the remaining single site with a cupola rotated 36° from the position it would occupy in the rhombicosidodecahedron. The parabigyrate rhombicosidodecahedron ($J_{73}$) requires both of the two decagonal faces of the parabidiminished rhombicosidodecahedron to be fused to cupolae which are rotated 36° from the position they would occupy in the rhombicosidodecahedron, while the metabigyrate rhombicosidodecahedron ($J_{74}$, Figure 150) requires two of the three decagonal faces of the tridiminished rhombicosidodecahedron to be fused to cupolae which are rotated 36° from the position they would occupy in the rhombicosidodecahedron, while the third decagonal face is fused to a cupola oriented in the same way as in the rhombicosidodecahedron. The trigyrate rhombicosidodecahedron ($J_{75}$, Figure 152) requires *all* of the three decagonal faces of the tridiminished rhombicosidodecahedron to be fused to cupolae which are rotated 36° from the position they would occupy in the rhombicosidodecahedron.

The first of the polyhedra to be discussed in this chapter is $J_{72}$, the gyrate rhombicosidodecahedron (Figure 146, below). It would be useful, in looking at this solid, to note that the original rhombicosidodecahedron is an Archimedean solid, or in other words, one with all vertices alike. Thus in the original rhombicosidodecahedron, all vertices are identical, of a 3•4•5•4 type; rotating the cupola (at the bottom in the views of Figure 146) produces a ring of 3•4•4•5 vertices, some of which can be seen in the front and side views, but which are best visualized in the bottom view. The rotation of the cupola at the bottom destroys all symmetry elements of the original rhombicosidodecahedron, except for one fivefold axis and the five mirror planes through that axis, producing a figure of $C_{5v}$ symmetry. Except for that ring of vertices, all the other vertices of the gyrate rhombicosidodecahedron are identical, and thus the dual polyhedron will have all faces congruent except for a single ring of ten faces, the duals of the 3•4•4•5 vertices of the gyrate rhombicosidodecahedron.

It will be noted that rotating a cupola in this manner does not change the position of the centroid of the polyhedron, so that the distortions found in comparing the duals of other Johnson solids to the duals of Platonic and Archimedean solids to which they were related do not occur in this case, and while the symmetry of this solid is quite different from that of the rhombicosidodecahedron, and the transitivity classes into which the faces divide are greater in number, even the dual will have many of the congruent parts that the Catalan solid dual to the rhombicosidodecahedron has.

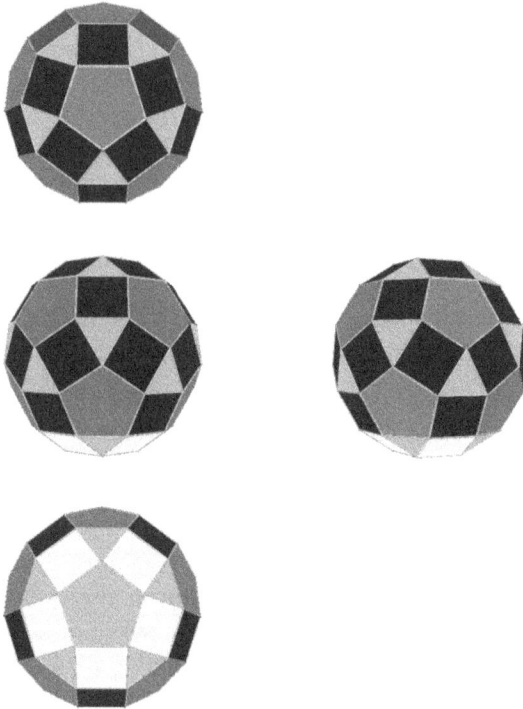

*Figure 146: Johnson solid 72: the gyrate rhombicosidodecahedron.*

The actual dual, obtained by polar reciprocation of the gyrate rhombicosidodecahedron, is illustrated below in Figure 147, and in fact, all the faces of the dual shown in Figure 147 (other than the faces in the ring surrounding the bottom five kite-quadrilaterals) are alike. It might be surprising that the five

*Gyrate, diminished, and gyrate diminished rhombicosidodecahedra: Johnson solids 72-83.*

kite-quadrilaterals at the bottom of Figure 147, corresponding to the vertices of the rotated cupola, are in fact congruent to the remaining forty kite-quadrilaterals in the figure; however, in fact, they are.

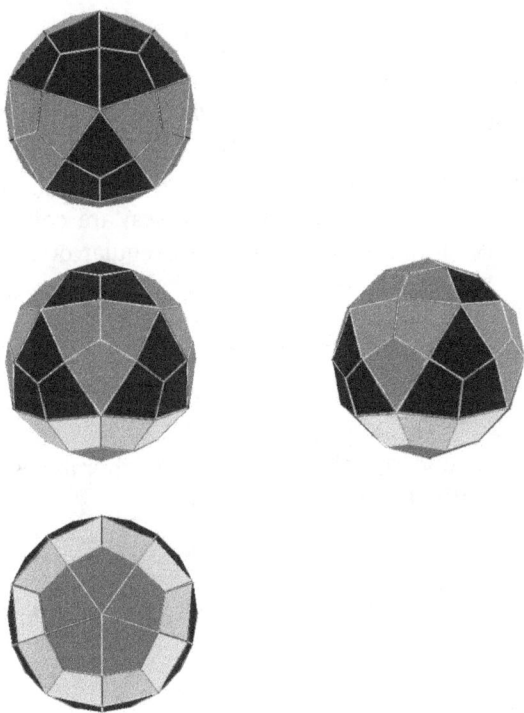

Figure 147: The dual of Johnson solid 72 (the gyrate rhombicosidodecahedron).

There are four sets of five kite-quadrilaterals that are required to be such by the $C_{5v}$ symmetry of the polyhedron, the two sets at top and bottom and two sets near the middle, clearly visible in the front and side views. But there are thirty other kite-quadrilateral faces that would not be required to be congruent to these twenty on the basis of symmetry alone, yet they are. And the vertex angles of the ten quadrilaterals dual to the 3•4•4•5 vertices are actually equal (but differently arranged) to the vertex angles of the fifty kite-quadrilaterals. Both the kite-quadrilaterals and the ten quadrilaterals dual to the 3•4•4•5 vertices of the gyrate rhombicosidodecahedron have an angle of

118.2687° (118° 16′ 7″), an angle of 67.7830° (67° 46′ 59″) and two angles of 86.9742° (86° 58′ 27″); however, in the kites, the 86.9742° angles are opposite each other, while in the ring quadrilaterals, they are at adjacent vertices. Interestingly, as well, all the 120 edges have exactly the same dihedral angle of 154.1214° (154° 7′ 17″).

Many of the Johnson solid duals described in this chapter show a large discrepancy between the number of *transitivity classes* of faces and the number of *different shapes* of faces. For example, the dual of $J_{72}$ has sixty faces, divided into *eight* different transitivity classes. However, in fact, fifty of the faces (falling into seven different transitivity classes) are congruent kite-quadrilaterals, and only one set of ten irregular quadrilaterals is different in shape. This is also seen for the duals of $J_{73}$ through $J_{75}$: each has the same two shapes of quadrilateral, though the number of transitivity classes is only four for the dual of $J_{73}$, no fewer than seventeen for the dual of $J_{74}$, and twelve for the dual of $J_{75}$.

Johnson solid 73, the parabigyrate rhombicosidodecahedron, is illustrated below in Figure 148.

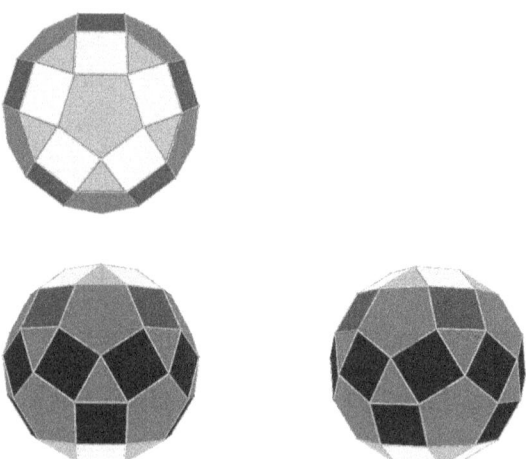

Figure 148: Johnson solid 73: the parabigyrate rhombicosidodecahedron.

Much of what was stated regarding $J_{72}$ previously can be

*Gyrate, diminished, and gyrate diminished rhombicosidodecahedra: Johnson solids 72-83.*

taken verbatim with regard to $J_{73}$. The symmetry is also $C_{5v}$, since both of the cupolae that are rotated 36° compared to their positions in the rhombicosidodecahedron are on the same fivefold axis. However, while $J_{72}$ can be considered as being derived from either of the two core polyhedra discussed in Chapter 14, only one ($J_{80}$) can be considered the core from which $J_{73}$ is derived: two pentagonal cupolae are fused to that core, both in the "gyrate" orientation. The symmetry of this polyhedron, however, is $S_{10v}$, as the twisting of two opposite cupolae generates a tenfold alternating axis.

The dual of the parabigyrate rhombicosidodecahedron is illustrated in Figure 149 below.

*Figure 149: The dual of Johnson solid 73 (the parabigyrate rhombicosidodecahedron).*

It shares the same $S_{10v}$ symmetry as its dual, of course. Because of this higher symmetry, the number of transitivity classes into which the faces can be divided is only four, the smallest of any of the Johnson solid duals discussed in this chapter. Even so, as was earlier mentioned, three of those four transitivity classes consist of identical kite-quadrilaterals.

Johnson solid 74, the metabigyrate rhombicosidodecahedron, is illustrated below in Figure 150. In this case, unlike $J_{73}$, which can be considered to be derived from a $J_{80}$ core, $J_{74}$ can only be derived from a $J_{83}$ core: in this case, two cupolae are fused in their standard orientations and one in the "gyrate" orientation. Unlike the parabigyrate rhombicosidodecahedron, the metabigyrate rhombicosidodecahedron has a very much lower $C_{2v}$ symmetry, which means that both it and its dual have exactly the same number of faces as the parabigyrate rhombicosidodecahedron and its dual, but the number of different transitivity classes is much larger. Where the parabigyrate rhombicosidodecahedron has faces that divide into seven transitivity classes, and its dual four, the metabigyrate rhombicosidodecahedron has faces that divide into *twenty-two* transitivity classes, and its dual seventeen.

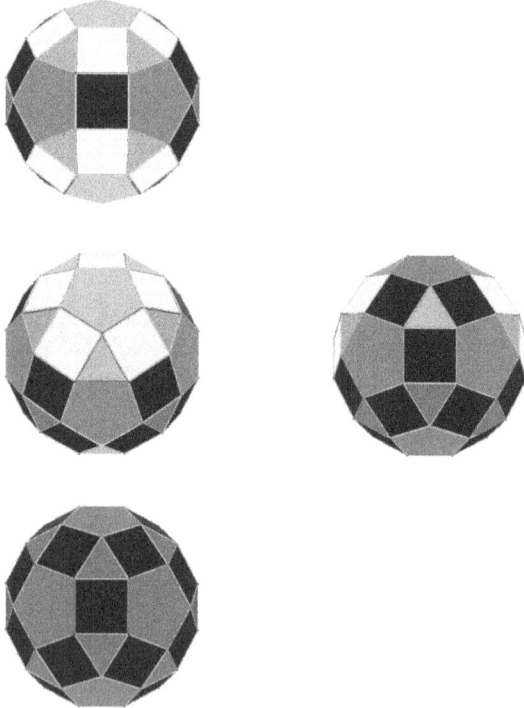

Figure 150: Johnson solid 74, the metabigyrate rhombicosidodecahedron.

The dual of the metabigyrate rhombicosidodecahedron is illustrated in Figure 151 below. As was noted, the faces divide into seventeen transitivity classes because of the lower symmetry, but all the kite-quadrilaterals are identical to the kite-quadrilaterals of the previous Johnson solid duals, as are the irregular quadrilaterals to those of the previous Johnson solid duals. The differences between faces in different transitivity classes have to do with more remote characteristics, such as the distances between two different faces, or the fact that some kites are adjacent to irregular quadrilaterals and others are not. Yet for all intents and purposes, the two types of quadrilateral, kite and irregular, are enough to characterize the faces, as shown in Table 33 below.

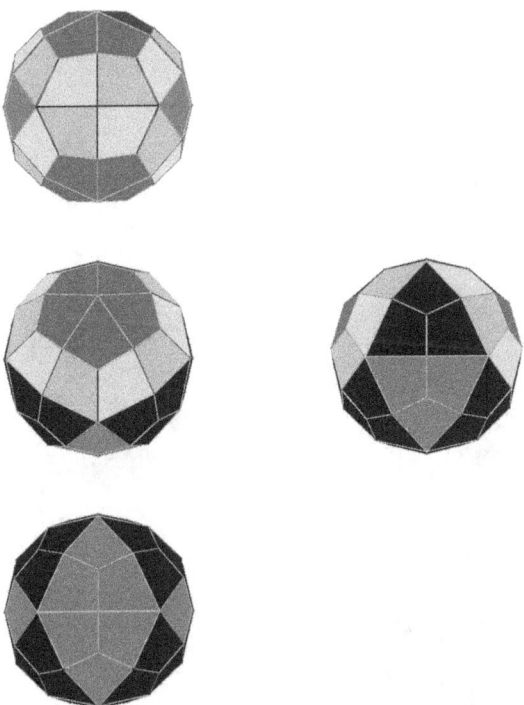

Figure 151: The dual of Johnson solid 74 (the metabigyrate rhombicosidodecahedron).

The trigyrate rhombicosidodecahedron, Johnson solid 75, is the last of the four Johnson solids in this chapter to retain all the faces of the original rhombicosidodecahedron, and all of what was common to the preceding three Johnson solids is characteristic of $J_{75}$ as well. The symmetry, however, is $C_{3v}$, which makes for a different set of transitivity classes in both $J_{75}$ and its dual from any of the others. Yet what was stated before about the relative irrelevance of transitivity classes to the characterization of the faces (and vertices) of the previous three Johnson solids and duals applies here as well.

It was noted in conjunction with a number of other Johnson solids that were derived by triple alteration of a solid of $I_h$ symmetry that the symmetry is not $S_{6v}$ or $D_{3h}$, but $C_{3v}$, with the "top" and "bottom" views looking very similar but not identical. The same is true here, as the three cupolae that are rotated are slightly above the center of the polyhedron.

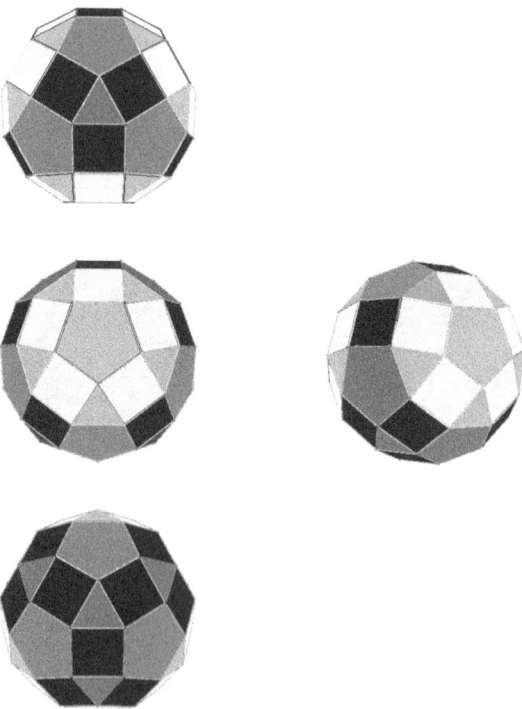

Figure 152: Johnson solid 75: the trigyrate rhombicosidodecahedron.

The dual of the trigyrate rhombicosidodecahedron is illustrated in Figure 153 below. It shares the same $C_{3v}$ symmetry as its dual, of course, and what was earlier stated on p. 328 about transitivity classes being greater in number than face shapes

*Gyrate, diminished, and gyrate diminished rhombicosidodecahedra: Johnson solids 72-83.*

applies here, as it did in the previous Johnson solid duals. As a result, Table 33 can have relatively simple entries for all four of these Johnson solids and duals. By contrast, this will not be true for many of the remaining Johnson solids and duals to be discussed in this chapter.

Unlike the trigyrate rhombicosidodecahedron itself, the appearance of the bottom is significantly different from the top. This is because the rotating of the cupolae produces the ring of irregular quadrilaterals replacing the typical kites, and they are nearer the top than the bottom in the orientation chosen in Figure 153.

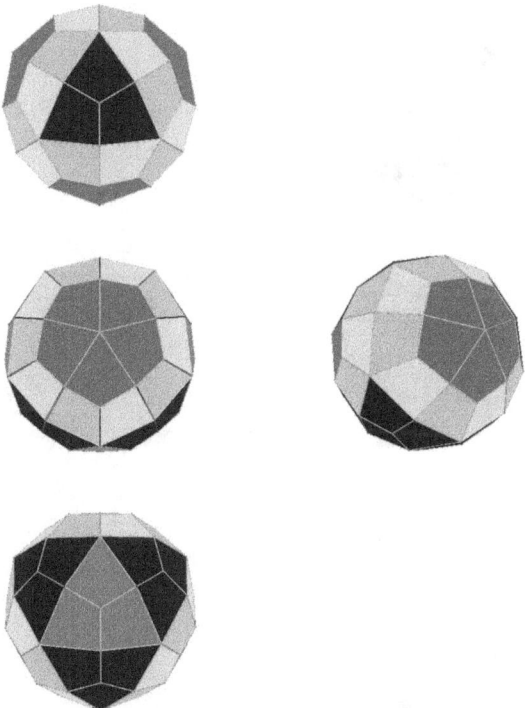

Figure 153: The dual of Johnson solid 75 (the trigyrate rhombicosidodecahedron).

In Chapter 15, the process of cutting a rhombicosidodecahedron by a plane to produce a diminished rhombicosidodecahedron and a pentagonal cupola was described. The pentagonal cupola, as was mentioned, is itself a Johnson solid, given (as was stated in the previous chapter) the designation of $J_5$ as described in Chapter 3. But so is the diminished rhombicosidodecahedron (Figure 154), and this is the next of the Johnson solids described in this chapter. The others are the various polyhedra produced by cutting any permissible number of pentagonal cupolae off, and rotating some of those cupolae to fuse them again, while discarding the rest of the cupolae.

The diminished rhombicosidodecahedron, Johnson solid 76, (Figure 154, below) retains one of the fivefold axes of the original rhombicosidodecahedron, and the five mirror planes through that axis, so its symmetry is $C_{5v}$.

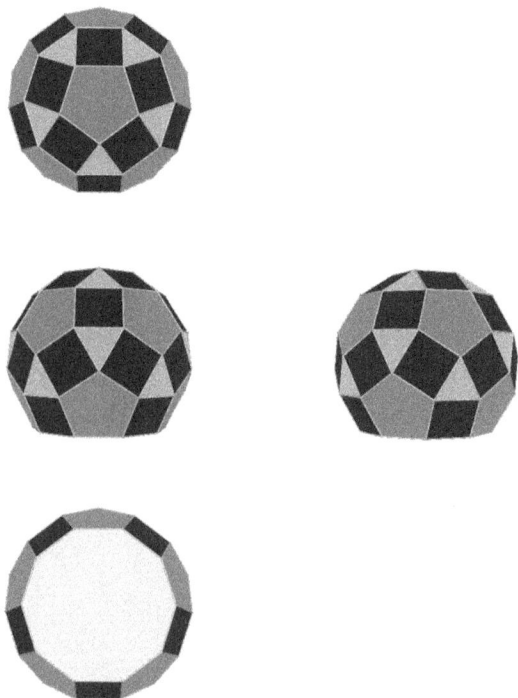

Figure 154: Johnson solid 76, the diminished rhombicosahedron.

While Johnson solids 72 to 75, which simply rotate one or more cupolae in a rhombicosidodecahedron, have the same 62 faces, 60 vertices, and 120 edges as the rhombicosidodecahedron itself, Johnson solid 76 (and 77 to 79) have replaced eleven faces by one (reducing the face count to 52), and eliminated five vertices (reducing the vertex count to 55) and fifteen edges (reducing the edge count to 105).

Of course, as is always the case, the dual of the diminished rhombicosidodecahedron (Figure 155 below) has the same $C_{5v}$ symmetry as the original diminished rhombicosidodecahedron, as dualization always preserves symmetry. However, the diminishing of the solid, unlike the rotating of a cupola as was done in the previous four Johnson solids, displaces the centroid from where it was in the original rhombicosidodecahedron. This has the effect that the quadrilateral faces are not all alike, as the kite-quadrilaterals were in the previous Johnson solids.

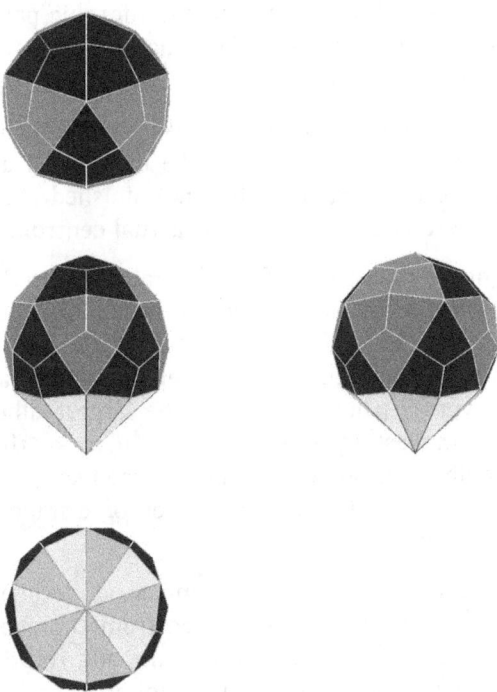

Figure 155: The dual of Johnson solid 76 (the diminished rhombicosahedron).

The faces fall into seven transitivity classes (six of quadrilaterals and one of triangles), and each of the seven has different face angles (as will be given in more detail below). It has a set of congruent triangular faces all coming together at the decavalent vertex that is dual to the decagonal face. However, these are not isosceles triangles; if the edge of the original Johnson solid is taken as 1, the edges meeting at the decagonal face are ~2.0543 and ~2.2286. And the third edges of these triangles are not all in a plane. So this polyhedron is not quite obtainable by fusing a pyramid to another solid. And although the several quadrilaterals constituting all the remaining faces are all of very similar dimensions, they are not identical; they fall, in fact, into six sets: three of five each and three of ten. Each of the sets of five are kite-quadrilaterals; however, the pairs of equal angles in the kites vary from ~86.8° to ~87.5°, the sets of ten have no particular symmetry in any single quadrilateral, but fall into enantiomorphic pairs. And yet, the smallest angles of all forty-five quadrilaterals, kite or not, only range from ~65.7° to ~69.4°, and the largest angles range from ~116.9° to ~119.2°. It is rather odd how close they are, but it can be assumed that they would be identical if a reciprocation center at the centroid of the undiminished rhombicosidodecahedron were used instead of the actual centroid of $J_{76}$ itself.

It might be noted that the dihedral angles, which are all equal in the rhombicosidodecahedron itself or in the various gyrate rhombicosidodecahedra, vary smoothly, with the largest values nearest to the decavalent vertex and the smallest opposite to that vertex (though, in fact, the largest dihedral angle exceeds the smallest by less than 4°!) To make this clear, in Table 33 the designations have been assigned so that the alphabetical sequence agrees with the order of distance from the decavalent vertex.

The paragyrate diminished rhombicosidodecahedron, $J_{77}$, is the first of the Johnson solids derived from the rhombicosidodecahedron that involve both diminishing and rotation of a cupola. For both this one and the next Johnson solid, the terms "meta-" and "para" relate to these two different alterations of the rhombicosidodecahedron; if the cupola removed from it and the one rotated by 36° are diametrically opposite, as in $J_{77}$, the "para-" prefix is used, and if they are at positions corresponding to $J_{74}$ (or $J_{81}$), as in $J_{78}$, the "meta-" prefix.

*Gyrate, diminished, and gyrate diminished rhombicosidodecahedra: Johnson solids 72-83.*

Because the diminishing and the rotation of the cupola are opposite each other, the fivefold axis that passes through both remains, so that like the diminished rhombicosidodecahedron ($J_{76}$) and the gyrate rhombicosidodecahedron ($J_{72}$), the paragyrate diminished rhombicosidodecahedron has $C_{5v}$ symmetry. Like the diminished rhombicosidodecahedron, the paragyrate diminished rhombicosidodecahedron has 52 faces, 55 vertices, and 105 edges, and in fact all the statistics of the two are very similar, which is why in Table 33 they are listed together (as were Johnson solids 72 to 75; Johnson solids 78 and 79 are combined with 76 and 77 as well).

Figure 156: Johnson solid 77: the paragyrate diminished rhombicosidodecahedron.

The dual of the paragyrate diminished rhombicosidodecahedron is illustrated in Figure 157 below. It is actually almost identical to the dual of the diminished rhombicosidodecahedron, with the same ten triangular faces meeting at one decavalent vertex, and six classes of quadrilaterals, some of which are kite-quadrilaterals and some near-kites, except that one class, which were near-kites in the dual of the diminished rhombicosidodecahedron, are completely irregular quadrilaterals in the dual of the paragyrate diminished rhombicosidodecahedron, though with exactly the same angles (though occurring in a different order). This identity can be considered a result of two factors: the fact that both $J_{76}$ and $J_{77}$ have a $C_{5v}$ symmetry, and (because rotating a cupola does not displace the centroid) both have the same centroid.

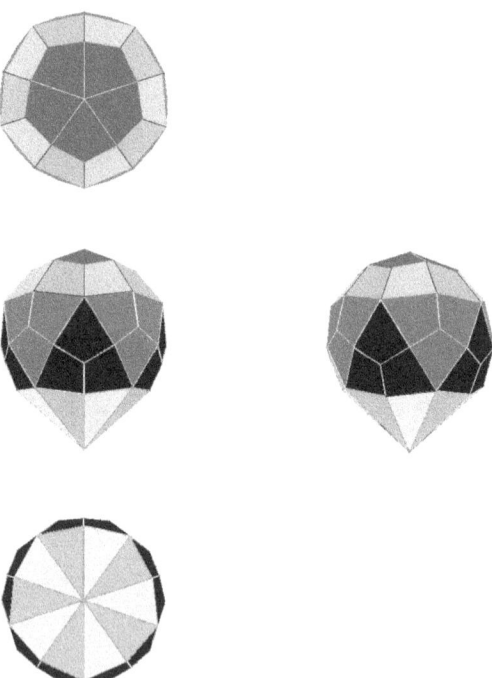

*Figure 157: The dual of Johnson solid 77 (the paragyrate diminished rhombicosidodecahedron).*

The metagyrate diminished rhombicosidodecahedron, $J_{78}$, is one of the least symmetric of the Johnson solids. It can be conceived of either as a rhombicosidodecahedron with one cupola removed and another, in a "meta" position relative to it as previously described, rotated 36°, which is the way Johnson thought of it in naming it, or as might make visualizing it a bit easier, as a fusion of $J_{83}$ with two cupolae, one in the "normal" orientation and one in the "gyrate" orientation, i. e., rotated 36° compared to its orientation in the rhombicosidodecahedron.

Figure 158: Johnson solid 78: the metagyrate diminished rhombicosidodecahedron.

When one starts with a rhombicosidodecahedron and either diminishes it (as in the case of $J_{76}$) or rotates a cupola 36° around (as in $J_{72}$), the fivefold axis through the altered cupola is preserved, so that the resulting polyhedron is of symmetry $C_{5v}$. If this is done to two opposite cupolae (designated "para"), since both are around the same fivefold axis, the resulting polyhedron still retains that one fivefold axis, regardless of which (diminution or rotation by 36°) is done to either, leaving a symmetry of $C_{5v}$ or even $S_{10v}$. If it is done to two cupolae in the "meta" position, however, things are different. If both are treated alike (both diminished, as will be seen in $J_{81}$, or both rotated 36°, as in $J_{74}$), then there will be a twofold axis and a mirror plane passing halfway between the two, so the symmetry will be found to be $C_{2v}$. However, when one cupola is diminished and the other rotated 36°, they are not equivalent, so that neither of those symmetry elements is present. The only symmetry still possessed when this construction is carried out is one mirror plane, passing through the center of the decagonal face and the center of the gyrate cupola, so the symmetry is merely $C_s$.

The dual of the metagyrate diminished rhombicosidodecahedron is illustrated in Figure 159 below.

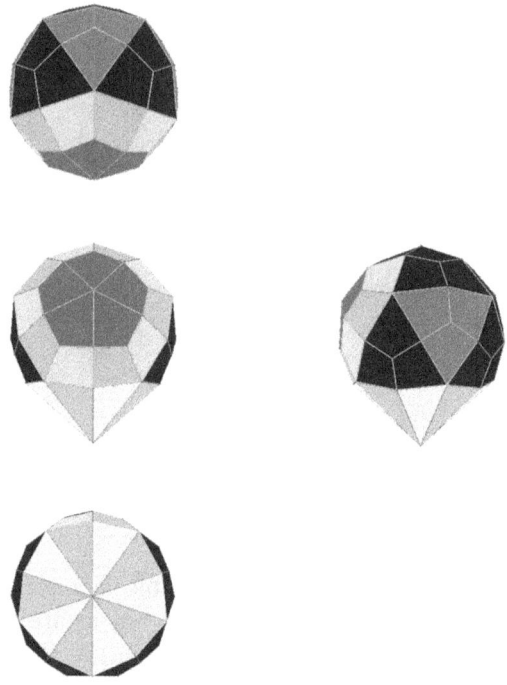

Figure 159: The dual of Johnson solid 78 (the metagyrate diminished rhombicosidodecahedron).

Although the triangular faces of this polyhedron are identical to the triangular faces of the duals of $J_{76}$ and $J_{77}$ (despite the fact that they divide into five different transitivity classes), the quadrilaterals have now changed, and there are so many different shapes as to make keeping track of them difficult. The combination of a large number of faces, in this case 55, with such a low symmetry leads to a very large number of transitivity classes; with $C_s$ symmetry, no more than two faces can belong to a transitivity class. So the faces could not divide into fewer than 28 transitivity classes, and in fact there are 29 altogether. A full specification of the faces according to transitivity classes in Table 33 would be [2+2+2+2+2](3) + [1+1+1+2+2+2+2+2+2+2+2+2+2+2+2+2+2+2+2+2+2+2+2+2+2+2](4), much too long to read comfortably.

And it would probably take excessive room to give all the face angles, as there are now fourteen different shapes of quadrilateral instead of only six, as well as the vast number of different dihedral angles involved (there are no fewer than twenty), so that for this polyhedron those numbers are omitted.

Johnson solid 79 can uniquely be described as the bigyrate diminished rhombicosidodecahedron. When one cupola has been diminished, there would be two places where another can be rotated, in the "para" position (seen in $J_{77}$) or the "meta" position (seen in $J_{78}$). But starting with $J_{77}$, there would not be another cupola in a position to rotate it without disturbing the first rotated cupola. So only a "meta" relative position is possible for the diminishing and the 36° rotation if another 36° rotation is to take place, and the second cupola must be in a "meta" position relative to both the diminishing and the first 36° rotation. This yields the polyhedron illustrated in Figure 160 below. Exactly like the diminished rhombicosidodecahedron, the paragyrate diminished rhombicosidodecahedron, and the metagyrate diminished rhombicosidodecahedron, this polyhedron has 52 faces (15 triangular, 25 square, 11 pentagonal, and 1 decagonal), 55 vertices (10 trivalent and 45 quadrivalent), and 105 edges, since the 36° rotation of a cupola does not affect the face/vertex/edge count.

It can be seen that the two rotated cupolae are, as in $J_{74}$, symmetric to each other across a mirror plane. (There is not a twofold axis between them, because the two cupolae in question are somewhat forward of the center of the polyhedron, as is visible in the top view in Figure 160 above.) Also, that same mirror plane bisects the decagonal face, so the rather low symmetry of $C_s$ is all that this polyhedron possesses, exactly like $J_{78}$.

Because all the polyhedra forming the faces of a Johnson solid must be regular by definition, so that all edges bounding a single face must be equal, it can be seen that all edges ultimately must be equal, so all faces of the same number of sides must be congruent. However, the $C_s$ symmetry requires that no more than two faces can belong to the same transitivity class, so the faces actually divide into 31 different transitivity classes. The differences relate to how far a given face is from the decagonal face and from the rotated cupolae.

Figure 160: Johnson solid 79, the bigyrate diminished rhombicosidodecahedron.

The dual of the bigyrate diminished rhombicosidodecahedron is illustrated in Figure 161 below. In common with the dual of the metagyrate diminished rhombicosidodecahedron, the presence of the rotated cupolae destroys the fivefold axis, but as far as the triangular faces are concerned, they retain the shapes they had in the dual of the diminished rhombicosidodecahedron. All four Johnson solid duals (the three just mentioned and the dual of the paragyrate diminished rhombicosidodecahedron) have ten identical triangular faces (identical to the others in the same solid and to those in the other three), even though, in the duals of the diminished rhombicosidodecahedron and of the paragyrate diminished rhombicosidodecahedron, all the triangles belong to a single transitivity class, while in the other two Johnson solid duals, they are divided into five different transitivity classes.

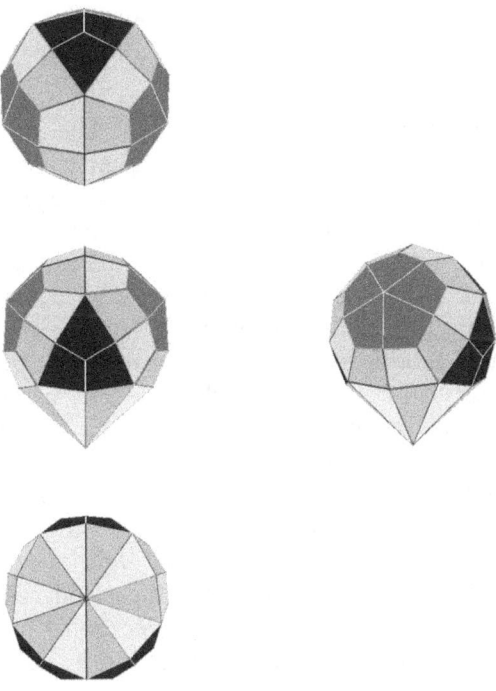

Figure 161: The dual of Johnson solid 79 (the bigyrate diminished rhombicosidodecahedron).

The combination of a large number of faces, in this case 55, with such a low symmetry leads to a very large number of transitivity classes; with $C_s$ symmetry, no more than two faces can belong to a transitivity class. So the faces could not divide into fewer than 28 transitivity classes, and in fact there are 29 altogether. A full specification of the faces according to transitivity classes in Table 33 would be [2+2+2+2+2](3) + [1+1+1+2+2+2+2+2+2+2+2+2+2+2+2+2+2+2+2+2+2+2](4), much too long to read comfortably.

Although the triangular faces of this polyhedron, as in the case of the dual of $J_{78}$, are identical to the triangular faces of the duals of $J_{76}$ and $J_{77}$ (despite the fact that they divide into five different transitivity classes), the quadrilaterals have now changed. And it would probably take excessive room to give all the face angles, as there are now, as in the case of the dual of $J_{78}$, fourteen different shapes of quadrilateral (in fact, most are identical to the faces of the dual of $J_{78}$) instead of only six, as well as the vast number of different dihedral angles involved (there are again twenty), so that for this polyhedron those numbers are omitted.

The parabidiminished rhombicosidodecahedron, $J_{80}$, is the next polyhedron to be discussed. It is unusual in being *bi/bigeneral*, in the terminology of my earlier book, *Polyhedra: a New Approach*. Most of the tectal polyhedra among the Platonic, Archimedean, and Johnson solids are *uni/unigeneral*: the principal axis of rotation cuts the surface in two polygons whose number of sides is equal to the order of that axis. In a few cases (the truncated tetrahedron, among the Archimedean solids, and the cupolae, rotunda, and their elongated derivatives, among the Johnson solids, described in Chapters 3 and 7) they are *uni/bigeneral*: the principal axis of rotation cuts the surface in two polygons, one of whose number of sides is equal to the order of that axis, and the other of whose number of sides is twice the order of the axis. But the fivefold principal axis of rotation of the parabidiminished rhombicosidodecahedron cuts the surface at the centers of the two decagonal faces, making this one of a small number of Johnson solids that is bi/bigeneral. (The bilunabirotunda, $J_{91}$, is another; see p. 397.)

The top and bottom are identical (except for a 36° rotation), leading to a tenfold alternating axis; thus the symmetry is $S_{10v}$.

On p. 334 it was noted that Johnson solids 72 to 75, which simply rotate one or more cupolae in a rhombicosidodecahedron, have the same 62 faces, 60 vertices, and 120 edges as the rhombicosidodecahedron itself, and Johnson solids 76 to 79 have replaced eleven faces by one (reducing the face count to 52), and eliminated five vertices (reducing the vertex count to 55) and fifteen edges (reducing the edge count to 105). The same reasoning makes it clear that Johnson solid 80 (and 81 and 82) has 42 faces, 50 vertices, and 90 edges.

Johnson solid 80 is one of the two core polyhedra described in Chapter 15 which can be used as a basis for the polyhedra of this chapter when one might prefer to treat them as cupola-augmented cores rather than as diminished rhombicosidodecahedra.

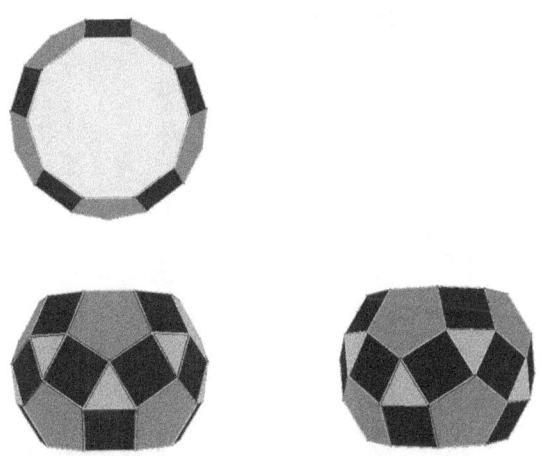

Figure 162: Johnson solid 80, the parabidiminished rhombicosidodecahedron, one of the two core polyhedra for describing the twelve Johnson solids of this chapter.

The dual of the parabidiminished rhombicosidodecahedron, illustrated in Figure 163 below, shares the $S_{10v}$ symmetry of its dual. Because the diminishings are symmetrically placed around the centroid of the original rhombicosidodecahedron, the centroid of the parabidiminished rhombicosidodecahedron remains in the same place, so the distortions that occcurred in what had been kite-quadrilaterals when the reciprocation center did not coincide with the centroid of the original rhombicosidodecahedron do not occur in this case, and the quadrilaterals in the central belt are perfect kites, identical to the faces of the dual to the original rhombicosidodecahedron (though they do fall into two transitivity classes, as the ones with an edge in common with the triangular faces differ in classification from the ones that do not have such an edge). The triangles at the two apices, however, are not isosceles, but alternate in chirality. A glance at Table 33 shows that they have their two largest angles equal to two angles of the kite-quadrilaterals; since the three angles of a triangle must add to 180° and the four angles of a quadrilateral must add to 360°, the third angle must also be related to angles in the kite-quadrilateral by

67.7830 + 86.9742 + 25.2428 = 180,

67.7830 + 2(86.9742) + 118.2687 = 360, therefore

25.2428 = 86.9742 + 118.2687 − 180.

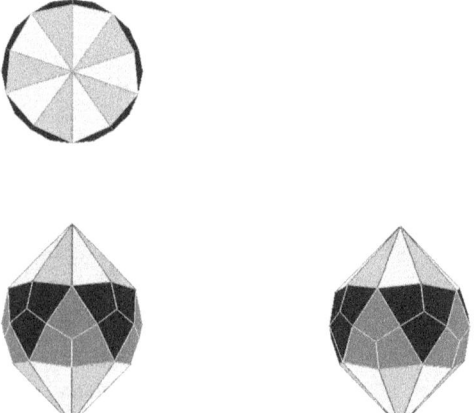

Figure 163: The dual of Johnson solid 80 (the parabidiminished rhombicosidodecahedron).

Johnson solid 81, the metabidiminished rhombicosidodecahedron, is illustrated in Figure 164 below. Although such properties as the face/vertex/edge count are identical to those of the parabidiminished rhombicosidodecahedron, any property dependent on symmetry, such as the number of transitivity classes of the faces, will be very different. The placement of the second diminishing in the "meta" position destroys the fivefold symmetry, but creates a mirror plane and twofold axis halfway

*Gyrate, diminished, and gyrate diminished rhombicosidodecahedra: Johnson solids 72-83.*

between the two diminishings, which shows as a vertical line in the top, front, and bottom views in Figure 164. There remains, in addition, a mirror plane that passes through the centers of the decagonal faces produced by the diminishings, which appears as a horizontal line in the top and bottom views and as a vertical line in the side view, so the overall symmetry is $C_{2v}$.

As it is the policy of this book to show all the solids in an aspect that locates the principal axis of rotation as a vertical line in front and side views, and the center of the top and bottom views, the orientation of Figure 164 does not correspond to that of Figure 162 above.

Figure 164: Johnson solid 81, the metabidiminished rhombicosidodecahedron.

The dual of the metabidiminished rhombicosidodecahedron, illustrated in Figure 165 below, clearly shows the effects of having a reciprocation center displaced from the centroid of the original rhombicosidodecahedron.

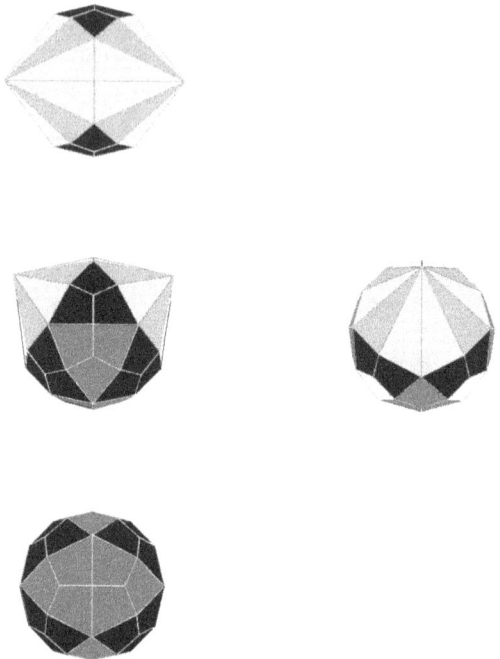

Figure 165: The dual of Johnson solid 81 (the metabidiminished rhombicosidodecahedron).

The centroid of the parabidiminished rhombicosidodecahedron is exactly at the same point as that of the original rhombicosidodecahedron; the centroid of the metabidiminished rhombicosidodecahedron is displaced slightly because the two diminishings remove vertices that, in the orientation of Figure 164, are both above the center, moving the actual centroid downward.

So unlike the case of the parabidiminished rhombicosidodecahedron, where all the kite-quadrilaterals of the original rhombicosidodecahedron remained as kites congruent to their original size and shape, in the metabidiminished rhombicosidodecahedron, there are differences among them. (In addition, the fact that the triangular faces range from nearer to further from the mirror plane that is halfway from the two decavalent

*Gyrate, diminished, and gyrate diminished rhombicosidodecahedra: Johnson solids 72-83.*

vertices causes them to differ as well.) Altogether the triangular faces fall into five transitivity classes, each of which is a different shape, and the quadrilaterals into nine transitivity classes, each of which is a different shape, though all the quadrilaterals are either kites or near-kites and all nearly congruent.

A full specification of the faces according to transitivity classes in Table 33 would be [4+4+4+4+4](3) + [2+2+2+4+4+4+4+4+4](4), much too long to read comfortably. With so many slightly different sets of angles, they too will not be included in Table 33 (nor will their dimensions be included in the text as was done for many of the duals in Chapter 14).

While the metabidiminished rhombicosidodecahedron has two mirror planes, one halfway between the two diminishings and one passing through the two centers of the decagonal faces that they create, Johnson solid 82, the gyrate bidiminished rhombicosidodecahedron, illustrated in Figure 168 below, only has one. The cupola that is rotated 36° is symmetrical with respect to the mirror plane that is halfway between the two diminishings, so that mirror plane remains, but it is not centered on an axis in the second plane, so that that plane ceases to be a mirror plane. As a result, the symmetry of $J_{82}$ is only $C_s$, making it one of only four Johnson solids of that low symmetry, the other three being $J_{78}$, $J_{79}$, and $J_{87}$.

The low symmetry is the cause of the large number of different transitivity classes into which the faces fall. While, in a polyhedron of $D_{5h}$ (or the $S_{10v}$ symmetry of $J_{80}$) symmetry, it is possible to have a transitivity class containing 10 different faces, if the symmetry is only $C_s$, no more than two faces can form a transitivity class, and any face that straddles the mirror plane necessarily forms a transitivity class by itself. So with 42 faces, there must be at least 21 transitivity classes; in fact, because several faces fall across the mirror plane, the actual number is 27, while with the same 42 faces, $J_{80}$, with its higher symmetry, has only 5. Yet it should be noted that all the triangular faces are congruent, as are all the square faces, and all the pentagonal, because this is a Johnson solid, requiring all polygons to be regular. With all polygons regular, no two edges of the polyhedron can be different, because *some* face would have two different sides, an impossibility.

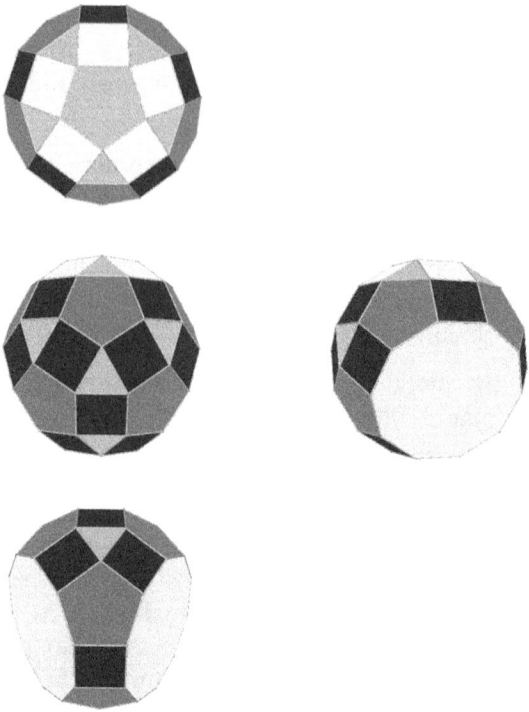

Figure 166: Johnson solid 82, the gyrate bidiminished rhombicosahedron.

The dual of the gyrate bidiminished rhombicosidodecahedron, illustrated in Figure 167 below, of course shares the $C_s$ symmetry of its dual. The combination of a large number of faces, in this case 50, with such a low symmetry leads to a very large number of transitivity classes; as was stated in the discussion of the gyrate bidiminished rhombicosidodecahedron itself, with $C_s$ symmetry, no more than two faces can belong to a transitivity class. So the faces could not divide into fewer than 25 transitivity classes, and in fact there are 27 altogether. A full specification of the faces according to transitivity classes in Table 33 would be [2+2+2+2+2](3) + [1+1+1+1+2+2+2+2+2+2+2+2+2+2+2+2+2+2+2+2+2](4), much too long to read comfortably.

*Gyrate, diminished, and gyrate diminished rhombicosidodecahedra: Johnson solids 72-83.*

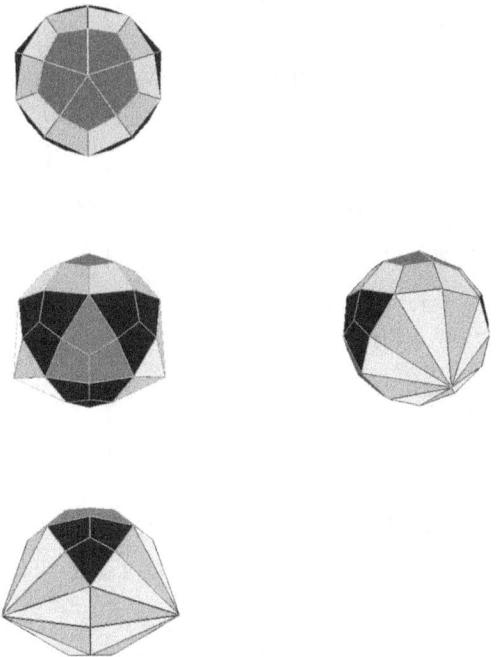

*Figure 167: The dual of Johnson solid 82 (the gyrate bidiminished rhombicosahedron).*

In Table 33, one difference will be made from the procedures which were employed in the other tables of this kind. In the other tables, the row headed "Faces" shows the faces broken down into transitivity classes. To do that in Table 33 would produce a rather messy table which would not be as enlightening as might be hoped. Since these solids are all derived from the rhombicosidodecahedron, most of such properties as the dihedral angles between faces of a particular number of sides will not vary from one transitivity class to another. The only reason that the polygons of any particular type are divided into a number of transitivity classes is more remote relationships in the solid in general, such as the distance of the face in particular from where a cupola was removed or rotated. So in Table 33, the "Faces" row only shows the numbers of polygons of each type. A more precise listing such as is in the other tables would show:

72: [5+5+10](3) + [5+5+5+5+10](4) + [1+1+5+5](5)

73: [10+10](3) + [10+10+10](4) + [2+10](5)

74: [2+2+2+2+4+4+4](3) + [1+1+2+2+4+4+4+4+4](4) + [2+2+2+2+4](5)

75: [1+1+3+3+6+6](3) + [3+3+3+3+6+6+6](4) + [3+3+3+3](5)

76: [5+5+5](3) + [5+5+5+10](4) + [1+5+5](5) + [1](10)

77: [5+5+5](3) + [5+5+5+10](4) + [1+5+5](5) + [1](10)

78: [1+1+1+2+2+2+2+2+2](3) + [1+1+1+2+2+2+2+2+2+2+2+2](4) + [1+1+1+2+2+2+2](5) + [1](10)

79: [1+1+1+2+2+2+2+2](3) + [ 1+1+1+2+2+2+2+2+2+2+2+2](4) + [1+1+1+2+2+2+2](5) + [1](10)

80: [10](3) + [10+10](4) + [10](5) + [2](10)

81: [2+2+2+4](3) + [1+1+2+4+4+4+4](4) + [2+2+6](5) + [2](10)

82: [1+1+1+1+2+2+2](3) + [1+1+1+1+2+2+2+2+2+2+2](4) + [1+1+1+1+2+2+2](5) + [2](10)

83: [1+1+3](3) + [3+3+3+6](4) + [3+3+3](5) + [3](10)

However, the data as listed in the tables are probably more useful.

*Gyrate, diminished, and gyrate diminished rhombicosidodecahedra: Johnson solids 72-83.*

| Johnson solid number | 72 – 75 | 76 – 79 | 80 – 82 |
|---|---|---|---|
| Name | 72: Gyrate rhombicosidodecahedron<br>73: Parabigyrate rhombicosidodecahedron<br>74: Metabigyrate rhombicosidodecahedron<br>75: Trigyrate rhombicosidodecahedron | 76: Diminished rhombicosidodecahedron<br>77: Paragyrate diminished rhombicosidodecahedron<br>78: Metagyrate diminished rhombicosidodecahedron<br>79: Bigyrate diminished rhombicosidodecahedron | 80: Parabidiminished rhombicosidodecahedron<br>81: Metabidiminished rhombicosidodecahedron<br>82: Gyrate bidiminished rhombicosidodecahedron |
| Symmetry* | 72: $C_{5v}$<br>73: $S_{10v}$<br>74: $C_{2v}$<br>75: $C_{3v}$ | 76, 77: $C_{5v}$<br>78: $C_s$<br>79: $C_s$ | 80: $S_{10v}$<br>81: $C_{2v}$<br>82: $C_s$ |
| Faces | [20](3) + [30](4) + [12](5) | [15](3) + [25](4) + [11](5) + [1](10) | [10](3) + [20](4) + [10](5) + [2](10) |
| Edges* | 120 | 105 | 90 |
| Vertices | 60 | 55 | 50 |
| Dihedral angles | 3-4: 159.0948° (159° 5′ 41″),<br>3-5: 153.9424° (153° 56′ 33″),<br>4-4: 153.4349° (153° 26′ 6″),<br>4-5: 148.2825° (148° 16′ 57″)<br>4-10: 121.7175° (121° 43′ 3″),<br>5-10: 116.5651° (116° 33′ 54″)♠ | | |
| Area | 59.30598284 | 58.11465078 | 56.92331871 |
| Volume | 41.61532378 | 39.29127846 | 36.96723315 |
| **Dual** | | | |
| Name | Unnamed | Unnamed | Unnamed |
| Faces | 60¤ | 76, 77: [10](3) + [5+5+5+10+10+10](4)<br>78, 79: 55◈ | 80: [20](3) + [10+20](4)<br>81, 82: 50◈ |
| Vertices | 62 | 52 | 42 |

| Johnson solid number | 72 − 75 | 76 − 79 | 80 − 82 |
|---|---|---|---|
| Face angles[†] | [2]86.9742° (86° 58′ 27″), 67.7830° (67° 46′ 59″), 118.2687° (118° 16′ 7″)[¤] | 76, 77: See list below[*][△] 78: See text (p. 341) 79: See text (p. 344) | 80: 3: 25.2428° (25° 14′ 34″), 67.7830° (67° 46′ 59″), 86.9742° (86° 58′ 27″); 4: [2]86.9742° (86° 58′ 27″), 67.7830° (67° 46′ 59″), 118.2687° (118° 16′ 7″); 81: See text 82: See text |

*Gyrate, diminished, and gyrate diminished rhombicosidodecahedra: Johnson solids 72-83.*

| Johnson solid number | 72 – 75 | 76 – 79 | 80 – 82 |
|---|---|---|---|
| Dihedral angles | 154.1214° (154° 7' 17")ᵅ | 76, 77: <br> 3-3: 155.6019° (155° 36' 7"), <br> 3-4a: 155.2944° (155° 17' 40"), <br> 4a-4a: 154.8632° (154° 51' 48"), <br> 4a-4b: 154.6429° (154° 38' 35"), <br> 4a-4c: 154.5262° (154° 31' 34"), <br> 4b-4d: 154.0140° (154° 0' 50"), <br> 4c-4d: 153.8104° (153° 48' 38"), <br> 4d-4d: 153.5114° (153° 30' 41"), <br> 4d-4e: 153.1154° (153° 6' 55"), <br> 4e-4e: 152.5583° (152° 33' 30"), <br> 4e-4f: 152.2703° (152° 16' 13"), <br> 4f-4f: 151.9150° (151° 54' 54")* <br> 78: See text (p. 341) <br> 79: See text (p. 344) | 80: 154.1214° (154° 7' 17"); <br> 81, 82: See text |
| Area | 59.76739510 | 43.84238259 | 80: 63.17055841 <br> 81: 42.13978817 <br> 82: 41.72884894 |
| Volume | 42.25536942 | 26.27684324 | 80: 44.66139570 <br> 81: 24.45873303 <br> 82: 24.10183131 |

Table 33: *Properties of Johnson solids 72 through 82 and their duals.*

*The symmetry and number of edges are always the same for each solid and its dual, so are not repeated in the "dual" section.

†Face angles are only given for the duals, as the Johnson solids have regular faces, whose angles can be found in Table 1 on p. 3.

♦Not all of the edge types in this list are found in all of these solids. However, all of the edges between polygons with the number of sides specified have the dihedral angles indicated, regardless of transitivity class.

□See text, p. 328. The number of transitivity classes ranges from four to seventeen, but all the faces have the same angles and all dihedral angles are equal.

°For the breakdown into transitivity classes of the faces of the dual of $J_{78}$, see p. 341; $J_{79}$, see p. 343; $J_{81}$, see p. 348; and $J_{82}$, see p. 350.

‡For both the duals of $J_{76}$ and $J_{77}$, the designation 4a refers to the ten near-kites sharing an edge with the triangular faces; 4b, to the five kite-quadrilaterals that share only a vertex with the triangular faces, stiuated between the 4a faces; 4c, to the five kite-quadrilaterals that share the remaining edges of the 4a faces; 4d, to the ten near-kites sharing the remaining edges of the 4b faces; 4e, to the ring of ten (near-kites in the dual of $J_{76}$, completely irregular in the dual of $J_{77}$) between the 4d faces and the last five faces around the fivefold axis; 4f, to those last five, which are also kite-quadrilaterals.

△The list of face angles of the duals of $J_{76}$ and $J_{77}$ is too large to fit in one cell. It is given here instead:

- 3: 26.6750° (26° 40′ 30″), 66.9231° (66° 55′ 23″), 86.4019° (86° 24′ 7″);

- 4a: 68.6502° (68° 39′ 1″), 85.3608° (85° 21′ 39″), 88.0865° (88° 5′ 11″), 117.9026° (117° 54′ 9″);

- 4b: [2]86.8265° (86° 49′ 36″), 69.3989° (69° 23′ 56″), 116.9480° (116° 56′ 53″);

- 4c: [2]87.5267° (87° 31′ 36″), 65.7344° (65° 44′ 4″), 119.2122° (119° 12′ 44″);

*Gyrate, diminished, and gyrate diminished rhombicosidodecahedra: Johnson solids 72-83.*

- 4d: 67.0999° (67° 6′ 0″), 85.7199° (85° 43′ 12″), 88.3075° (88° 18′ 27″), 118.8727° (118° 52′ 22″);

- 4e: 68.7605° (68° 45′ 38″), 85.8528° (85° 51′ 10″), 87.7107° (87° 42′ 39″), 117.6760° (117° 40′ 34″);

- 4f: [2]87.1534° (87° 9′ 12″), 66.9891° (66° 59′ 21″), 118.7042° (118° 42′ 15″)

*The Johnson Solids and Their Duals*

Johnson solid 83, the tridiminished rhombicosidodecahedron, was mentioned earlier as one of the two "core polyhedra" from which the others in this chapter may be derived by augmentation. It is illustrated in Figure 166 below.

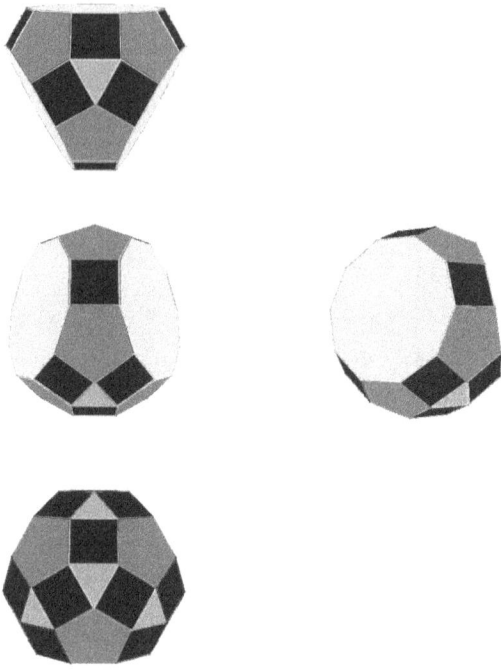

*Figure 168: Johnson solid 83, the tridiminished rhombicosahedron, one of the two core polyhedra for describing the twelve Johnson solids of this chapter.*

The symmetry is $C_{3v}$, and the face/vertex/edge count can be predicted from the trends in the earlier cases discussed in this chapter. On p. 334 it was noted that Johnson solids 72 to 75, which simply rotate one or more cupolae in a rhombicosidodecahedron, have the same 62 faces, 60 vertices, and 120 edges as the rhombicosidodecahedron itself; Johnson solids 76 to 79 have replaced eleven faces by one (reducing the face count to 52), and eliminated five vertices (reducing the vertex count to 55) and fifteen edges (reducing the edge count to 105); and Johnson solid 80 to 82 have 42 faces, 50 vertices, and 90 edges.

The same reasoning makes it clear that and Johnson solid 83 has 32 faces, 45 vertices, and 75 edges.

The dual of the tridiminished rhombicosidodecahedron, illustrated in Figure 169 below, shares the same $C_{3v}$ symmetry as its dual. Unlike some of the earlier-described duals in this chapter, the five different transitivity classes of triangular face are really different in shape, although very similar; there are four transitivity classes of quadrilateral face, also all different but very similar; three of those four are kite-quadrilaterals, but the quadrilaterals in the fourth class are only near-kites.

Figure 169: The dual of Johnson solid 83 (the tridiminished icosahedron).

Because of the three mirror planes that are part of the $C_{3v}$ symmetry description, the only quadrilaterals that are not forced to be kites are the six that are designated "4c" in Table 34 below. The three mirror planes that pass through the threefold axis cut the three quadrilaterals that surround each end of that axis diagonally, making them necessarily kite-quadrilaterals; in addition there are three quadrilaterals that are adjacent to two triangles where the pyramid-like arrays of triangles diverge from each other, and they too are each bisected by one of the three mirror planes, making them necessarily kite-quadrilaterals.

*Gyrate, diminished, and gyrate diminished rhombicosidodecahedra: Johnson solids 72-83.*

| Johnson solid number | 83 |
|---|---|
| Name | Tridiminished rhombicosidodecahedron |
| Symmetry* | $C_{3v}$ |
| Faces | [5](3) + [15](4) + [9](5) + [3](10) |
| Edges* | 75 |
| Vertices | 45 |
| Dihedral angles | 3-4: 159.0948° (159° 5' 41"), 4-5: 148.2825° (148° 16' 57"), 4-10: 121.7175° (121° 43' 3"), 5-10: 116.5651° (116° 33' 54")♣ |
| Area | 55.73198664 |
| Volume | 34.64318783 |
| **Dual** | |
| Name | Unnamed |
| Faces | [6+6+6+6+6](3) + [3+3+3+6](4) |
| Vertices | 32 |

| Johnson solid number | 83 |
|---|---|
| Face angles[†] | 3a: 24.5154° (24° 30' 56"), 67.9673° (67° 58' 2"), 87.5172° (87° 31' 2"); 3b: 24.8893° (24° 53' 21"), 68.6685° (68° 40' 7"), 86.4422° (86° 26' 32"); 3c: 25.4856° (25° 29' 8"), 66.7423° (66° 44' 32"), 87.7722° (87° 46' 20"); 3d: 26.0657° (26° 3' 56"), 68.1354° (68° 8' 7"), 85.7990° (85° 47' 56"); 3e: 26.4130° (26° 24' 47"), 66.7495° (66° 44' 58"), 86.8374° (86° 50' 15"); 4a: [2]87.0390° (87° 2' 20"), 67.4677° (67° 28' 4"), 118.4543° (118° 27' 16"); 4b: [2]86.9153 (86° 54' 55"), 68.1217° (68° 7' 18"), 118.0478° (118° 2' 52"); 4c: 67.6459° (67° 38' 45"), 86.3285° (86° 19' 43"), 87.6188° (87° 37' 8"), 118.4068° (118° 24' 24"); 4d: [2]86.8374° (86° 50' 15"), 68.8012° (68° 48' 4"), 117.5239° (117° 31' 26")[*] |

*Gyrate, diminished, and gyrate diminished rhombicosidodecahedra: Johnson solids 72-83.*

| Johnson solid number | 83 |
|---|---|
| Dihedral angles | 3a-3a: 155.3059° (155° 18′ 21″), <br> 3a-3b: 155.1449° (155° 8′ 42″), <br> 3b-3b: 154.9567° (154° 57′ 24″), <br> 3b-3c: 154.6957° (154° 41′ 44″), <br> 3c-3c: 154.3588° (154° 21′ 32″), <br> 3c-3d: 154.0783° (154° 4′ 42″), <br> 3d-3e, 3d-4d: 153.5215° (153° 31′ 17″), <br> 3e-3e: 153.2937° (153° 17′ 37″), <br> 3a-4a: 155.4153° (155° 24′ 55″), <br> 3e-4c, 4c-4d: 153.0895° (153° 5′ 22″), <br> 4a-4a: 155.5146° (155° 30′ 52″), <br> 4b-4b: 152.5650° (152° 33′ 54″), <br> 4b-4c: 152.7146° (152° 42′ 53″), <br> 4c-4c: 152.8503° (152° 51′ 1″)* |
| Area | 50.95300520 |
| Volume | 31.98713404 |

Table 34: *Properties of Johnson solid 83 and its dual.*

°The symmetry and number of edges are always the same for each solid and its dual, so are not repeated in the "dual" section.

†Face angles are only given for the duals, as the Johnson solids have regular faces, whose angles can be found in Table 1 on p. 3.

✦All of the edges between polygons with the number of sides specified have the dihedral angles indicated, regardless of transitivity class.

*The designation 3a refers to the triangles, two in each bundle of ten, that share an edge with the three kite-quadrilaterals that are isolated at one end of the threefold axis; 3b refers to the triangles, two in each bundle of ten, adjacent to the 3a triangles, which also share an edge with identical triangles of another bundle of ten triangles; 3c refers to the triangles, two in each bundle of ten, adjacent to the 3b triangles, which also share an edge with identical triangles of another bundle of ten triangles; 3d refers to the triangles, two in each bundle of ten, adjacent to the 3c triangles as well as to the three kite-quadrilaterals at the outer corners of the complex of twelve around the other end of the threefold axis from the one where the three isolated kites are; 3e refers to the remaining triangles, two in each bundle of ten, adjacent to the 3d triangles, to each other, and to the only quadrilaterals that are not kites; 4a refers to the three kite-quadrilaterals that are isolated at one end of the threefold axis; 4b refers to set of three kite-quadrilaterals at the other end of the threefold axis; 4c refers to the quadrilaterals, which are irregular but nearly kites, surrounding the 4b faces; and 4d refers to the remaining kite-quadrilaterals, each one adjacent to two different bundles of ten triangular faces.

*Gyrate, diminished, and gyrate diminished rhombicosidodecahedra: Johnson solids 72-83.*

This completes the set of twelve Johnson solids related to the rhombicosidodecahedron, and in fact completes the set of Johnson solids that can easily be grouped into sets of related polyhedra, which prompted the organization of this book into chapters. The next chapter is basically a treatment of all the remaining Johnson solids, which must be considered as individuals, since there is no easy way to classify them, as the first eighty-three were classified in this book.

# Chapter 17: Miscellany: Johnson solids 84-92.

The last nine of the Johnson solids do not seem to be easy to classify. It is clear that Norman Johnson left them for last for just that reason.

$J_{84}$ was termed the "snub disphenoid" by Johnson. It has twelve faces, and any polyhedron with twelve faces can be termed a *dodecahedron*, although the term usually refers to the Platonic solid with twelve regular pentagons as faces (regular dodecahedron or pentagon-dodecahedron), and when not used in that context, usually means the Catalan (Archimedean dual) solid with twelve rhombi as faces (rhombic dodecahedron). When it is desired to reference this solid by the term *dodecahedron*, it has been called a "Siamese dodecahedron" to distinguish it. All of its faces are triangular, but they fall into two transitivity classes. Given the $S_{4v}$ symmetry of the polygon, the fourfold alternating axis pierces two edges; the four faces adjacent to those edges form one transitivity class; the remaining eight, forming a zigzag band between them, constitute another.

*Figure 170: Johnson solid 84: the snub disphenoid, dodecadeltahedron, or Siamese dodecahedron.*

The $S_{4v}$ symmetry of this solid is rare among the Johnson solids; it has been seen in $J_{26}$ previously and will be met with again later in this chapter in the discussion of $J_{90}$, but those are the only three with $S_{4v}$ symmetry among the ninety-two Johnson solids.

Since it has only equilateral triangles as faces, it is also a member of the set of polyhedra called *deltahedra*. (See also p. 61 in Chapter 5.) It is illustrated in Figure 170. Since there are twelve faces, it has been called (e. g., in Pugh's book, *Polyhedra: a Visual Approach*) a *dodecadeltahedron*. The snub disphenoid is the fifth and last of the Johnson solids that are deltahedra, the others being $J_{12}$, $J_{13}$, $J_{17}$, and $J_{51}$.

The dual polyhedron, of course, shares the same $S_{4v}$ symmetry, but its faces include pentagons of a form described as "isosceles" in my earlier book, Polyhedra: *a New Approach*, and isosceles trapezoids. It is illustrated in Figure 171.

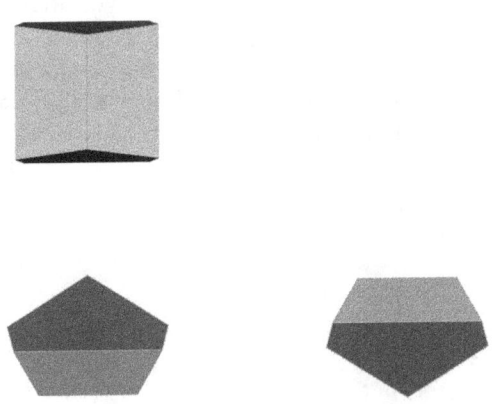

Figure 171: The dual of Johnson solid 84 (the snub disphenoid): an as-yet unnamed polyhedron with trapezoidal and pentagonal faces.

The pentagons actually have four equal angles, more regular than required for the designation "isosceles," but there is no special name to describe them. All four of the isosceles trapezoids are identical, as are all four pentagons.

Johnson gave the name "snub square antiprism" to his

solid 85, illustrated in Figure 172 below. The name appears somewhat inappropriate, as it is not obtained by performing the same operation on a square antiprism that generates a snub cube from a cube, or a snub dodecahedron from a regular dodecahedron. However, since this was the name Johnson chose to give it, this is the name by which it is generally known. The solid can, in fact, best be thought of as a generalization of a regular icosahedron, with the threefold axis of the icosahedron replaced by a fourfold axis, thus converting two opposite triangular faces to square ones.

One can consider the regular icosahedron and the snub square antiprism ($J_{85}$) as the triangular and square members of a family (see my *Encyclopedia of Polyhedron Families*). In fact, the *n* = 2 member of this family is the snub disphenoid just described. In the same way as Johnson's list only includes the pentagonal member of the family of rotundae (and all families based on the rotunda like elongated and gyroelongated rotundae, birotundae, and cupolarotundae), it only includes this one member of this family (at least by name; two members if you include $J_{84}$ as the digonal member), but (as is covered in more detail in my planned book, *Encyclopedia of Polyhedron Families*) a complete family which, in accordance with Johnson's naming, would be called the family of snub antiprisms, can be constructed.

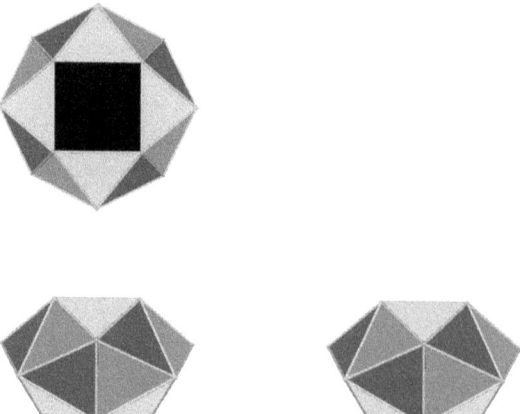

*Figure 172: Johnson solid 85, named by him the snub square antiprism.*

## Miscellany: Johnson solids 84-92.

It was noted that Johnson solid 26, although named "gyrobifastigium" rather than "digonal gyrobicupola" by Johnson, was placed in the appropriate place in the sequence before the other bicupolae. The placement of Johnson solids 84 and 85 is a sign of similar thought; $J_{84}$ was not named "snub digonal antiprism" (or even "snub tetrahedron," as a digonal antiprism is a tetrahedron) to show that it is the $n = 2$ member of the family whose $n = 3$ member is the regular icosahedron and whose $n = 4$ member is $J_{85}$, but placing it immediately before $J_{85}$ in the sequence points to the likelihood that Johnson recognized that this family existed.

As the snub antiprisms are generalizations of the regular icosahedron, clearly the duals would be generalizations of the dual of the icosahedron, namely the regular dodecahedron, and in particular, the dual of $J_{85}$ is a figure obtained by replacing one threefold axis of the dodecahedron by a fourfold axis, as illustrated in Figure 173 below. While the family to which the regular dodecahedron and this solid both belong had not been named, to the best of my knowledge, prior to my earlier book, *Polyhedra: a New Approach*, in that book I named these polyhedra *pentagonized globoids*. They have the property of having only pentagons as faces, and if the principal axis of rotational symmetry is *n*-fold, they have 4*n* faces (falling into two sets of 2*n* congruent isosceles pentagons, though except for the special case of the regular dodecahedron, the two sets are not congruent to each other). When discussing these two sets of pentagons, one can refer to the *n* faces meeting at each of the two apices as *apical pentagons* and the remaining 2*n* faces as *lateral pentagons*. In Table 35, the symbols 5a and 5L will be used to distinguish them in the portion of the table referring to the dual of Johnson solid 85. (It might be noted, in Table 35, that the lateral pentagons have *four* equal angles, not just two sets of two, which would be required by the symmetry of an isosceles pentagon.) As was just noted, all the apical pentagons are congruent to each other, and all the lateral pentagons as well, though as can be seen in Table 35, the lateral and apical pentagons are distinctly different.

Like the family of snub antiprisms, the family of pentagonized globoids is discussed in greater detail in my *Encyclopedia of Polyhedron Families*, planned to come out shortly after this book.

In discussing the dual of Johnson solid 85, it is a little difficult to come up with terminology to allow us to refer to specific sets of faces. All are pentagons — in fact all are *isosceles* pentagons, in the terminology of my earlier book, *Polyhedra: a New Approach* — so just referring to them by shape is useless. The apical/lateral distinction mentioned above helps, but in addition, it is often necessary to add one more qualifier. One set of four apical pentagons can easily be seen to be "upper" if the polyhedron is oriented as in Figure 173, and the remaining four apical pentagons are consequently "lower," and the terms "upper" and "lower" can be extended to the lateral pentagons as well. Each lateral pentagon has two long shared edges (each of length $\sim$0.4979) with the apical pentagons, and one much shorter edge (of length $\sim$0.2056) shared with an apical pentagon from the other side. It is natural to describe as "upper" the lateral pentagons whose longer shared edges are with upper apical pentagons and whose shorter shared edge is with a lower apical pentagon, and analogously for the term "lower." We then also can refer to those longer edges as "same-side" apical/lateral edges, and the shorter ones as "opposite-side" apical/lateral edges.

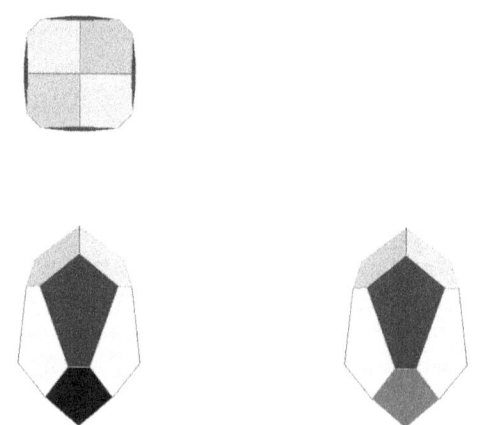

Figure 173: The dual of Johnson solid 85 (the snub square antiprism): the square pentagonized globoid.

Johnson solid 86, the sphenocorona, can be derived from the icosahedron. The bottom view of Figure 174, in fact, is similar enough to what one would see viewing an icosahedron from a point along one of its twofold axes that it would take careful measurements to distinguish them. The top view looks much like a cube viewed from along its twofold axis, except that the leftmost and rightmost vertices are pulled apart from where they would be in a cube (directly below the two vertices in the center of the top view). If one assumes that the edges of the icosahedron and the sphenocorona are equal and represented by $e$, the two vertices at the leftmost and rightmost points of the bottom view would be approximately $0.789428e$ to the left and right of the vertical mirror plane (that is, vertical in the bottom view of the sphenocorona), while for an icosahedron they would be approximately $0.809017e$ (or more exactly $\tau e/2$, where $\tau$ represents the golden ratio) to the left and right of the corresponding mirror plane; the four vertices at the corners of the bottom view of Figure 174 would be $0.852727e$ above and below the horizontal mirror plane (that is, horizontal in the bottom view of the sphenocorona), while for an icosahedron they would be the same distance given before of approximately $0.809017e$, or more exactly $\tau e/2$, above and below the corresponding mirror plane. The difference comes from the fact that in an icosahedron there are four more vertices directly below the four nearest to the horizontal mirror plane, but in the sphenocorona there are only two unseen vertices, the two shown in the center of the top view, which must be $0.5e$ to the left and right of the vertical mirror plane, and they have the effect of pulling together the leftmost and rightmost vertices, while that in turn separates the four vertices in the corners in order to keep all the triangular faces equilateral.

Explanations of the name "sphenocorona" are hard to come by, but the term "spheno-" is supposed to refer to the pair of squares shown in the top view of Figure 174, while at least one site explains a "corona" as a polygon surrounded by triangles. If one surrounds the *pair* of squares by triangles, they can be packed so as to completely join up, needing no more faces to make a polyhedron, and the result is the polyhedron that Johnson named the sphenocorona. So this may be the source of the name.

*The Johnson Solids and Their Duals*

In effect, one is attempting to combine six of the eight vertices of a cube with eight of the twelve vertices of a regular icosahedron; they do not fit exactly, but by distorting the icosahedron and the cube, they can be made to fit. Although this book has not, in general, tried to present detailed constructions of the Johnson solids, the combination of the unfamiliarity of the sphenocorona to most readers and the special nature of this construction (combining "almost-fitting" pieces of two polyhedra) seems to justify the inclusion of this one here, so it is given now.

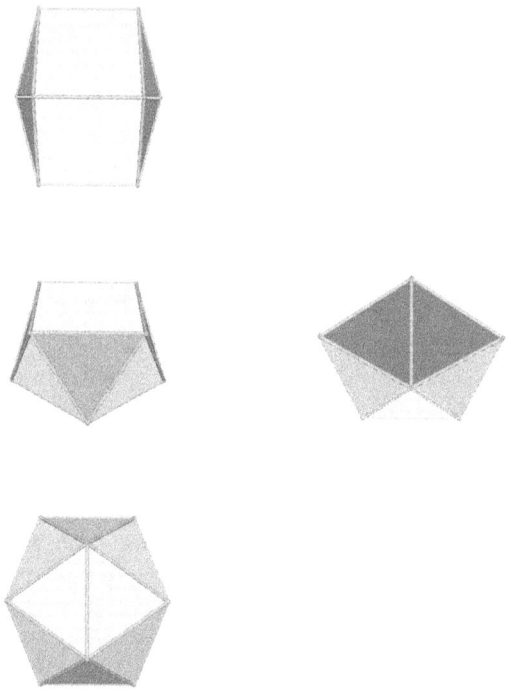

Figure 174: Johnson solid 86: the sphenocorona.

Suppose one takes a cube, with edge 1, and orients it so that the z-axis is along a twofold axis of the cube. If this is done, the eight vertices of the cube fall into three sets, each in a plane perpendicular to the z-axis:

$$\mathbf{A}_{1\text{-}2}\colon (\pm 1/2,\ 0,\ 1/2\sqrt{2}),$$

## Miscellany: Johnson solids 84-92.

$$B_{1\text{-}4}: (\pm\tfrac{1}{2}, \pm\tfrac{1}{2}\sqrt{2}, 0),$$

$$C_{1\text{-}2}: (\pm\tfrac{1}{2}, 0, -\tfrac{1}{2}\sqrt{2}),$$

or, in decimal form, approximately:

$$A_{1\text{-}2}: (\pm 0.5, 0, 0.70710678),$$

$$B_{1\text{-}4}: (\pm 0.5, \pm 0.70710678, 0),$$

$$C_{1\text{-}2}: (\pm 0.5, 0, -0.70710678).$$

If an icosahedron is similarly oriented, the twelve vertices fall into five similar sets:

$$A_{1\text{-}2}: (0, \pm\tfrac{1}{2}, \tau/2),$$

$$B_{1\text{-}2}: (\pm\tau/2, 0, \tfrac{1}{2}),$$

$$C_{1\text{-}4}: (\pm\tfrac{1}{2}, \pm\tau/2, 0),$$

$$D_{1\text{-}2}: (\pm\tau/2, 0, -\tfrac{1}{2}),$$

$$E_{1\text{-}2}: (0, \pm\tfrac{1}{2}, -\tau/2),$$

where $\tau$ represents the golden ratio and is equivalent to $\tfrac{1}{2}(1 + \sqrt{5})$; or, in decimal form, approximately:

$$A_{1\text{-}2}: (0, \pm 0.5, 0.80901699),$$

$$B_{1\text{-}2}: (\pm 0.80901699, 0, 0.5),$$

$$C_{1\text{-}4}: (\pm 0.5, \pm 0.80901699, 0),$$

$$D_{1\text{-}2}: (\pm 0.80901699, 0, -0.5),$$

$$E_{1\text{-}2}: (0, \pm 0.5, -0.80901699).$$

The B vertices of the cube form a rectangle that is slightly smaller than the rectangle formed by the C vertices of the icosahedron, and if a compromise arrangement were made including the A vertices of the cube, a rectangle halfway between the two rectangles in question, and the D and E vertices of the icosahedron, a figure with ten vertices would be encountered:

$$A_{1\text{-}2} \text{ (from } A_{1\text{-}2} \text{ of the cube)}: (\pm 0.5, 0, 0.70710678),$$

$B_{1\text{-}4}$ (from $B_{1\text{-}4}$ of the cube and $C_{1\text{-}4}$ of the icosahedron): ($\pm 0.5$, $\pm 0.75806189$, 0),

$C_{1\text{-}2}$ (from $D_{1\text{-}2}$ of the icosahedron): ($\pm 0.80901699$, 0, $-0.5$),

$D_{1\text{-}2}$ (from $E_{1\text{-}2}$ of the icosahedron): (0, $\pm 0.5$, $-0.80901699$).

It would, in fact, require some adjustment to make the polygons all regular, creating a Johnson solid. The resulting polyhedron has coordinates:

$A_{1\text{-}2}$: ($\pm 0.5$, 0, $0.52235693$),

$B_{1\text{-}4}$: ($\pm 0.5$, $\pm 0.85272694$, 0),

$C_{1\text{-}2}$: ($\pm 0.78942763$, 0, $-0.43484297$),

$D_{1\text{-}2}$: (0, $\pm 0.5$, $-0.79093850$).

The cubical faces $A_1A_2B_2B_1$ and $A_1A_2B_3B_4$ become the faces $A_1A_2B_2B_1$ and $A_1A_2B_3B_4$ of the new polyhedron, and the icosahedral faces $D_1E_1E_2$, $D_2E_1E_1$, $C_1D_1E_1$, $C_2D_2E_1$, $C_3D_2E_2$, $C_4D_1E_2$, $C_1C_2E_1$, and $C_3C_4E_2$ become the faces $C_1D_1D_2$, $C_2D_1D_2$, $B_1C_1D_1$, $B_2C_2D_1$, $B_3C_2D_2$, $B_4C_1D_2$, $B_1B_2D_1$, and $B_3B_4D_2$ of the new polyhedron. There are two cubical faces $A_1B_1C_1B_4$ and $A_2B_2C_2B_3$, which break up. The cubical vertices $C_1$ and $C_2$ are no longer present, but the vertices $C_1$ and $C_2$ of the new polyhedron are in approximately the same area, however the sets $A_1B_1C_1B_4$ and $A_2B_2C_2B_3$ would not be planar; they are replaced by four triangular faces $A_1B_1C_1$, $A_1C_1B_4$, $A_2B_2C_2$, and $A_2C_2B_3$. This leaves a total of fourteen faces. (The final four faces can also be derived from four of the icosahedral faces, $B_1C_1D_1$, $B_1C_4D_1$, $B_1C_2D_2$, and $B_1C_3D_2$, by noting that the A vertices of the cube approximately replace the B vertices of the icosahedron.)

Through all these operations, the twofold axis along the z-axis remained a twofold axis, though of course the other twofold axes (three others for the cube, fifteen others for the icosahedron) were destroyed. The mirror planes through that twofold axis (the xz- and yz-planes) also remained, though all other mirror planes also were destroyed, so the resulting symmetry is $C_{2v}$. As was mentioned, there are two square faces ($A_1A_2B_2B_1$ and $A_1A_2B_3B_4$) and twelve triangular faces, which divide into four transitivity classes: $A_1B_1C_1$, $A_1C_1B_4$, $A_2B_2C_2$, and $A_2C_2B_3$ form one transitivity class; $B_1B_2D_1$, and $B_3B_4D_2$ form the second; $B_1C_1D_1$,

## Miscellany: Johnson solids 84-92.

$B_2C_2D_1$, $B_3C_2D_2$, $B_4C_1D_2$ form the third; and $C_1D_1D_2$ and $C_2D_1D_2$ the fourth.

Of course, since all faces of a Johnson solid are regular polygons, it is always the case that all edges of the polyhedron are equal, and that all faces of the same number of sides are congruent polygons. Thus, all twelve of the triangles are equilateral triangles which are identical in size. However, under the $C_{2v}$ symmetry of the whole polyhedron, the triangles fall into four distinct transitivity classes:

1. Two that appear in the center of the bottom view, which share an edge that is bisected by the twofold axis. These can be termed the *bottom triangles*;

2. Four that "enclose" the bottom triangles. Two of these appear in the front view as well as in the bottom view; two more, visible in the bottom view, are in back; similarly, two of these appear in the right side view, while two more are invisible from the right side, as they are on the left side of the polyhedron (using the terms "left" and "right" to denote *your own* left and right as you look at the polyhedron); these can be termed the *lower foursome*;

3. Another set of four that are so oriented that the two rightmost are seen most clearly in the right side view; two more are in corresponding positions on the left side. There are two of them just barely visible in the front view; all four can be seen in the top view, a little more clearly than in the front view, but still only barely. They are the triangles that share a vertex with *both* of the square faces, though each has an edge in common with only one square face; these can be termed the *upper foursome*;

4. The remaining two (one of which is visible almost head-on in the lower half of the front view; the other is in a corresponding position in back) share one edge with the square and the remaining two with triangles in the lower foursome. They are bisected by the same mirror plane that cuts both square faces, and are referred to in this discussion as the *lower front* and *lower back* face.

One mirror plane cuts through the centers of both square faces, the lower front, and the lower back face, and includes the edge between the two bottom faces. It appears in Figure 174 as a vertical line in the top, front, and bottom views. The other mirror plane, perpendicular to the first, includes the edge between the two square faces, as well as the edge between front and back triangles in the upper foursome, and passes through the centers of the two bottom triangles. It appears as a vertical line in the side view and a horizontal line in the top and bottom views.

The dual of the sphenocorona is illustrated in Figure 175 below. The faces fall into four transitivity classes: two classes of quadrilateral and two classes of pentagon.

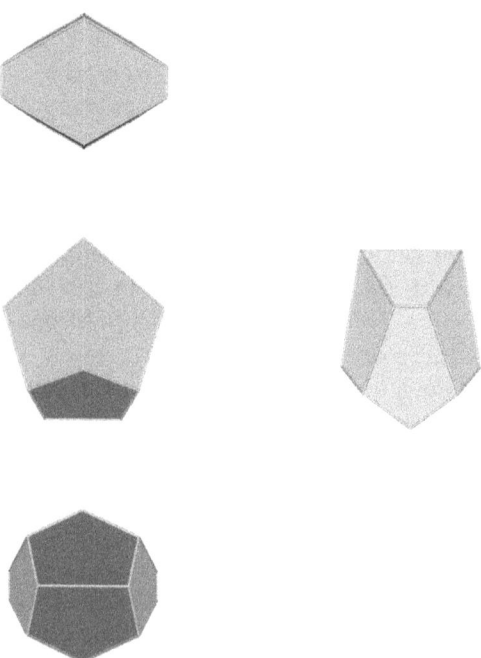

Figure 175: The dual of Johnson solid 86 (the sphenocorona).

Each of these classes has only two faces, except that there are four irregular quadrilaterals of the type designated 4a in

Table 35. This makes a total of ten faces.

Since every polyhedron and its dual have the same symmetry, the dual of the sphenocorona has the same $C_{2v}$ symmetry as the sphenocorona itself, with two mirror planes and a twofold axis where the mirror planes intersect. One of those mirror planes is viewed as a vertical line through the top, front, and bottom views; the other runs horizontally in the top and bottom views and vertically in the side view of Figure 175.

If one looks at Table 35 below, it is interesting to note that the quadrilaterals designated 4a form dihedral angles of 114.1351° (114° 8′ 6″), 114.4569° (114° 27′ 25″), and 114.8378° (114° 50′ 16″) with the three other neighboring faces, almost equal. (The exception is the dihedral angle between the two 4a faces themselves; at 119.8984° [119° 53′ 54″] it is significantly different.) All the other three types of face are bisected by one or the other of the two mirror planes of the polyhedron, so both types of pentagonal face are isosceles (defined as in my earlier book, *Polyhedra: a New Approach*) and the 4i faces are isosceles trapezoids.

There are a lot of other angles that are nearly equal in the tabulation of the properties of the dual of the sphenocorona which is given in Table 35. This may be related to the fact that the sphenocorona itself is almost, but not exactly, a fit of parts of two Platonic solids together, as was described earlier.

## The Johnson Solids and Their Duals

| Johnson solid number | 84 | 85 | 86 |
|---|---|---|---|
| Name | Snub disphenoid | Snub square antiprism | Sphenocorona |
| Symmetry* | $S_{4v}$ | $S_{8v}$ | $C_{2v}$ |
| Faces | [4+8](3) | [8+16](3) + [2](4) | [2+2+4+4](3) + [2](4) |
| Edges* | 18 | 40 | 22 |
| Vertices | 8 | 16 | 10 |
| Dihedral angles | 3a-3a: 96.1983° (96° 11′ 54″), 3a-3b: 121.7432° (121° 44′ 35″), 3b-3b: 166.4406° (166° 26′ 26″)♦ | 3a-3b: 144.1436° (144° 8′ 37″), 3b-3b: 164.2574° (164° 15′ 27″), 3b-3c: 114.6452° (114° 38′ 43″), 3a-4: 145.4406° (145° 26′ 26″)△ | 3a-3a: 159.8924° (159° 53′ 33″), 3a-3c: 118.8922° (118° 53′ 32″), 3b-3b: 131.4416° (131° 26′ 30″), 3b-3c: 143.4787° (143° 28′ 43″), 3c-3d: 135.9915° (135° 59′ 29″), 3a-4: 109.5240° (109° 31′ 27″), 3d-4: 97.4555° (97° 27′ 20″), 4-4: 117.0190° (117° 1′ 9″)‡ |
| Area | 5.19615242 = $3\sqrt{3}$ | 12.39230485 = $2 + 6\sqrt{3}$ | 7.19615242 = $2 + 3\sqrt{3}$ |
| Volume | 0.85949365 | 3.60122201 | 1.51535164 |
| Dual |||| 
| Name | Unnamed | Square pentagonized globoid※ | Unnamed |
| Faces | [4](4) + [4](5) | [8+8](5) | [2+4](4) + [2+2](5) |
| Vertices | 12 | 26 | 14 |

— 378 —

Miscellany: Johnson solids 84-92.

| Johnson solid number | 84 | 85 | 86 |
|---|---|---|---|
| Face angles[†] | 4: [2]80.2690° (80° 16′ 8″), [2]99.7310° (99° 43′ 52″), 5: [4]106.9003° (106° 54′ 1″), 112.3986° (112° 23′ 55″) | 5a: [2]117.1808° (117° 10′ 51″), [2]118.1556° (118° 9′ 20″), 69.3271° (69° 19′ 38″); 5L: [4]110.0571° (110° 3′ 26″), 99.7715° (99° 46′ 17″)[#] | 4a: 59.9317° (59° 55′ 54″), 97.2870° (97° 17′ 13″), 97.7577° (97° 45′ 28″), 105.0236° (105° 1′ 25″); 4i: [2]69.7547° (69° 45′ 17″), [2]110.2453° (110° 14′ 34″); 5a: [2]109.6064° (109° 36′ 23″), [2]109.7651° (109° 45′ 55″), 101.2569° (101° 15′ 25″); 5b: [2]102.7968° (102° 47′ 49″), [2]110.4836° (110° 29′ 1″), 113.4393° (113° 26′ 21″)[*] |

| Johnson solid number | 84 | 85 | 86 |
|---|---|---|---|
| Dihedral angles | 4-4: 114.9396° (114° 56′ 23″), 4-5: 104.8424° (104° 50′ 33″), 5-5: 84.7037° (84° 42′ 12″) | 5a-5a: 118.5654° (118° 33′ 55″), 5a-5L: 127.5535° (127° 33′ 12″)# | 4a-4a: 119.8984° (119° 53′ 54″), 4a-4i: 114.4569° (114° 27′ 25″), 4i-4i: 105.9601° (105° 57′ 36″), 4a-5a: 114.8378° (114° 50′ 16″), 4a-5b: 114.1351° (114° 8′ 6″), 4i-5a: 106.5835° (106° 35′ 0″), 5a-5b: 106.0693° (106° 4′ 9″), 5b-5b: 104.8844° (104° 53′ 4″)* |
| Area | 1.44634184 | 6.78898855 | 4.10264058 |
| Volume | 0.12370036 | 1.44567217 | 0.64354985 |

*Table 35: Properties of Johnson solids 84, 85, and 86 and their duals.*

\*The symmetry and number of edges are always the same for each solid and its dual, so are not repeated in the "dual" section.

†Face angles are only given for the duals, as the Johnson solids have regular faces, whose angles can be found in Table 1 on p. 3.

#The designation 3a refers to the four triangles which include the two edges whose midpoints determine the twofold axis; the designation 3b refers to the remaining eight triangles.

△The designation 3a refers to those triangular faces that are adjacent to a square face; 3b and 3c are both used for the other triangular faces, with the 3b/3c distinction only used when it is necessary to distinguish faces on the same side from faces on opposite sides of the solid (*i. e.*, if one considers those 3a-faces adjacent to the *same* square face as defining one side, two 3b-faces which are adjacent to 3a-faces on the same side are considered to be on the same side themselves).

*The designation 3a refers to the four triangular faces that are adjacent to both the square faces and to another triangular face of the same type; the designation 3b refers to the two triangular faces that are opposite the square faces; the designation 3c refers to the four triangular faces that share edges with the 3b-faces (as well as with the 3a-faces); and the designation 3d refers to the two remaining faces, which are adjacent each to *one* square face and *two* 3c-faces. In this case, because of the five different 3-3 dihedral angles and two different 3-4 dihedral angles, each transitivity class needs a separate designation, unlike the cases in nearly all other listings in these tables.

%Name introduced in my earlier book, *Polyhedra: a New Approach*.

#The dihedral angles are the same between apical and lateral pentagonal faces, regardless of whether they are same-side or opposite-side.

°The designation 4a to the four irregular quadrilaterals, two in front and two in back in the orientation of Figure 175; the designation 4i to the two isosceles trapezoidal faces at the top of the polyhedron; 5a to the two isosceles pentagons that share an edge with the isosceles trapezoidal faces; 5b to the two isosceles pentagons at the front and back, which have no common edges with the isosceles trapezoidal faces.

Johnson solid 87, the augmented sphenocorona, is simply obtained, as its name suggests, by fusing a square pyramid ($J_1$) to one of the square faces of the sphenocorona ($J_{86}$) just described. This destroys the $C_{2v}$ symmetry, and leaves only one mirror plane as the sole symmetry element, so the symmetry is $C_s$. It is illustrated in Figure 176 below.

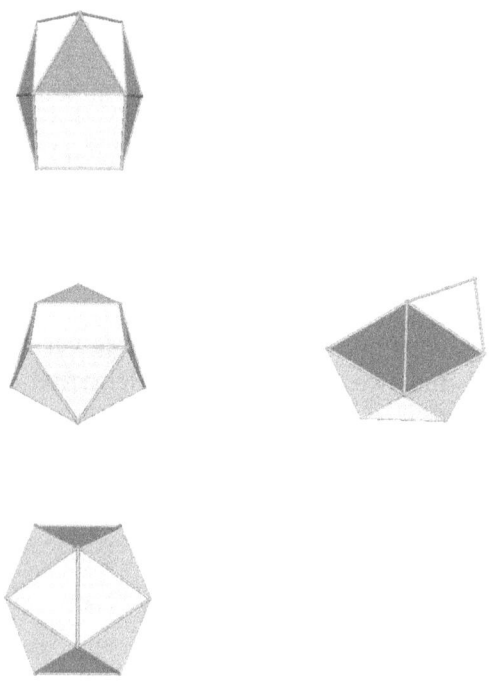

*Figure 176: Johnson solid 87: the augmented sphenocorona*

The dual of the augmented sphenocorona is illustrated in Figure 177 below. Like the augmented sphenocorona itself, this polyhedron has symmetry $C_s$. With such an asymmetrical arrangement, it is notable that there are eleven faces, but they divide into *seven* different transitivity classes! Three of the faces (one quadrilateral and two pentagonal faces) form single-member transitivity classes, and the other eight faces are paired, forming four more transitivity classes. As there is only one mirror plane, which passes through the three faces described as forming single-member transitivity classes, those three are required by the overall $C_s$ symmetry of the polyhedron to have mirror lines, so the one quadrilateral in question is a kite-quadrilateral, and the two pentagonal faces are isosceles pentagons (using the definition in my earlier book, *Polyhedra: a New Approach*).

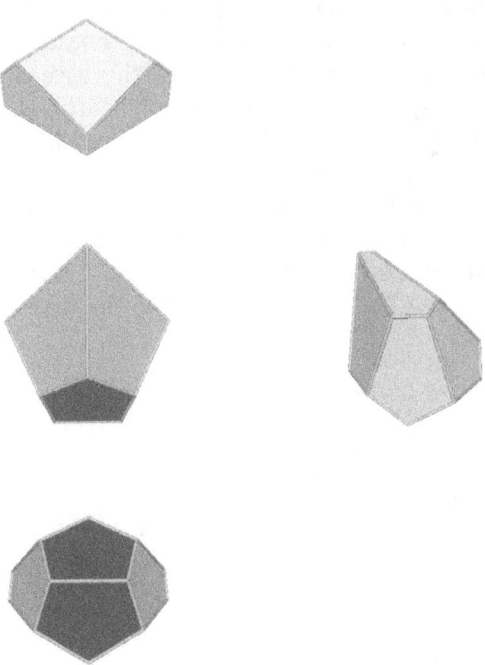

Figure 177: *The dual of Johnson solid 87 (the augmented sphenocorona).*

In the views of Figure 177, the two pentagons at the center of the bottom view (which are *not* congruent) are the isosceles pentagons, and the sole kite-quadrilateral is only visible in the top view, as it is in the back of the polyhedron as it is oriented in the figure.

As can be noted in Table 36 below, one of the two isosceles pentagons (the one designated 5a in the table) is very close to regular, with no angle differing by more than 1° from the 108° of a regular pentagon. The sides, too, are nearly equal, although not quite as close as the angles, with two of ~0.3647, two of ~0.3085, and one of ~0.4040.

Johnson solid 88, the sphenomegacorona, is also a derivative of the sphenocorona ($J_{86}$) described earlier in this chapter. In looking at Figure 174 (particularly the side view), one can see that the triangles of the upper foursome (as defined on p. 374) form two pairs of adjacent triangles. Replacing each pair by a set of four (resembling a pyramid with a rhombic base, but in fact the four vertices that would form the base of the "pyramid" are not truly in a plane), a new polyhedron is obtained. It has eighteen faces, sixteen of which are equilateral triangles and the remaining two, squares. Its symmetry is $C_{2v}$, and it can best described by looking the top view of Figure 178 below. There are two mirror planes, which intersect along the twofold axis. One mirror plane contains the edge between the two square faces; the other is perpendicular to the first one.

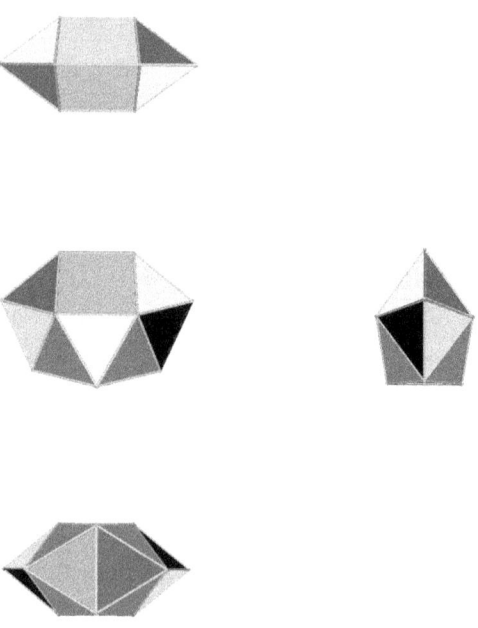

*Figure 178: Johnson solid 88: the sphenomegacorona.*

Two vertices, the leftmost and rightmost in all views except the side view, can be thought of as octahedral vertices, where four equilateral triangles come together. At all the remaining vertices that do not involve the square faces, five equilateral triangles come together, as in an icosahedron. Altogether, there are four kinds of vertex, which can be described as $3^4$ (two), $3^5$ (four), $3^2 \cdot 4^2$ (two), and $3^4 \cdot 4$ (four), accounting for the twelve vertices.

The dual of the sphenomegacorona is illustrated in Figure 179 below. All of its faces are either quadrilateral or pentagonal, and because the $C_{2v}$ symmetry provides for two perpendicular mirror planes, which between them bisect all the quadrilaterals and half of the eight pentagons, all the quadrilaterals are isosceles trapezoids and half the pentagons are isosceles (as defined in my earlier book, *Polyhedra: a New Approach*).

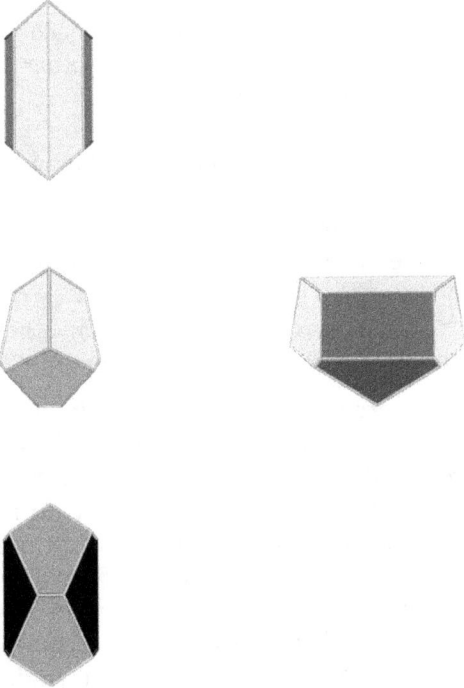

Figure 179: The dual of Johnson solid 88 (the sphenomegacorona).

The only completely irregular pentagons are the two at

the top of the front view in Figure 179 and two corresponding ones in the back (or, stating it a little differently, all four of the pentagonal faces visible in the bottom view are isosceles, and the four that are not are the irregular ones). The two isosceles trapezoidal faces that take up most of the top view are actually close to 60°-60°-120°-120° isosceles trapezoids such as one obtains by cutting off part of an equilateral triangle by a line parallel to a side, but not precisely; and the other two isosceles trapezoids are nearly rectangular, with none of their angles differing by as much as one degree from a right angle (see Table 36), but again not precisely.

The bottom view shows four isosceles pentagons, two of which share an edge and the other two completing the vertices at the ends of that edge. This somewhat resembles the configuration found in part of a regular dodecahedron (though there the four pentagons would be regular, not merely isosceles).

Johnson solid 89, the hebesphenomegacorona, is another singular shape, not clearly related to any other of the Johnson solids. One might, by its name, see it as related to $J_{88}$, the sphenomegacorona, and there are some resemblances: its symmetry is $C_{2v}$, as is the symmetry of the sphenomegacorona, and most of its faces are triangular. All together, eighteen of its twenty-one faces are equilateral triangles, and three are squares, and one of the three square faces is different from the other two in that the twofold axis passes through its center. The other two squares are related to each other by the symmetries of the solid. A picture of $J_{89}$ is in Figure 180 below.

This polyhedron is actually related closely to the regular icosahedron. If one looks at the bottom view of Figure 180, an icosahedron would look identical to it when viewed from both top and bottom. Suppose that one starts with an icosahedron and considers that the top view has a vertical edge like that one in the center of the bottom view of Figure 180. Now split the two vertices at the ends of that edge into two pairs, and move them apart, so that the single edge becomes a square. The two triangles corresponding to the ones shown in light color in the bottom view of Figure 180 would each have one vertex split, and when the split parts have reached a distance sufficient to equal all the other edges of the polyhedron, those triangles will become squares, producing a configuration such as the top view of Figure 180. (One might think that there are enough edges in an icosahedron that this process could be repeated, but apparently it introduces enough distortion that, within the Johnson requirement that all faces be regular, no additional applications of the process are possible.)

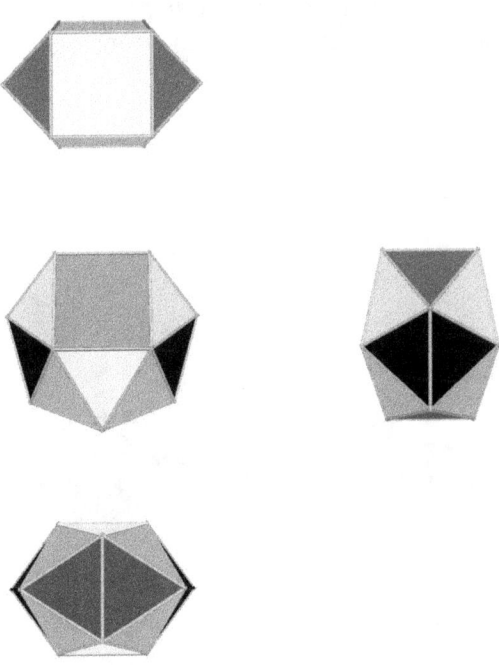

*Figure 180: Johnson solid 89: the hebesphenomegacorona.*

Figure 181 shows the dual of the hebesphenomegacorona. It can best be constructed with reference to the procedure, earlier described, where the hebesphenomegacorona is generated by splitting two adjacent vertices of a regular icosahedron, making the edge that joins them into a square, and making two of the triangular faces (one at each of the two vertices that were split) into squares. The four irregular quadrilateral faces seen in the top view of Figure 181 similarly derive from two pentagonal faces of a regular dodecahedron, obtained by splitting them along new edges created in this process (appearing as vertical lines in that top view). The edge which separated those two pentagonal faces is thus split into two, which meet the two new edges in a new vertex (at the exact center of the top view). As the vertices of the dodecahedron are all trivalent, the additional edges create two new quadrivalent vertices in addition to the one in the center of the top view of Figure 181, each where one of the new edges meets three of the edges of the original dodecahedron. (These new vertices are at the top and bottom of the top view of Figure 181, and one of them also appears in the approximate center of the front view.)

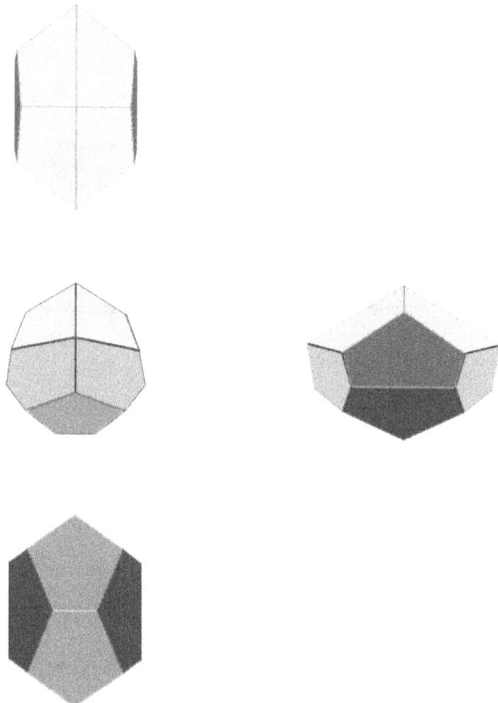

Figure 181: The dual of Johnson solid 89 (the hebesphenomegacorona).

The bottom view of Figure 181 resembles the bottom view of Figure 179, with four isosceles pentagons, two of which share an edge (which is also in a mirror plane of the polyhedron), and the other two completing the vertices at the ends of that edge. Two other pentagons are also isosceles as a result of the mirror plane just mentioned, which bisects them, but the remaining four pentagonal faces, cut by neither of the two mirror planes of the polyhedron, are totally irregular.

## The Johnson Solids and Their Duals

| Johnson solid number | 87 | 88 | 89 |
|---|---|---|---|
| Name | Augmented sphenocorona | Sphenomegacorona | Hebesphenomegacorona |
| Symmetry* | $C_s$ | $C_{2v}$ | $C_{2v}$ |
| Faces | [1+1+1+1+2+2+2+2+2+2](3) + [1](4) | [2+2+4+4+4](3) + [2](4) | [2+2+2+4+4+4](3) + [1+2](4) |
| Edges* | 26 | 28 | 33 |
| Vertices | 11 | 12 | 14 |
| Dihedral angles | 3a-3a: 131.4416° (131° 26′ 30″), 3a-3b: 143.4787° (143° 28′ 43″), 3b-3c: 135.9915° (135° 59′ 29″), 3b-3d: 118.8922° (118° 53′ 32″), 3c-3e: 152.1911° (152° 11′ 28″), 3d-3d: 159.8924° (159° 53′ 33″), 3d-3f: 164.2596° (164° 15′ 35″), 3e-3f: 109.4712° (109° 28′ 16″), 3c-4: 97.4555° (97° 27′ 20″), 3d-4: 109.5240° (109° 31′ 27″), 3e-4: 171.7546° (171° 45′ 17″)✦ | 3a-3a: 86.7268° (86° 43′ 37″), 3a-3d: 129.4446° (129° 26′ 40″), 3b-3c: 143.7383° (143° 44′ 18″), 3c-3d: 171.6457° (171° 38′ 45″), 3c-3e: 117.3556° (117° 21′ 20″), 3d-3d: 86.7268° (86° 43′ 37″), 3e-3e: 161.4828° (161° 28′ 58″), 3a-4: 154.7223° (154° 43′ 20″), 3b-4: 137.2401° (137° 14′ 24″), 4-4: 72.9730° (72° 58′ 23″)⊞ | 3a-3a: 111.7348° (111° 44′ 5″), 3a-3d: 157.1481° (157° 8′ 53″), 3b-3b: 149.5648° (149° 33′ 53″), 3b-3d: 128.4960° (128° 29′ 46″), 3c-3d: 141.3411° (141° 20′ 28″), 3d-3d: 128.4960° (128° 29′ 46″), 3c-4: 152.9756° (152° 58′ 32″), 3d-4: 133.9728° (133° 58′ 22″), 4-4: 102.5238° (102° 31′ 26″)⊛ |
| Area | 7.92820323 = 1 + 4√3 | 8.92820323 = 2 + 4√3 | 10.79422863 = 3 + (9/2)√3 |
| Volume | 1.75105390 | 1.94810823 | 2.91291041 |
| Dual | | | |
| Name | Unnamed | Unnamed | Unnamed |
| Faces | [1+2](4) + [1+1+2+2](5) | [2+2](4) + [2+2+4](5) | [4](4) + [2+2+4](5) |
| Vertices | 16 | 16 | 18 |

— 390 —

Miscellany: Johnson solids 84-92.

| Johnson solid number | 87 | 88 | 89 |
|---|---|---|---|
| Face angles† | See list below^Δ# | 4a: [2]89.0295° (89° 1' 46"), [2]90.9705° (90° 58' 14"); 4b: [2]56.7849° (56° 47' 6"), [2]123.2151° (123° 12' 54"); 5a: [2]110.4140° (110° 24' 50"), [2]111.7769° (111° 46' 37"), 95.6183° (95° 37' 6"); 5b: [2]96.8730° (96° 52' 23"), [2]122.2086° (122° 12' 31"), 101.8368° (101° 50' 12"); 5c: 71.9827° (71° 58' 58"), 105.7267° (105° 43' 36"), 114.5801° (114° 34' 48"), 123.5319° (123° 31' 55"), 124.1786° (124° 10' 43")* | 4: 64.0914° (64° 5' 29"), 75.2084° (75° 12' 30"), 108.4731° (108° 28' 23"), 112.2271° (112° 13' 37"); 5a: [2]106.3403° (106° 20' 25"), [2]108.3607° (108° 21' 38"), 110.5980° (110° 35' 53"); 5b: [2]108.7352° (108° 44' 7"), [2]112.0560° (112° 3' 22"), 98.4175° (98° 25' 3"); 5c: [2]107.0683° (107° 4' 6"), [2]107.2200° (107° 13' 12"), 111.4233° (111° 25' 24"); 5d: 76.9797° (76° 58' 47"), 113.2503° (113° 15' 1"), 116.1906° (116° 11' 26"), 116.7689° (116° 46' 8"), 116.8105° (116° 48' 38")* |

## The Johnson Solids and Their Duals

| Johnson solid number | 87 | 88 | 89 |
|---|---|---|---|
| Dihedral angles | 4a-4a: 123.2854° (123° 17′ 7″), 4a-5b: 119.9860° (119° 59′ 10″), 4a-5c: 118.4880° (118° 29′ 17″), 4a-5e: 115.4496° (115° 26′ 59″), 4k-5d: 124.3093° (124° 18′ 33″), 4k-5e: 121.9099° (121° 54′ 35″), 5a-5b: 112.3696° (112° 22′ 11″), 5a-5c: 109.7462° (109° 44′ 46″), 5a-5d: 111.9617° (111° 57′ 42″), 5b-5c: 113.3503° (113° 21′ 1″), 5c-5d: 112.9668° (112° 58′ 0″), 5c-5e: 105.4850° (105° 29′ 6″), 5d-5d: 114.8079° (114° 48′ 28″), 5d-5e: 108.2500° (108° 15′ 0″), 5e-5e: 98.1052° (98° 6′ 19″)# | 4a-4b: 131.3369° (131° 20′ 13″), 4b-4b: 122.0725° (122° 4′ 21″), 4a-5b: 131.5776° (131° 34′ 39″), 4a-5c: 130.5969° (130° 35′ 49″), 4b-5c: 114.8436° (114° 50′ 37″), 5a-5a: 113.4199° (113° 25′ 12″), 5a-5b: 119.4664° (119° 27′ 59″), 5a-5c: 108.0985° (108° 5′ 54″), 5b-5c: 116.1977° (116° 11′ 52″), 5c-5c: 100.6586° (100° 39′ 31″)* | 4-4: 126.3873° (126° 23′ 14″), 4-5c: 126.8206° (126° 49′ 14″), 4-5d: 121.0060° (121° 0′ 22″), 5a-5b: 125.8108° (125° 48′ 35″), 5a-5c: 126.8986° (126° 53′ 55″), 5a-5d: 121.1458° (121° 8′ 45″), 5b-5b: 125.0100° (125° 0′ 36″), 5b-5d: 119.8609° (119° 51′ 39″), 5c-5d: 121.7444° (121° 44′ 40″), 5d-5d: 110.9773° (110° 58′ 38″)* |
| Area | 2.33094575 | 2.66733911 | 4.27497152 |
| Volume | 0.27316111 | 0.32798012 | 0.71777431 |

*Table 36: Properties of Johnson solids 87, 88, and 89 and their duals.*

\*The symmetry and number of edges are always the same for each solid and its dual, so are not repeated in the "dual" section.

†Face angles are only given for the duals, as the Johnson

*Miscellany: Johnson solids 84-92.*

solids have regular faces, whose angles can be found in Table 1 on p. 3.

✦Although the triangular faces fall into ten different transitivity classes, it is only necessary to distinguish six sets of triangular faces for this tabulation: 3a refers to the two which share an edge in the mirror plane; 3b to the four others which share an edge with the two 3a-faces; 3c to the two triangles which share a vertex with the 3a-faces and which are bisected by the mirror plane; 3d to those four triangles which share a vertex with the 3a-faces but are not near the mirror plane; 3e to the two triangles derived from the pyramidal augment that are bisected by the mirror plane, and 3f to the other two triangular faces of the pyramidal augment.

⌗The designation 3a refers to the four triangles that are adjacent to the square faces and to each other; the designation 3b refers to the two other triangles adjacent to the square faces (opposite each other); the designation 3c refers to the four triangles adjacent to the 3b-faces; the designation 3d refers to the four other triangles which share a vertex with the square faces; and the designation 3e refers to the remaining two triangles, farthest from the square faces (the only triangles that do not even share a vertex with either of the square faces).

⊛Although the triangular faces fall into *six* transitivity classes and the square faces into *two*, it is only necessary to divide the triangles into four groups: if one refers to the *first* mirror plane as the one that bisects all three square faces and the *second* as the one, perpendicular to the first, that bisects only the middle square, 3a designates the four faces which lie in pairs with their edges lying in the second mirror plane; 3b designates the two triangles furthest from the three square faces, with a common edge lying in the first mirror plane; 3c designates the two triangles that are adjacent with the two outer squares along the edges opposite the 4-4 edges; and 3d refers to all the remaining ten triangular faces. There is no particular reason to distinguish in this listing between the two types of square faces.

°The designation 4a refers to the near-rectangular isosceles trapezoidal faces bisected by one mirror plane but not adjacent

— 393 —

to the other; 4b to the isosceles trapezoidal faces whose longest sides, sharing an edge, lie along one of the mirror planes of the polyhedron; 5a to the isosceles pentagons with no common edge with any of the isosceles trapezoidal faces; 5b to the isosceles pentagons that share an edge with the 4a faces; 5c to the irregular pentagons which share edges with both classes of isosceles trapezoidal faces.

*For this purpose, the "first mirror plane" is the one that includes both two of the edges between two quadrilaterals and an edge between two pentagons, at opposite ends of the polyhedron. The "second mirror plane" is perpendicular to it, and includes two of the edges between two quadrilaterals, as well as the edges between two irregular pentagonal faces adjacent to those quadrilaterals, but bisects the isosceles pentagons opposite to the four quadrilateral faces. The designation 5a refers to the two isosceles pentagonal faces bisected by the first mirror plane which are not adjacent to any quadrilateral faces; 5b refers to the two isosceles pentagonal faces with a shared edge in the first mirror plane, opposite the four quadrilateral faces; 5c refers to the two isosceles pentagonal faces bisected by the first mirror plane which are each adjacent to two quadrilateral faces; 5d refers to the irregular pentagonal faces that are adjacent to quadrilateral faces, but each with an edge in the second mirror plane separating them from another 5d face.

#The designation 4a refers to the two irregular quadrilaterals flanking the mirror plane of the polyhedron; 4k to the single kite-quadrilateral bisected by the mirror plane; 5a to the near-regular isosceles pentagon farthest from all quadrilateral faces; 5b to the isosceles pentagon which is adjacent to the two irregular quadrilaterals; 5c to the two irregular pentagons that share edges with both the 5b and 5a faces; 5d to the two irregular pentagons flanking the mirror plane between the 4k and the 5a faces; 5e to the two irregular pentagons that share edges with both the 4a and 4k quadrilaterals.

$^\triangle$The list of face angles of the dual of $J_{87}$ is too large to fit in one cell. It is given here instead:

- 4a: 55.3957° (55° 23′ 45″), 94.4822° (94° 28′ 56″), 101.9760° (101° 58′ 34″), 108.1461° (108° 8′ 46″);

- 4k: [2]91.2439° (91° 14′ 38″), 78.9260° (78° 55′ 34″), 98.5862° (98° 35′ 10″);
- 5a: [2]107.5935° (107° 35′ 37″), [2]107.9178° (107° 55′ 4″), 108.9773° (108° 58′ 38″);
- 5b: [2]102.7248° (102° 43′ 29″), [2]110.5304° (110° 31′ 49″), 113.4896° (113° 29′ 23″);
- 5c: 103.6227° (103° 37′ 22″), 106.2265° (106° 13′ 36″), 106.5864° (106° 35′ 11″), 110.9152° (110° 54′ 55″), 112.6491° (112° 38′ 57″);
- 5d: 99.5312° (99° 31′ 52″), 102.9887° (102° 59′ 19″), 104.9425° (104° 56′ 33″), 115.8707° (115° 52′ 15″), 116.6669° (116° 40′ 1″);
- 5e: 73.5313° (73° 31′ 53″), 109.5717° (109° 34′ 18″), 114.6003° (114° 36′ 1″), 119.5939° (119° 35′ 36″), 122.7032° (122° 42′ 12″).

Another singular member of the list of Johnson solids is the *disphenocingulum*, $J_{90}$. The name which Johnson gave it implies two (*di-*) roofs (*spheno-*) and a belt (*cingulum*): the "roof" part refers to the two squares at an angle to each other at the top and bottom as the figure is shown in Figure 182 below, and has nothing to do with the term "roof" in the sense I have been using it in this book, as a face diametrically opposite to one designated as the *base*. It has $S_{4v}$ symmetry, with a fourfold alternating axis bisecting two of its edges. As was noted earlier in this chapter, this $S_{4v}$ symmetry is rather rare among Johnson solids, only $J_{26}$ and $J_{84}$ sharing this symmetry. Four of the faces are squares, the two which are adjacent at the top and two similarly adjacent at the bottom; it is the edges between these pairs of squares which have, at their midpoints, the ends of the fourfold alternating axis. All the other faces of this solid are equilateral triangles; there are twenty of them, which can be divided into three sets with different symmetry properties. The first set, consisting of four triangles, is the four which are adjacent to the square faces opposite to the edges where the squares border each other. The second and third sets consist of eight each; the second set is the eight which have an edge in common with the square faces, other than the ones included in the first set; the third set consists of the eight triangles that only border other triangles.

*Figure 182: Johnson solid 90: the disphenocingulum.*

The dual of the disphenocingulum is illustrated in Figure 183 below. Since duals always share symmetry, this figure also has $S_{4v}$ symmetry, and from the nature of the duality relationship, the fourfold alternating axis also bisects two edges. The faces adjacent to those edges are isosceles trapezoids; there are four of these. Adjacent to the opposite edges of those isosceles trapezoids are four pentagons, of the sort that I referred to as *isosceles* in my earlier book, *Polyhedra: a New Approach*. There are eight more pentagonal faces, all irregular but all of the same shape (except for mirror reflection), making a total of sixteen faces.

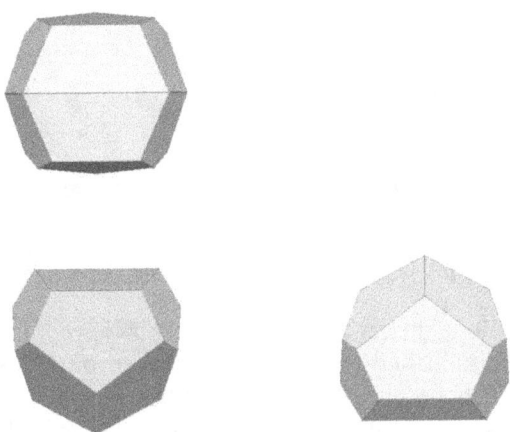

Figure 183: *The dual of Johnson solid 90 (the disphenocingulum).*

The *bilunabirotunda*, $J_{91}$, is in a category of its own as well. It has $D_{2h}$ symmetry, and the "birotunda" part of its name comes from the two pentagons and two triangles that occur together on both sides of the polyhedron. It is illustrated in Figure 184 below. It was earlier noted (in Chapter 11) that Johnson solid 55 (the parabiaugmented hexagonal prism) is the only other Johnson solid with $D_{2h}$ symmetry among the entire set of ninety-two.

The top view shows a complex of a square and four equilateral triangles, and the bottom is identical in form. If one considers the square and the two triangles which extend forward and back from the top as one base of a birotunda and the corresponding faces of the bottom as the other base, with the pentagons and the remaining triangles as the lateral faces of the birotunda, the name is explained. However, it is sufficiently different from the pentagonal birotunda ($J_{34}$) that Johnson did not place it in the sequence near that solid.

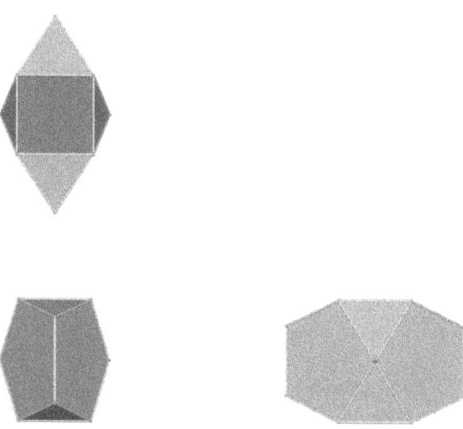

Figure 184: Johnson solid 91: the bilunabirotunda.

The bilunabirotunda is unusual in being *bi/bigeneral*, as is the parabidiminished rhombicosidodecahedron ($J_{80}$, see p. 344). In fact, it can be classified in a number of ways, because the principal axis of rotation is not uniquely defined. The bilunabirotunda has three twofold axes, any of which can be considered the principal axis. And if the principal axis is taken as the axis which is vertical, when the solid is in the position implied by Figure 184, the top view shows that the axis pierces the surface in the center of a square (a 4-gon). The bottom would show the same. If the twofold axis chosen were the one that runs from *left to right* in the position implied by Figure 184, the polyhedron would be *diapical*, in the terminology of my earlier book, *Polyhedra: a New Approach*: the principal axis cuts the surface at two vertices of the polyhedron. And if the remaining twofold axis were chosen, running from front to back, it would cut the surface at the midpoint of two edges, which can be considered the center of digonal faces, so with that choice of principal axis, the polyhedron is uni/unigeneral. That a polyhedron can be considered to fall in more than one class, depending on which axis is taken as principal, is common. For most of the Johnson solids, however, there is a unique choice of principal axis implied by the symmetry of the polyhedron.

The dual of the bilunabirotunda is illustrated in Figure 185 below.

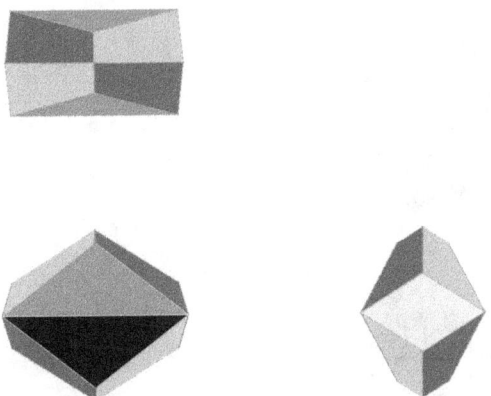

*Figure 185: The dual of Johnson solid 91 (the bilunabirotunda).*

I find puzzling to me the system Norman Johnson used to arrive at the name he gave to the final member of the list of Johnson solids, the *triangular hebesphenorotunda* ($J_{92}$). The term *"triangular"* leads to a discussion similar to what was found when discussing Johnson solid 6 (the pentagonal rotunda). In both cases, Johnson gave the solid a name which implies a whole family of polyhedra, but only one could be made with the faces all regular polygons, so as to qualify as a Johnson solid. Rotundae other than pentagonal, and hebesphenorotundae other than triangular, simply require the faces other than those perpendicular to the axis of rotational symmetry to be either isosceles or, in the terminology of my earlier book, *Polyhedra: a New Approach*, strombic. However, while from the pentagonal rotunda can be derived (mostly by fusion) other Johnson solids such as the pentagonal birotunda, the elongated and gyroelongated pentagonal rotunda and birotunda, and the pentagonal cupolarotunda (and its elongated and gyroelongated derivatives), the triangular hebesphenorotunda does not lead to any similar derivatives, which, I suppose, is why it came last in the list of Johnson solids. (The dihedral angles at the base of the triangular hebesphenorotunda are quite obtuse, at 138.1897° = 138° 11′ 23″. An attempt to fuse a prism, an antiprism, or any of the other solids that are fused to any of the cupolae or to a rotunda in the various Johnson solids would not yield a convex polyhedron.)

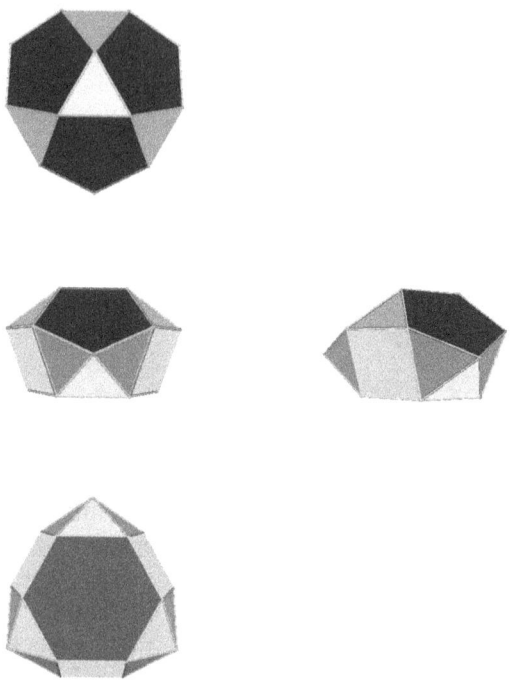

*Figure 186: Johnson solid 92: the triangular hebesphenorotunda.*

But even more than Johnson's naming the triangular hebesphenorotunda as a member of a family, I wonder about the name "hebesphenorotunda." The only other Johnson solid with "hebespheno-" in its name is $J_{89}$, the hebesphenomegacorona. The term there seems to have been used to refer to the configuration of three squares in the top view of Figure 180 (only the central one of which clearly appears, but the other two are greatly foreshortened). In general, Johnson's nomenclature does not use the same term for the digonal case of a polyhedron and those with higher-order axes of rotation; compare $J_{26}$ (named by him the *gyrobifastigium*) with $J_{29}$ (the *square gyrobicupola*) and $J_{31}$ (the *pentagonal gyrobicupola*). But the only part of $J_{92}$ that resembles that configuration of three squares is the configuration of three squares and a hexagon in the bottom view of Figure 186. This is in fact the part that, in the case of $n = 2$, would be the three squares of Figure 180. And this would be the one case where Johnson used the same word for the $n = 2$ and some $n > 2$ case of a family.

The term "rotunda" also seems to be stretched in the name of this polyhedron. While $J_{92}$ resembles a rotunda in having an $n$-gonal roof and a $2n$-gonal base, in neither portion is the configuration like that in a rotunda. In a rotunda, each side of the roof is an edge shared with a triangular face, and each vertex of the roof is common to two triangles and a pentagon as well as the roof; in $J_{92}$, the roles of the triangles and pentagons are reversed.

The base is even more unlike the rotunda base: the sides of the $2n$-gon in a rotunda alternate between an edge shared with a triangular face and one shared with a pentagonal face; in $J_{92}$ there is an alternation, but between triangles and *squares*.

It is interesting that if one takes the triangle at the top, the three pentagons which share an edge with it, and the three triangles which share an edge with *two* of the pentagons, the seven polygons together are arranged exactly as they would be in an icosidodecahedron, an Archimedean solid of $I_h$ symmetry, so that all the dihedral angles between those pentagons and triangles are equal, despite the fact that the triangles belong to two different transitivity classes. The outer nine edges of this set of seven polygons do not form a plane nonagon, so it is somewhat surprising that the remaining thirteen faces (nine triangles in three different transitivity classes, three squares, and a hexagon) form a set that fits the first set of seven precisely!

There is more in the way of connections between the triangular hebesphenorotunda and the icosidodecahedron. The triangular hebesphenorotunda has a height of

$$\sqrt{[(3 + \sqrt{5})/6]} + \tfrac{1}{3}\sqrt{3} = [\cos(\pi/5) + \sin(\pi/6)]/\cos(\pi/6) \cong 1.5115,$$

exactly half the height of the icosidodecahedron measured between two opposite triangular faces, so that if two triangular hebesphenorotundae are fused together at their hexagonal faces in the "gyro" orientation, although the resulting polyhedron is not convex, both of those sets of triangles and pentagons will be exactly where they would be in an icosidodecahedron. The icosidodecahedron itself would have six vertices in the same plane as the fusion plane, forming a hexagon whose sides are τ (the golden ratio).

The hebesphenorotundae (considered as a family, including $J_{92}$ and its non-Johnson generalizations) *do* resemble the rotundae in being uni/bigeneral tectal polyhedra. The faces of $J_{92}$ include triangles, squares, pentagons, and a hexagon, and no other Johnson solid includes all four types of polygon. The symmetry of $J_{92}$ is $C_{3v}$.

The dihedral angles between the pentagonal faces and the adjacent triangular faces are 142.6226° (142° 37′ 21″) and 100.8123° (100° 48′ 44″). It was noted in the discussion of $J_2$ that the dihedral angle between a triangular face and the pentagonal base is 37.3774° (37° 22′ 39″). So the attempt to augment a pentagonal face of $J_{92}$ by fusing a pentagonal pyramid to that face would lead to a 180° dihedral angle with some of the adjacent triangular faces, the ones where the dihedral angle in $J_{92}$ is 142.6226°. This means that augmenting $J_{92}$ at any pentagonal face does not create a Johnson solid, but one allows all three pentagonal faces to be augmented by pyramids that are *shorter* than $J_2$, the resulting polyhedron is in fact the same (in its arrangements of polygon types and vertex valencies) as the threefold member of a family whose twofold member is $J_{89}$, providing a relationship between these two Johnson solids.

The last polyhedron which this book will present is the dual of the triangular hebesphenorotunda, which is illustrated in Figure 187 below. It has, of course, the same $C_{3v}$ symmetry as $J_{92}$ itself.

While Johnson solid 92 contains four different kinds of polygons as faces, namely triangles, squares, pentagons, and a hexagon, as stated above, the faces of its dual are all quadrilaterals, though of four different shapes. Two of those types are kite-quadrilaterals and the other two are irregular, in pairs of alternating chirality. One set of kites is centered in the top view of Figure 187; the other has one member which is viewed nearly head-on in the front view.

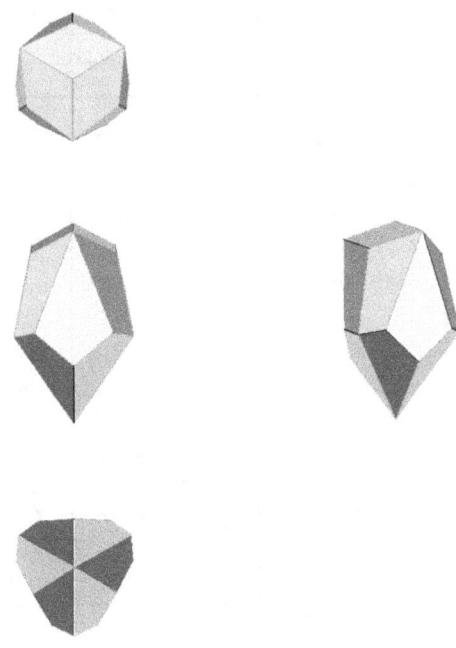

*Figure 187: The dual of Johnson solid 92 (the triangular hebesphenorotunda).*

## The Johnson Solids and Their Duals

| Johnson solid number | 90 | 91 | 92 |
|---|---|---|---|
| Name | Disphenocingulum | Bilunabirotunda | Triangular hebesphenorotunda |
| Symmetry* | $S_{4v}$ | $D_{2h}$ | $C_{3v}$ |
| Faces | [4+8+8](3) + [4](4) | [4+4](3) + [2](4) + [4](5) | [1+3+3+6](3) + [3](4) + [3](5) + [1](6) |
| Edges* | 38 | 26 | 36 |
| Vertices | 16 | 14 | 18 |
| Dihedral angles | 3a-3a: 124.7019° (124° 42′ 7″), 3a-3b: 148.4340° (148° 26′ 2″), 3b-3b: 166.8114° (166° 48′ 41″), 3b-3c: 133.5912° (133° 35′ 28″), 3a-4: 136.3359° (136° 20′ 9″), 3c-4: 154.4188° (154° 25′ 8″), 4-4: 100.1939° (100° 11′ 38″)♦ | 3a-4: 159.0948° (159° 5′ 41″), 3b-4: 110.9052° (110° 54′ 19″), 3a-5: 100.8123° (100° 48′ 44″), 3b-5: 142.6226° (142° 37′ 21″), 5-5: 63.4349° (63° 26′ 6″)△ | 3b-3c: 138.1897° (138° 11′ 23″), 3a-4: 110.9052° (110° 54′ 19″), 3b-4: 159.0948° (159° 5′ 41″), 3a-5: 142.6226° (142° 37′ 21″), 3b-5: 100.8123° (100° 48′ 44″), 3c-6: 138.1897° (138° 11′ 23″), 4-6: 110.9052° (110° 54′ 19″)* |
| Area | 12.66025404 = 4 + 5√3 | 12.34601122 | 16.38867354 |
| Volume | 3.77764534 | 3.09371765 | 5.10874597 |
| Dual | | | |
| Name | Unnamed | Unnamed | Unnamed |
| Faces | [4](4) + [8+4](5) | [4](3) + [8+2](4) | [3+3+6+6](4) |
| Vertices | 24 | 14 | 20 |

*Miscellany: Johnson solids 84-92.*

| Johnson solid number | 90 | 91 | 92 |
|---|---|---|---|
| Face angles[†] | 4: [2]70.9719° (70° 58' 19"), [2]109.0281° (109° 1' 41"); 5a: [3]114.2843° (114° 17' 3"), 80.4820° (80° 28' 55"), 116.6651° (116° 39' 54"); 5b: [3]105.4936° (105° 29' 37"), [2]111.7596° (111° 45' 34")[¤] | 3: [2]40.8934° (40° 53' 36"), 98.2132° (98° 12' 48"); 4a: 73.9549° (73° 57' 18"), 106.0451° (106° 2' 42"), 58.6528° (58° 39' 10"), 121.3472° (121° 20' 50"); 4r: [2]63.4349° (63° 26' 6"), [2]116.5651° (116° 33' 54")[⊛] | 4a: 53.9620° (53° 57' 43"), 76.1880° (76° 11' 17"), 111.5211° (111° 31' 16"), 118.3289° (118° 19' 44"); 4b: 35.2681° (35° 16' 5"), 83.0026° (83° 0' 9"), 120.1523° (120° 9' 8"), 121.5770° (121° 34' 37"); 4k: [2]65.1174° (65° 7' 3"), 110.2141° (110° 12' 51"), 119.5511° (119° 33' 4"); 4t: [2]107.0220° (107° 1' 19"), 44.4950° (44° 29' 42"), 101.4610° (101° 27' 40")[∗] |

— 405 —

| Johnson solid number | 90 | 91 | 92 |
|---|---|---|---|
| Dihedral angles | 4-4: 131.2857° (131° 17′ 8″), 4-5a: 127.7343° (127° 44′ 4″), 4-5b: 130.4506° (130° 27′ 2″), 5a-5a: 121.6737° (121° 40′ 25″), 5a-5b: 126.3896° (126° 23′ 23″)¤ | 3-3: 138.1897° (138° 11′ 23″), 3-4a: 134.6983° (134° 41′ 54″), 4a-4a: 124.5354° (124° 32′ 7″), 4a-4r: 117.7323° (117° 43′ 56″)⊗ | 4a-4a: 136.5015° (136° 30′ 5″), 4a-4b: 134.2483° (134° 14′ 54″), 4a-4k: 132.5996° (132° 35′ 59″), 4a-4t: 139.6258° (139° 37′ 33″), 4b-4b: 130.6597° (130° 39′ 35″), 4b-4t: 138.7442° (138° 44′ 39″), 4k-4k: 121.8672° (121° 52′ 2″)* |
| Area | 8.20382436 | 4.99779387 | 6.87839497 |
| Volume | 1.95134429 | 0.86890628 | 1.45275326 |

Table 37: Properties of Johnson solids 90, 91, and 92 and their duals.

\*The symmetry and number of edges are always the same for each solid and its dual, so are not repeated in the "dual" section.

†Face angles are only given for the duals, as the Johnson solids have regular faces, whose angles can be found in Table 1 on p. 3.

♦The designation 3a refers to those eight triangular faces that are adjacent to square faces on the edges which are *adjacent* to the 4-4 edges; the designation 3b to the triangular faces that are adjacent only to *other triangular faces* (including each other); the designation 3c to those four triangular faces that are adjacent to square faces on the edges which are *opposite* to the 4-4 edges.

ΔThe definition of the 3a-3b distinction is rather hard to express, since both are surrounded by two pentagons and a square. It is best to refer to the mirror plane which includes the two 5-5 edges. The four triangular faces bisected by this mirror plane are designated as 3a; the four that do not contact this mirror plane are designated as 3b.

*The designation 3a refers to the single triangle opposite the hexagonal base and the three others sharing a vertex with it; 3b refers to all those triangles sharing only one vertex with the hexagonal base, and 3c to the three triangles sharing an edge with the hexagonal base.

"The designation 5a refers to the irregular pentagons that constitute most of the faces; 5b to the four isosceles pentagons, each of which has its base along the same edge that constitutes the short side of one of the four isosceles trapezoidal faces.

°The designation 4a refers to the eight irregular quadrilaterals; 4r to the two rhombi.

⁺The designation 4a to the six irregular quadrilaterals adjacent to the kite-quadrilaterals that form an antibipyramid-like apex; 4b to the six irregular quadrilaterals meeting at the hexavalent apex; 4k refers to the three (near-rhombus) kite-quadrilaterals forming an antibipyramid-like apex; 4t to the three kite-quadrilaterals with their smallest angles each sharing a quinquevalent vertex with two 4k and two 4a faces.

# Chapter 18: The Johnson solids and polyhedron fusion.

This chapter will explore some aspects of the creation of composite polyhedra by the fusion of simple polyhedra together. Because it is concerned with the use of Johnson solids either as the simple polyhedra involved in the fusion or as the composite polyhedra resulting from the fusion, the only polyhedra with which this chapter is concerned are those with regular-polygonal faces, but unlike the other chapters of the book, the Platonic and Archimedean solids, the regular-polygon-faced prisms, and the regular-polygon-faced antiprisms, will specifically be included.

For the purposes of this chapter, an *element* will mean a polyhedron that is used as one of the pieces fused together to produce a composite polyhedron. The elements that will be used are:

1. Pyramids — only triangular, square, and pentagonal pyramids are possible if the restriction to regular polygons as faces is to be maintained. They will be abbreviated $Py_3$, $Py_4$, and $Py_5$ in the recipes that follow. A pyramid can only be the first or the last element in a fusion, and if first, it must be oriented with base toward the next element, while if last, it must be oriented with base toward the preceding element.

2. Prisms — no restriction in the order is forced, but because of other conditions, only a few orders will actually be found. It will be understood that the bases of the prisms are the faces involved in the fusion, except as otherwise noted. They will be abbreviated $Pr_3$, $Pr_4$, $Pr_5$,... Although, as noted below, a triangular prism in which a lateral face is the site of a fusion can be treated as a digonal cupola, and will be when this facilitates the discussion, it will in other cases be more feasible to simply write $Pr_n(L)$ for an *n*-gonal prism where a lateral face is fused with whatever follows. If

more than one lateral face is fused, relative locations are cited, thus $Pr_n(L1,3)$ means that two lateral faces that are fused are separated by one that is not, and so forth; one will always be numbered 1 and the other fused faces will be numbered in sequence.

3. Antiprisms — the same comments apply as those made regarding the prisms. They will be abbreviated $A_3$, $A_4$, $A_5$,... Technically speaking, $A_2$ would be a regular tetrahedron, but there is no way of fusing one at the digonal base, so the notation $A_2$ will not occur here.

4. Cupolae — as with the pyramids, only triangular, square, and pentagonal cupolae are possible if the restriction to regular polygons as faces is to be maintained. They will be abbreviated $C_3$, $C_4$, and $C_5$ in the recipes that follow. The symbol $C_2$ will be used for a triangular prism fused to some other polyhedron at one of its square faces, since, in fact, a digonal cupola is a triangular prism if the definitions are properly construed.

5. Rotundae — actually, only the pentagonal rotunda is possible if the restriction to regular polygons as faces is to be maintained, but the notation $R_5$ is used even though the subscript 5 would not be needed. (Note that any combination that has two C's, a C and an R, or two R's, there are two possible orientations. If the orientation is "ortho," i. e., the roofs of the two are aligned, the second one will be accompanied by an "o"; if "gyro," i. e., the roofs of the two are staggered, the second one will be accompanied by a "g." An exception to this is when an antiprism intervenes. In that case the two are forced into an orientation that is halfway between an ortho and a gyro orientation, so neither an "o" nor a "g" is appropriate.)

6. Dodecahedron — this is really an element, though more complex in appearance than any of the ones mentioned earlier, as no plane can be made to cut

it into two smaller polyhedra with all regular faces. It will be abbreviated D, and when more than one pyramid is fused to it, the relative positions will not be described, but pictures will make the differences clear. Although it is the fivefold member of a family of polyhedra, which was named the pentagonized globoids in my earlier book, *Polyhedra: a New Approach*, we do not put the subscript 5 on the D because no other member of this family can appear in a Johnson solid.

7. Truncated tetrahedron — in fact, like the dodecahedron, this is a member of a family (in this case the threefold member of what could be termed the fully truncated pyramids), but no other member can appear in a Johnson solid. So it is symbolized TT. It only occurs in one Johnson solid, anyway.

8. Truncated cube — yet another case, like the two previous, of a polyhedron which could be considered a member of a family. It is the fourfold member of what could be termed the fully truncated prisms, but no other member can appear in a Johnson solid. So it is symbolized TC.

9. Truncated dodecahedron — this is much like the dodecahedron itself; not only is it the only member of its family that can appear in a Johnson solid, but when more than one cupola is fused to it, it would be too difficult to develop a code to describe the relative positions, but pictures will make the differences clear. It is symbolized TD.

10. Tridiminished icosahedron (see p. 255, Chapter 13) — this can not really be broken down any further, and thus must be considered an element in itself. It is represented by $J_{63}$ below.

Similarly to $J_{63}$, one must consider $J_{80}$ and $J_{83}$ as elements. The formation of other Johnson solids by fusion of cupolae to those two was summarized in Table 32 (p. 321, Chapter 15) in more detail than is done here, so this information will not be

*The Johnson solids and polyhedron fusion.*

repeated here.

With this symbolism, the Platonic solids can be represented as follows (note that some can be represented in more than one way, and some are elements by themselves in the above list):

- Tetrahedron — $Py_3$ (also, as mentioned above, $A_2$, but nothing is gained by this alternative designation).
- Cube — $Pr_4$.
- Octahedron — $Py_4 + Py_4$, or $A_3$.
- Dodecahedron — D.
- Icosahedron — $Py_5 + A_5 + Py_5$, or $J_{63} + 3Py_5$.

The Archimedean solids can be similarly treated:

- Truncated tetrahedron — TT.
- Cuboctahedron — $C_3 + C_3g$.
- Truncated cube — TC.
- Truncated octahedron — this would be an element by itself, but because it does not figure in any of the combinations discussed here, no symbol is provided.
- Rhombicuboctahedron — $C_4 + Pr_4 + C_4o$.
- Great rhombicuboctahedron (sometimes called, incorrectly, the truncated cuboctahedron) — this would be an element by itself, but because it does not figure in any of the combinations discussed here, no symbol is provided.
- Snub cube (or snub cuboctahedron) — this would be an element by itself, but because it does not figure in any of the combinations discussed here, no symbol is provided.
- Icosidodecahedron — $R_5 + R_5g$.
- Truncated dodecahedron — TD.
- Truncated icosahedron — this would be an element by itself, but because it does not figure in any of the

combinations discussed here, no symbol is provided.

- Rhombicosidodecahedron — this would be an element by itself, but because it does not figure in any of the combinations discussed here, no symbol is provided.

- Great rhombicosidodecahedron (sometimes called, incorrectly, the truncated icosidodecahedron) — this would be an element by itself, but because it does not figure in any of the combinations discussed here, no symbol is provided.

- Snub dodecahedron (or snub icosidodecahedron) — this would be an element by itself, but because it does not figure in any of the combinations discussed here, no symbol is provided.

And many of the Johnson solids can be treated the same way, but not all. Note that some are elements in their own right:

- $J_1$ — $Py_4$ (note that $Py_3$ is a Platonic solid).
- $J_2$ — $Py_5$.
- $J_3$ — $C_3$.
- $J_4$ — $C_4$.
- $J_5$ — $C_5$.
- $J_6$ — $R_5$.
- $J_7$ — $Py_3 + Pr_3$.
- $J_8$ — $Py_4 + Pr_4$.
- $J_9$ — $Py_5 + Pr_5$.
- $J_{10}$ — $Py_4 + A_4$ (note that $Py_3 + A_3$ is not constructable because a dihedral angle would be 180°).
- $J_{11}$ — $Py_5 + A_5$.
- $J_{12}$ — $Py_3 + Py_3$.
- $J_{13}$ — $Py_5 + Py_5$ (note that $Py_4 + Py_4$ is a Platonic solid).
- $J_{14}$ — $Py_3 + Pr_3 + Py_3$.

- $J_{15}$ — $Py_4 + Pr_4 + Py_4$.
- $J_{16}$ — $Py_5 + Pr_5 + Py_5$.
- $J_{17}$ — $Py_4 + A_4 + Py_4$ (note that $Py_3 + A_3 + Py_3$ is not constructable because a dihedral angle would be 180°, and $Py_5 + A + Py_5$ is a Platonic solid).
- $J_{18}$ — $C_3 + Pr_3$.
- $J_{19}$ — $C_4 + Pr_4$.
- $J_{20}$ — $C_5 + Pr_5$.
- $J_{21}$ — $R_5 + Pr_5$.
- $J_{22}$ — $C_3 + A_3$.
- $J_{23}$ — $C_4 + A_4$.
- $J_{24}$ — $C_5 + A_5$.
- $J_{25}$ — $R_5 + A_5$.
- $J_{26}$ — $C_2 + C_2g$ (note that an ortho orientation is not constructable because a dihedral angle would be 180°).
- $J_{27}$ — $C_3 + C_3o$ (note that a gyro orientation would give an Archimedean solid).
- $J_{28}$ — $C_4 + C_4o$.
- $J_{29}$ — $C_4 + C_4g$.
- $J_{30}$ — $C_5 + C_5o$.
- $J_{31}$ — $C_5 + C_5g$.
- $J_{32}$ — $C_5 + R_5o$.
- $J_{33}$ — $C_5 + R_5g$.
- $J_{34}$ — $R_5 + R_5o$ (note that a gyro orientation would give an Archimedean solid).
- $J_{35}$ — $C_3 + Pr_3 + C_3o$.
- $J_{36}$ — $C_3 + Pr_3 + C_3g$.
- $J_{37}$ — $C_4 + Pr_4 + C_4g$ (note that an ortho orientation would give an Archimedean solid).

## The Johnson Solids and Their Duals

- $J_{38} - C_5 + Pr_5 + C_5o$.
- $J_{39} - C_5 + Pr_5 + C_5g$.
- $J_{40} - C_5 + Pr_5 + R_5o$.
- $J_{41} - C_5 + Pr_5 + R_5g$.
- $J_{42} - R_5 + Pr_5 + R_5o$.
- $J_{43} - R_5 + Pr_5 + R_5g$.
- $J_{44} - C_3 + A_3 + C_3g$.
- $J_{45} - C_4 + A_4 + C_4$.
- $J_{46} - C_5 + A_5 + C_5$.
- $J_{47} - C_5 + A_5 + R_5$.
- $J_{48} - R_5 + A_5 + R_5$.
- $J_{49} - Py_4 + C_2$, or $Pr_3(L) + Py_4$.
- $J_{50} - Pr_3(L1,2) + 2Py_4$.
- $J_{51} - Pr_3(L1,2,3) + 3Py_4$.
- $J_{52} - Pr_5(L) + Py_4$.
- $J_{53} - Pr_5(L1,3) + 2Py_4$.
- $J_{54} - Pr_6(L) + Py_4$.
- $J_{55} - Pr_6(L1,3) + 2Py_4$.
- $J_{56} - Pr_6(L1,2) + 2Py_4$.
- $J_{57} - Pr_6(L1,3,5) + 3Py_4$.
- $J_{58} - D + Py_5$.
- $J_{59} - D + 2Py_5$.
- $J_{60} - D + 2Py_5$.
- $J_{61} - D + 3Py_5$.
- $J_{62} - J_{63} + P_5$
- $J_{63}$ – an element in itself (see above).
- $J_{64} - J_{63} + P_3$

*The Johnson solids and polyhedron fusion.*

- $J_{65} - C_3 + TT$
- $J_{66} - C_4 + TC$
- $J_{67} - C_4 + TC + C_4$
- $J_{68} - C_5 + TD$
- $J_{69} - 2C_5 + TD$
- $J_{70} - 2C_5 + TD$
- $J_{71} - 3C_5 + TD$
- $J_{72}$ to $J_{83}$ — Please refer to Table 32 for the decompositions.

The remaining Johnson solids — $J_{84}$ to $J_{92}$ — do not decompose, with the exception of $J_{87}$ which is $Py_4 + J_{86}$.

Of course, from some of these can be derived other relationships among the Johnson solids; for example, since $J_9$ = $Py_5 + Pr_5$ and $J_{16} = Py_5 + Pr_5 + Py_5$, one can conclude that $J_{16} = J_9 + Py_5$. A very nice set of diagrams illustrating many of these relationships was produced by Allison Chen and is viewable on her blog at

http://chenallison.blogspot.com/2013/04/johnson-solids-wip.html#!/2013/04/johnson-solids-wip.html

as well as on her personal Website, at

http://allisonychen.com/

# Index

**A**

Acute angle.................30, 92, 109, 118, 139, 140, 143, 160, 184, 187
Akisation................................................................................................36
Alternating axis......................3, 120, 127, 147, 149, 161, 165, 289, 366
Alternating chirality.79, 80, 83, 86, 92, 98, 99, 152, 163, 182, 270, 272, 303, 346, 403
Antiapex........................19, 20, 29, 30, 79, 98, 182, 183, 259, 279, 286
Antiapical star.........................................................284, 285, 287, 290, 310
Antiapical-star fill-in triangle....................284, 285, 286, 287, 290, 310
Antiapical-star fill-in-opposed triangle............................285, 286, 290
Antibipyramid.................................................................19, 20, 40, 51, 174
Antibipyramid-like antiapex................................................................98
Antibipyramid-like apex. 20, 21, 29, 30, 31, 51, 79, 83, 85, 98, 99, 129, 131, 151, 407
Antihermaphrodite............................................................................258
Antiprism.1, 15, 36, 37, 40, 51, 57, 65, 78, 96, 106, 112, 144, 170, 172, 174, 179, 180, 182, 185, 186, 190, 193, 400, 409
Apex 5, 10, 14, 15, 19, 20, 29, 41, 42, 46, 47, 51, 57, 63, 64, 65, 66, 79, 81, 95, 145, 152, 160, 169, 172, 186, 195, 196, 197, 199, 200, 205, 207, 214, 215, 225, 243, 256, 258, 261, 265, 272, 275, 279, 346, 369
Apical kite......30, 92, 95, 110, 132, 137, 140, 160, 163, 169, 182, 184, 186, 187, 190
Apical pentagon...............................................................................369, 370
Apical/lateral kite rule.................................30, 92, 132, 139, 160, 184, 187
Apicobasal polyhedron..........................................................................20, 42
Archimedean dual solid.............................................................118, 259, 366
Archimedean solid......1, 12, 23, 28, 35, 38, 83, 87, 113, 119, 127, 131, 138, 148, 149, 150, 152, 193, 255, 266, 267, 270, 280, 286, 316, 344, 401
Asymmetric bipyramid..........................................................................40
Augmentation.....36, 37, 51, 194, 200, 201, 205, 207, 209, 214, 215, 216, 225, 229, 230, 231, 232, 234, 237, 239, 245, 251, 253, 261, 265, 266, 273, 315, 358
Augmented cube..................................................................................37

*Index*

Augmented dodecahedron................................................15, 232, 236, 318
Augmented hexagonal prism.....................................................194, 214
Augmented pentagonal antiprism......................................................37
Augmented pentagonal prism..............................................37, 194, 207
Augmented Platonic.......................................................................43, 44, 59
Augmented polyhedron.................................................................................40
Augmented prism..............................................................................37, 192, 194
Augmented sphenocorona.......................................................................382
Augmented square antiprism...................................................................37
Augmented triangular prism..........................................................37, 194
Augmented tridiminished icosahedron...........................................260
Augmented truncated cube.........................................................270, 271
Augmented truncated dodecahedron.............280, 281, 282, 290, 309
Augmented truncated dodecahedron dual........................................310
Augmented truncated tetrahedron.........................266, 267, 269, 270
Augmenting pyramid....................................................................................59
Axis (of a prism)...........................................................................................207
Axis of rotational symmetry.....................2, 17, 18, 42, 75, 218, 225, 399
B
Basal dihedral angle.............................................................................15, 37
Basally augmented antiprism..................................................................38
Basally augmented cupola........................................................................81
Basally augmented pentagonal antiprism.............................53, 252
Basally augmented triangular cupola..............................................145
Basally augmented triangular prismoid..............................................40
Basally biaugmented antiprism..............................................................67
Basally biaugmented pentagonal antiprism................................252, 319
Basally biaugmented prism.......................................................................67
Basally monoaugmented prism.............................................37, 39, 192
Basally monoaugmented triangular prism......................................38
Base (of a polyhedron)..17, 18, 21, 22, 24, 25, 28, 30, 31, 34, 35, 36, 43, 46, 50, 51, 52, 57, 59, 67, 68, 69, 75, 78, 92, 96, 97, 99, 101, 106, 112, 113, 116, 117, 119, 121, 126, 130, 131, 144, 151, 163, 170, 182, 185, 186, 197, 256, 265, 266, 267, 268, 270, 271, 272, 316, 396, 397, 398, 400, 401, 402, 407
Base (of a prism).37, 38, 59, 65, 66, 82, 92, 146, 161, 162, 163, 193, 194,

195, 197, 198, 200, 205, 209, 225, 408
Base (of a pyramid) 7, 11, 12, 14, 38, 40, 66, 81, 193, 197, 207, 229, 233, 268, 384, 408
Base (of an antiprism) .................................. 37, 65, 182, 184, 185, 186, 409
Base (of an isosceles polygon) .......................................................... 407
Base (of an isosceles triangle) ........................................... 11, 272, 275
Base-augmented antiprism ................................................................ 51
Base-augmented prism ...................................................................... 51
Bi/bigeneral tectal polyhedron .................................................. 344, 398
Biaugmented antiprism ...................................................................... 65
Biaugmented hexagonal prism ................................................... 199, 216
Biaugmented pentagonal prism .......................................................... 199
Biaugmented prism ............................................................................ 65
Biaugmented triangular prism ........................................ 197, 198, 199, 200
Biaugmented truncated cube ........................................... 273, 274, 289
Bicupola ................................... 58, 113, 115, 128, 131, 133, 144, 158, 184
Bicupola dual .................................................................................... 132
Bifrustum of antibipyramid .................................................................. 67
Bifrustum of bipyramid ........................................................................ 67
Bifrustum of pentagonal bipyramid ..................................................... 70
Bifrustum of square bipyramid ..................................................... 68, 69
Bifrustum of triangular bipyramid .................................................. 67, 68
Bigyrate diminished rhombicosidodecahedron .......................... 341, 342
Bilunabirotunda ........................................................ 216, 344, 397, 398, 399
Bipyramid .......... 14, 15, 40, 51, 58, 59, 62, 82, 147, 174, 195, 201, 215, 260
Birotunda ........................... 58, 113, 133, 138, 140, 144, 158, 184, 397
C
Catalan solid ................................ 44, 118, 139, 150, 259, 286, 287, 326, 366
Center of symmetry ........................................................................... 28
Centroid 5, 8, 51, 131, 132, 140, 141, 184, 209, 210, 214, 219, 220, 235, 237, 239, 246, 291, 303, 335, 336, 345, 347
Chiral polyhedron ..................................................................... 170, 181
Combinatorial duality ......................................................................... 44
Component polyhedron (in a fusion) ....................................... 146, 163, 183
Convex polygon ................................................................................ 192
Convex polyhedron ........................................................................... 302

*Index*

Conway, John Horton..................................................................39
Core polyhedron.................319, 320, 321, 323, 324, 329, 345, 358
Corona...............................................................................371
Cromwell, Peter R............................................................148
Cube................................44, 62, 230, 270, 271, 368, 372, 373, 374
Cuboctahedron.......................................................87, 113, 119, 127, 128
Cumulation.........................................................................36
Cupola 17, 19, 21, 25, 28, 30, 35, 82, 87, 96, 97, 98, 106, 107, 113, 115, 117, 120, 121, 124, 129, 130, 131, 133, 137, 147, 157, 159, 161, 169, 170, 172, 175, 179, 180, 181, 182, 184, 190, 266, 267, 268, 280, 288, 303, 316, 317, 319, 321, 324, 325, 327, 329, 330, 336, 337, 338, 340, 341, 344, 358, 400, 409
Cupola dual...........................................................................20, 29, 184
Cupola-dual star...281, 284, 286, 287, 290, 291, 292, 294, 295, 303, 310
Cupola-dual-star fill-in triangle.....284, 285, 286, 287, 290, 293, 294, 297, 300, 307, 310
Cupola-dual-star fill-in-opposed triangle.....284, 285, 290, 293, 297, 307
Cupolarotunda.....................................113, 114, 129, 131, 133, 158, 184

D

Decagon.....28, 92, 106, 112, 130, 163, 266, 280, 288, 289, 302, 314, 316, 319, 324, 325, 336, 340, 341, 342, 344, 347
Decagonal antiprism...............................................................182, 186
Decagonal prism........................................................................153, 158, 162
Decagonal pyramid..................................................................93
Decavalent antiapex..................................................................284
Decavalent vertex..........................................................108, 336, 337, 348
Deltahedron..............................................................58, 60, 61, 74, 200, 367
Deltohedron..............................................................................40
Deltoidal hexecontahedron.......................................................287
Deltoidal icositetrahedron.................................................150, 259
Deltoido-triangular icosahedron................................................31
Diapical polyhedron..........................................19, 20, 42, 78, 172, 186, 398
Digon..............................................................................124, 398, 409
Digonal cupola.......................................................................113, 129, 409
Digonal gyrobicupola...........................................................2, 113, 114
Dihedral angle...4, 11, 12, 14, 15, 21, 24, 25, 28, 34, 35, 38, 39, 53, 61, 73,

74, 76, 80, 83, 85, 87, 91, 92, 95, 106, 107, 112, 113, 118, 119, 124, 125, 129, 137, 138, 139, 150, 169, 173, 175, 192, 193, 197, 205, 212, 217, 223, 229, 230, 233, 265, 268, 270, 273, 275, 285, 286, 287, 288, 291, 295, 296, 297, 298, 299, 300, 301, 305, 306, 307, 309, 328, 336, 341, 344, 351, 377, 381, 402

Diminished icosahedron .................................................... 53, 234, 251, 252, 318
Diminished rhombicosidodecahedron ..... 316, 319, 333, 334, 335, 337, 338, 341, 343
Dipyramid ........................................................................................................ 58
Disphenocingulum ..................................................................................... 396
Dodecadeltahedron ................................................................................... 367
Dodecahedron ................... 230, 231, 232, 234, 236, 237, 245, 251, 289, 409
Doubly truncated icosahedron .............................................................. 238
Dual polyhedron..4, 5, 6, 8, 10, 12, 19, 21, 22, 25, 30, 31, 32, 34, 40, 41, 42, 44, 46, 51, 55, 59, 61, 62, 67, 68, 70, 74, 75, 79, 81, 83, 85, 86, 92, 93, 97, 98, 100, 101, 102, 108, 114, 115, 117, 120, 121, 122, 125, 126, 128, 131, 133, 137, 139, 145, 149, 151, 153, 159, 161, 163, 164, 165, 172, 173, 174, 183, 184, 186, 195, 197, 199, 201, 202, 205, 209, 214, 216, 223, 234, 237, 239, 245, 251, 254, 259, 260, 261, 273, 281, 282, 287, 290, 292, 303, 309, 325, 326, 328, 329, 330, 332, 335, 337, 338, 340, 342, 343, 345, 347, 359, 376, 377, 382, 385, 387, 399, 402
Duality ........................................................................................................... 397
Dualization .................................................................................................. 116
E
Edge (of a polyhedron)...4, 5, 6, 11, 28, 30, 35, 38, 40, 44, 59, 60, 66, 73, 74, 76, 83, 87, 95, 97, 98, 106, 113, 117, 121, 128, 130, 131, 137, 138, 139, 144, 150, 153, 160, 163, 173, 182, 184, 192, 197, 205, 212, 223, 244, 254, 256, 265, 267, 273, 282, 286, 288, 289, 291, 294, 341, 346, 349, 356, 358, 363, 366, 370, 372, 375, 388, 393, 394, 397, 401, 406
Eightfold alternating axis ........................................................................ 149
Elongated bicupola ........................................................ 144, 154, 161, 164, 173
Elongated bicupola dual ............................................................ 145, 151, 152
Elongated bipyramid ...................... 14, 15, 65, 66, 67, 70, 144, 147, 192
Elongated birotunda ............................................................................. 144, 368
Elongated cupola ................................................................... 78, 81, 85, 86, 87
Elongated cupolarotunda ................................................................ 161, 163, 368

*Index*

Elongated dipyramid..................................................................65, 192
Elongated gyrobicupola.....................................................................144
Elongated gyrobicupola dual............................................................152
Elongated orthobicupola...................................................................144
Elongated pentagonal bipyramid..................................................69, 70
Elongated pentagonal birotunda................................................165, 399
Elongated pentagonal cupola.....................................78, 84, 85, 91
Elongated pentagonal cupolarotunda...............................................145
Elongated pentagonal gyrobicupola...............................152, 153, 154
Elongated pentagonal gyrobirotunda.............................163, 164, 165
Elongated pentagonal gyrocupolarotunda....................158, 160, 161
Elongated pentagonal orthobicupola.....................151, 152, 153, 162
Elongated pentagonal orthobirotunda....................162, 163, 164, 165
Elongated pentagonal orthocupolarotunda.............158, 159, 160, 162
Elongated pentagonal pyramid..................................................45, 46
Elongated pentagonal rotunda..........................................91, 92, 399
Elongated polyhedron........................................................36, 37, 65, 78
Elongated pyramid.........................14, 15, 39, 43, 51, 55, 82, 192
Elongated rotunda..............................................................................368
Elongated square bipyramid............................................68, 69, 198, 230
Elongated square cupola................................................................78, 83
Elongated square gyrobicupola.............................148, 149, 152, 153
Elongated square orthobicupola......................................................148
Elongated square pyramid...............................................43, 44, 194, 230
Elongated triangular bipyramid...................................................68, 69
Elongated triangular cupola....................................78, 79, 83, 84, 86
Elongated triangular gyrobicupola...............................148, 152, 153
Elongated triangular orthobicupola........................144, 145, 148, 151
Elongated triangular pyramid........................................38, 41, 43, 44
Elongation...............................................................................................15
Equilateral triangle....7, 9, 14, 17, 24, 28, 41, 44, 49, 58, 59, 60, 61, 74, 185, 193, 194, 200, 235, 237, 239, 246, 249, 258, 260, 265, 267, 269, 270, 316, 367, 371, 375, 384, 386, 396, 397
Euler formula.......................................................................58, 66, 97, 117
F
Face (of a polyhedron)...4, 5, 6, 7, 11, 21, 25, 28, 30, 32, 36, 39, 40, 41,

42, 43, 44, 45, 47, 50, 51, 59, 60, 65, 66, 69, 75, 81, 82, 83, 91, 93, 95, 96, 97, 106, 108, 112, 115, 116, 119, 121, 124, 130, 138, 139, 140, 150, 152, 153, 156, 157, 159, 164, 169, 171, 172, 173, 175, 182, 185, 192, 193, 194, 197, 199, 200, 205, 209, 212, 213, 214, 215, 218, 219, 223, 224, 228, 229, 230, 231, 233, 235, 237, 238, 239, 240, 243, 244, 245, 249, 251, 253, 254, 258, 260, 261, 265, 266, 267, 268, 270, 273, 274, 275, 278, 285, 286, 289, 291, 296, 302, 303, 305, 310, 314, 316, 319, 324, 325, 326, 327, 328, 336, 340, 341, 344, 349, 358, 366, 370, 371, 375, 376, 377, 381, 382, 383, 384, 385, 387, 393, 394, 397, 398, 399, 401, 402, 406, 407
Face angle..................................................................341, 344
Fivefold axis 10, 15, 28, 45, 74, 75, 86, 126, 138, 140, 142, 184, 209, 230, 231, 233, 235, 236, 292, 317, 323, 325, 329, 334, 337, 339
Four-pyramoid..................................................................258, 260
Fourfold alternating axis..................................................366, 396, 397
Fourfold axis 10, 15, 22, 43, 44, 51, 68, 74, 75, 83, 86, 119, 120, 149, 271, 368, 369
Frustum of pyramid..................................................................197
Frustum of square pyramid..................................................................197
Frustum of triangular 4-pyramoid..................................................258, 260
Frustum of triangular antihermaphrodite..................................................258
Further ring..................................................................231
Furthest vertex..................................................................97
Fusion.31, 36, 38, 39, 67, 81, 86, 96, 116, 121, 124, 127, 129, 130, 137, 138, 153, 162, 163, 182, 183, 185, 186, 192, 198, 199, 229, 250, 252, 271, 288, 399
Fusion plane..59, 61, 114, 131, 138, 140, 152, 159, 163, 169, 173, 184, 289, 302
Fusion-elongation-dual rule............146, 148, 151, 154, 160, 161, 163, 183
Fusion-gyroelongation-dual rule..................................................183, 186

G

Golden ratio..................................................................371, 373
Great rhombicosidodecahedron..................................................266
Great rhombicuboctahedron..................................................266
Group..................................................................4
Gyrate bidiminished rhombicosidodecahedron..................................349, 350
Gyrate diminished rhombicosidodecahedron..................................319

## Index

Gyrate rhombicosidodecahedron ............... 316, 319, 325, 326, 327, 337
Gyrobicupola ............................................................. 113, 117, 127, 128
Gyrobifastigium ..................................................... 2, 113, 114, 129, 401
Gyroelongated bicupola ......................... 144, 170, 172, 177, 181, 186
Gyroelongated bifrustum of pentagonal bipyramid ..................... 237
Gyroelongated bipyramid .................................................. 15, 65, 144
Gyroelongated birotunda ................................................ 144, 181, 368
Gyroelongated cupola .................................................................. 78, 98
Gyroelongated cupolarotunda ............................................... 181, 368
Gyroelongated dipyramid ................................................................. 65
Gyroelongated pentagonal bicupola ...................................... 175, 176
Gyroelongated pentagonal bipyramid ............................. 74, 252, 319
Gyroelongated pentagonal birotunda ............................ 185, 186, 399
Gyroelongated pentagonal cupola ............................... 101, 102, 108
Gyroelongated pentagonal cupolarotunda ................................... 186
Gyroelongated pentagonal pyramid .................................. 52, 53, 252
Gyroelongated pentagonal rotunda ........................................ 108, 399
Gyroelongated polyhedron .................................... 36, 37, 65, 78
Gyroelongated pyramid ........................... 15, 38, 50, 51, 54, 55
Gyroelongated rotunda .............................................................. 368
Gyroelongated square bicupola ............................................ 174, 175
Gyroelongated square bipyramid .................................................. 74
Gyroelongated square cupola ......................... 99, 100, 101, 102
Gyroelongated square pyramid ................................... 50, 51, 52
Gyroelongated triangular bicupola ............................................. 174
Gyroelongated triangular bipyramid ...................................... 15, 74
Gyroelongated triangular cupola ........... 96, 97, 99, 100, 101, 102
Gyroelongated triangular pyramid ........................................ 15, 50

## H

Hebesphenomegacorona ........................................ 386, 387, 400
Hebesphenorotunda ..................................................... 399, 402
Heccaideltahedron .................................................................. 74
Hermaphrodite ........................................................................ 39
Hexagon ............... 78, 96, 97, 106, 218, 220, 266, 267, 268, 401, 402, 407
Hexagonal antiprism ............................................................... 96
Hexagonal prism ................................... 214, 215, 216, 220, 225

Hexagonal pyramid..................................................................145, 268
Hexavalent vertex.......................................................................224

I

Icosahedral symmetry....................................................230, 318, 321
Icosahedron................15, 31, 53, 234, 237, 251, 252, 319, 371, 372, 373, 374
Icosidodecahedron....................................................................113
Icosidodecahedron..........................28, 35, 87, 131, 138, 139, 140, 401
Isosceles pentagon. 28, 51, 57, 75, 76, 173, 255, 369, 370, 377, 383, 385, 386, 388, 397, 407
Isosceles polygon.......................................................................5, 399
Isosceles trapezoid...5, 41, 42, 47, 49, 67, 68, 70, 73, 118, 120, 124, 125, 128, 131, 133, 137, 139, 140, 142, 234, 235, 237, 239, 240, 244, 246, 249, 251, 261, 265, 367, 377, 385, 386, 397, 407
Isosceles triangle......7, 11, 17, 19, 20, 28, 41, 42, 46, 49, 79, 80, 83, 115, 197, 209, 212, 213, 215, 216, 217, 219, 235, 239, 243, 244, 246, 249, 254, 256, 258, 259, 265, 269, 272, 274, 275, 279, 284, 296, 336

J

Johnson solid 66 dual upper half.......................................272, 275
Johnson solid dual..........................30, 145, 308, 329, 331, 333, 343
Johnson, Norman W...1, 17, 58, 59, 74, 113, 158, 192, 229, 323, 366, 367

K

Kite-quadrilateral.5, 18, 29, 30, 31, 32, 34, 42, 51, 57, 79, 80, 81, 83, 90, 92, 95, 98, 107, 109, 112, 120, 125, 129, 132, 137, 140, 150, 152, 157, 159, 169, 172, 173, 182, 186, 187, 200, 205, 209, 213, 215, 216, 219, 220, 224, 228, 255, 259, 260, 265, 269, 272, 274, 275, 279, 282, 283, 286, 292, 294, 301, 303, 326, 327, 331, 336, 338, 345, 346, 356, 359, 360, 363, 364, 383, 403

L

Lateral dihedral angle.................................................................15
Lateral edge..................................................................................11
Lateral face (of a polyhedron)7, 9, 11, 12, 17, 18, 25, 27, 28, 30, 36, 66, 69, 70, 99, 101, 114, 115, 117, 120, 121, 124, 137, 170, 182, 186, 193, 229, 234, 256, 268, 270, 288, 317, 398
Lateral face (of a prism).14, 15, 37, 59, 60, 66, 92, 193, 194, 198, 200, 208, 220, 225, 409
Lateral kite.....30, 92, 95, 110, 132, 137, 140, 160, 163, 169, 182, 184, 187,

*Index*

190
Lateral pentagon..................................................................369, 370
Lateral vertex.......................................................................201, 209
Laterally augmented hexagonal prism.........................194, 214
Laterally augmented pentagonal prism.......................194, 207
Laterally augmented prism..................................................192
Laterally augmented triangular prism.................................194
Laterally biaugmented hexagonal prism.............................198
Laterally biaugmented pentagonal prism...........................198
Laterally biaugmented triangular prism.............................198
Laterally monoaugmented prism....................................37, 194
Laterally monotruncated pentagonal bipyramid..............207
Laterally monotruncated triangular bipyramid................208
Lesser dome..............................................................................21

M
MathWorld................................................................................xii
Metabiaugmented dodecahedron......231, 236, 237, 238, 239, 245, 254
Metabiaugmented hexagonal prism............................218, 220
Metabiaugmented truncated dodecahedron 289, 291, 292, 308, 309
Metabidiminished icosahedron........................252, 253, 254, 258
Metabidiminished rhombicosidodecahedron.........323, 346, 347, 349
Metabigyrate rhombicosidodecahedron..............323, 325, 329, 330
Metagyrate diminished rhombicosidodecahedron.....323, 338, 340, 341, 342
Miller, Jeffrey Charles Percy................................................148
Miller's solid..........................................................148, 149, 155
Mirror line............................................................................5, 383
Mirror plane 2, 28, 42, 51, 52, 65, 67, 68, 97, 98, 117, 119, 120, 126, 138, 140, 142, 147, 152, 159, 161, 165, 172, 205, 207, 209, 214, 218, 225, 233, 244, 245, 253, 268, 269, 271, 273, 289, 293, 294, 303, 305, 325, 341, 347, 349, 359, 371, 374, 375, 377, 382, 384, 385, 388, 393, 394, 406
Monoaugmented antiprism..............................................37, 40
Monoaugmented prism.....................................................37, 40
Monofrustum of antibipyramid............................................40
Monofrustum of asymmetric triangular dipyramid.........40
Monofrustum of bipyramid...............................................40, 51

Monotruncated icosahedron ............................................................ 234
Monotruncated triangular antihermaphrodite ............................. 258
**N**
Nearer ring ............................................................................................. 231
Nonagon ................................................................................................ 401
**O**
Obtuse angle ........................... 30, 92, 109, 119, 139, 140, 143, 160, 184, 187
Obtuse triangle ..................................................................................... 41
Octagon ................................................ 106, 121, 152, 266, 270, 271, 273
Octahedron ......................................................................................... 15, 58
Orthobicupola ....................................................... 113, 117, 127, 128
Orthocupolarotunda ......................................................................... 133
**P**
Parabiaugmented dodecahedron ..................... 231, 236, 237, 238, 245
Parabiaugmented hexagonal prism ........................... 216, 220, 397
Parabiaugmented truncated dodecahedron ........ 289, 290, 291, 292
Parabidiminished icosahedron ........................................................ 252
Parabidiminished rhombicosidodecahedron. 320, 323, 324, 325, 344, 345, 346, 348, 398
Parabigyrate rhombicosidodecahedron ................... 323, 328, 329, 330
Parabitruncated icosahedron .......................................................... 237
Paragyrate diminished rhombicosidodecahedron 323, 336, 337, 338, 341, 343
Parallelogram ................................................. 114, 115, 125, 128, 131, 133, 137
Pentagon...24, 28, 35, 42, 47, 51, 52, 75, 78, 91, 92, 95, 112, 117, 126, 137, 151, 164, 169, 172, 173, 182, 185, 187, 190, 209, 230, 233, 235, 237, 240, 243, 244, 245, 249, 252, 253, 254, 255, 258, 259, 260, 265, 266, 280, 290, 291, 310, 316, 319, 324, 341, 349, 367, 369, 370, 376, 382, 383, 385, 386, 387, 389, 394, 397, 398, 401, 402, 406, 407
Pentagon-dodecahedron ........................................................ 31, 118, 366
Pentagonal antiprism ................................................... 108, 185, 252
Pentagonal bicupola ............................................................... 131, 184
Pentagonal bipyramid ................................................ 14, 58, 60, 61, 209
Pentagonal birotunda ........................................... 131, 184, 398, 399
Pentagonal cupola...17, 20, 23, 91, 108, 113, 126, 127, 129, 130, 153, 182, 280, 287, 288, 289, 302, 314, 316, 323, 329, 333, 409

## Index

Pentagonal cupolarotunda..................................................158, 399
Pentagonal dipyramid...............................................................58
Pentagonal gyrobicupola..................................................127, 128, 401
Pentagonal gyrobirotunda..........................................113, 114, 131, 134, 139
Pentagonal gyrocupolarotunda..........................................113, 132, 133
Pentagonal orthobicupola..................................................126, 127, 128
Pentagonal orthobirotunda..........................................113, 119, 126, 138, 139
Pentagonal orthocupolarotunda..........................................113, 129, 131, 132, 133
Pentagonal prism....................................................................60, 62, 91
Pentagonal pyramid....5, 9, 10, 14, 15, 38, 233, 234, 245, 250, 252, 258, 402
Pentagonal rotunda 25, 28, 29, 30, 31, 32, 35, 87, 91, 108, 113, 129, 138, 158, 162, 182, 185, 186, 399, 409
Pentagonized globoid..................................................................369
Peritruncated bipyramid................................................................201
Peritruncated triangular bipyramid.....................................................201
Plane........................................................41, 128, 174, 184, 197, 260, 273
Platonic solid....1, 7, 12, 38, 43, 58, 59, 60, 70, 193, 230, 266, 344, 366, 377
Polar reciprocation.....5, 12, 25, 44, 51, 131, 132, 140, 164, 171, 184, 209, 219, 220, 235, 237, 239, 246, 261, 326
Polytimoid....................................................................40, 56
Principal axis of rotation...9, 12, 18, 19, 20, 22, 25, 43, 45, 69, 80, 85, 86, 98, 114, 131, 172, 184, 194, 200, 237, 273, 291, 344, 347, 369, 398
Prism..1, 14, 15, 36, 38, 39, 40, 45, 51, 59, 62, 65, 66, 69, 78, 81, 82, 92, 144, 146, 157, 159, 161, 162, 163, 174, 193, 194, 195, 200, 201, 205, 209, 214, 220, 225, 400, 408, 409
Prismoid.............................................................................40
Pseudodeltoidal icositetrahedron.....................................................150
Pseudopyramid-like antiapex....................................................20, 29, 32
Pseudopyramid-like apex..............................................................20
Pseudorhombicuboctahedron......................................................148, 150
Pugh, Anthony.......................................................................200
Pyramid.1, 5, 7, 11, 14, 15, 19, 20, 24, 37, 38, 40, 43, 44, 45, 51, 55, 57, 61, 66, 69, 70, 78, 81, 82, 93, 95, 115, 146, 147, 154, 176, 193, 194, 197, 199, 200, 201, 205, 207, 209, 214, 215, 220, 225, 229, 234, 236, 245, 258, 260,

261, 265, 266, 268, 319, 336, 384, 393, 408, 409
Pyramid-like antiapex ............................................................... 20, 79, 83, 86
Pyramid-like apex ..................................................................... 20, 259, 261
Pyramoid ............................................................................................... 258
Q
Quadrilateral 5, 40, 42, 44, 46, 47, 49, 51, 57, 79, 80, 81, 83, 86, 93, 95, 108, 115, 124, 143, 150, 151, 157, 159, 160, 163, 164, 169, 197, 200, 205, 209, 210, 212, 213, 215, 217, 223, 224, 225, 228, 239, 244, 254, 255, 265, 269, 272, 274, 275, 279, 282, 283, 286, 291, 292, 294, 295, 296, 300, 301, 303, 304, 328, 336, 340, 344, 346, 359, 360, 364, 376, 377, 382, 383, 385, 387, 394, 402, 407
Quadrivalent vertex .................................. 6, 40, 44, 58, 74, 283, 341, 388
Quinquevalent apex .................................................................................. 213
Quinquevalent vertex ........................................................................ 74, 295
R
Reciprocation center .... 5, 7, 8, 51, 116, 131, 140, 164, 171, 184, 209, 219, 220, 235, 239, 303, 336, 345, 347
Rectangle ............................................................... 17, 60, 196, 373, 386
Rectangular pyramid ........................................................................... 194
Rectification .......................................................................................... 255
Regular decagon ................................................................................... 280
Regular dodecahedron ..... 15, 31, 44, 74, 75, 76, 118, 233, 235, 239, 250, 280, 288, 366, 368, 369, 386, 387
Regular hexagon ................................................................................... 267
Regular icosahedron ....................... 15, 31, 61, 74, 235, 368, 369, 372, 387
Regular octagon ................................................................................... 270
Regular octahedron ............................................................ 15, 60, 61, 62, 119
Regular pentagon .. 10, 28, 53, 57, 118, 237, 240, 255, 256, 316, 366, 386
Regular polygon ...... 1, 17, 42, 82, 130, 138, 192, 197, 201, 229, 268, 349, 374, 399, 409
Regular tetrahedron .7, 11, 12, 14, 15, 24, 38, 39, 44, 45, 50, 59, 61, 119, 193, 197, 256, 260, 267, 268, 270, 409
Regular-polygon-faced antiprism ......................................................... 38
Regular-polygon-faced prism .............................................................. 38
Rhombic dodecahedron ................................................... 31, 119, 128, 366
Rhombic triacontahedron .................................................................... 140

Rhombicosidodecahedron 23, 87, 251, 280, 287, 315, 316, 317, 319, 320, 321, 323, 324, 325, 326, 329, 331, 333, 335, 336, 339, 346, 348, 351, 358, 365
Rhombicuboctahedron .................................................. 83, 87, 148, 151
Rhombus 16, 42, 92, 109, 118, 120, 124, 132, 139, 140, 142, 160, 184, 187, 200, 215, 216, 225, 366, 384, 407
Right angle .................................................................................. 11, 92
Right prism ....................................................................................... 38
Roof (of a polyhedron) ................. 17, 28, 96, 97, 117, 182, 185, 401
Roof polygon .................................................................................... 17
Rossiter, Adrian ............................................................................. 280
Rotunda. 17, 28, 92, 112, 113, 120, 129, 130, 131, 132, 133, 137, 141, 159, 160, 161, 163, 164, 169, 181, 182, 184, 186, 190, 344, 368, 399, 400, 401, 409
Rotunda dual ......................................................... 29, 95, 110, 182, 184

## S

Scalene triangle ........................... 19, 20, 25, 29, 212, 244, 272, 279
Schönflies notation ............................................................ 2, 3, 230
Siamese dodecahedron ............................................................ 366
Side (of a polygon) ........................ 44, 45, 52, 282, 283, 291, 294, 296
Sixfold alternating axis ..................................................... 147, 317
Sixfold axis ................................................................................. 225
Small rhombicuboctahedron ................................................... 148
Snub antiprism .................................................................. 368, 369
Snub cube ................................................................................... 368
Snub disphenoid ........................................................................ 366
Snub dodecahedron .................................................................. 368
Snub square antiprism ........................................................ 367, 368
Sphenocorona .............................................. 371, 372, 376, 377, 382
Sphenomegacorona ..................................................... 384, 385, 386
Square...4, 7, 10, 17, 24, 39, 42, 43, 45, 47, 49, 51, 53, 65, 68, 69, 75, 78, 83, 91, 92, 95, 96, 97, 106, 117, 119, 124, 137, 144, 147, 156, 164, 169, 171, 175, 182, 193, 196, 197, 200, 205, 209, 213, 214, 215, 216, 217, 220, 223, 225, 228, 230, 266, 267, 270, 272, 280, 288, 289, 316, 317, 319, 324, 341, 349, 368, 375, 381, 382, 384, 386, 387, 393, 396, 397, 398, 401, 402, 406
Square antiprism ........................................................................ 368

Square bipyramid..................................................................58, 60, 70
Square cupola........17, 20, 21, 22, 83, 113, 119, 121, 130, 153, 270, 273, 409
Square gyrobicupola..........................................120, 121, 122, 127, 128
Square orthobicupola..................................................119, 120, 121, 126
Square pyramid.................................................7, 10, 38, 193, 194, 197, 382
Square rotunda...............................................................................130, 158
Straight line.......................................................................................... 44
Strombic polygon..................................................................................399
Strombo-triangular icosahedron......................................................31, 33
Symmetry element................................................................................325
Symmetry group......................................................................................2
Symmetry transformation.......................................................................4
T
Tectal polyhedron.......................................................................17, 42, 344
Tenfold alternating axis................127, 236, 237, 289, 291, 319, 329, 344
Tetracaidecadeltahedron......................................................................200
Tetrahedron...................................................................................58, 260
Three-four-five trend rule 12, 20, 21, 22, 25, 45, 46, 55, 61, 62, 69, 70, 78, 84, 85, 102, 115, 119, 128, 152, 153, 154, 175, 176
Threefold axis......22, 41, 44, 58, 65, 67, 68, 75, 83, 86, 97, 117, 147, 172, 196, 225, 245, 246, 260, 268, 269, 302, 303, 311, 317, 319, 359, 363, 364, 368, 369
Topological duality................................................................................ 44
Transitivity class 4, 5, 28, 46, 50, 54, 58, 78, 95, 98, 106, 108, 112, 156, 157, 169, 170, 173, 182, 185, 190, 205, 209, 212, 223, 233, 235, 238, 254, 258, 265, 278, 284, 285, 289, 291, 292, 296, 297, 300, 301, 304, 307, 309, 314, 328, 340, 342, 343, 344, 349, 351, 356, 359, 363, 366, 374, 375, 376, 381, 382, 393, 401
Trapezo-rhombic dodecahedron..........................................................118
Trapezohedron................................................................................19, 40
Trapezoid.............................................................................................. 92
Triakis icosahedron...........................................................286, 287, 291, 296
Triangle. 4, 7, 11, 14, 19, 28, 32, 35, 40, 41, 42, 43, 44, 46, 47, 49, 50, 57, 63, 65, 67, 68, 74, 78, 83, 91, 92, 95, 96, 97, 106, 112, 113, 115, 116, 117, 124, 137, 144, 151, 156, 164, 169, 170, 171, 175, 179, 182, 185, 190, 194, 196, 197, 200, 205, 209, 212, 214, 215, 216, 217, 220, 223, 224, 227, 233, 237,

239, 251, 252, 253, 255, 258, 259, 260, 265, 266, 267, 269, 270, 278, 280, 285, 286, 288, 289, 290, 291, 296, 305, 314, 317, 324, 340, 341, 343, 344, 346, 349, 359, 360, 363, 364, 366, 368, 371, 375, 376, 380, 381, 384, 387, 393, 401, 402, 407

Triangular antiprism..................................................................50
Triangular bipyramid...........................14, 58, 59, 61, 195, 230, 260
Triangular cupola.17, 18, 19, 20, 22, 96, 113, 116, 130, 145, 153, 267, 278, 409
Triangular cupolapyramid...................................................145
Triangular dipyramid............................................................58
Triangular gyrobicupola.........................................113, 119, 127
Triangular hebesphenorotunda...................................399, 402
Triangular orthobicupola.....................116, 117, 119, 120, 121, 126
Triangular prism........................38, 59, 62, 113, 129, 197, 200, 225, 409
Triangular pyramid.......................7, 38, 50, 58, 70, 197, 256, 257, 310
Triangular rotunda...........................................................130, 158
Triaugmented dodecahedron........................................232, 245, 246
Triaugmented hexagonal prism...............................................225
Triaugmented triangular prism...............................................200
Triaugmented triangular prism...............................................200
Triaugmented truncated dodecahedron...................302, 303, 308
Tridiminished icosahedron...............................255, 258, 260, 410
Tridiminished rhombicosidodecahedron.........320, 324, 325, 358, 359
Trigyrate rhombicosidodecahedron....................325, 331, 332, 333
Trivalent vertex......................6, 40, 44, 58, 59, 196, 283, 341, 388
Truncated cube.............................................................266, 270, 273
Truncated dodecahedron...........266, 280, 286, 288, 289, 291, 302, 314
Truncated icosahedron..................................................234, 266
Truncated octahedron.............................................................266
Truncated tetrahedron....................................266, 267, 268, 278, 344
Truncation...............................................40, 51, 201, 216, 234, 251, 255, 261
Twofold axis.....58, 207, 214, 218, 225, 236, 238, 253, 273, 291, 346, 374, 375, 377, 380, 384, 398

U

Uni/bigeneral tectal polyhedron..............................17, 344, 402
Uni/unigeneral tectal polyhedron...............................344, 398

**V**
Valency............................................................................................................44
Vertex (of a polygon)..................................................................................92, 237
Vertex (of a polyhedron)4, 5, 14, 15, 17, 24, 25, 28, 40, 42, 44, 58, 60, 66, 75, 83, 96, 116, 117, 121, 128, 130, 131, 140, 153, 163, 172, 182, 184, 197, 205, 215, 234, 235, 237, 239, 251, 255, 256, 260, 267, 268, 270, 273, 280, 282, 289, 290, 291, 304, 316, 319, 325, 341, 358, 371, 372, 373, 375, 384, 387, 388, 393, 398, 401, 407
Vertex angle..............................................................................................52, 327
**W**
Wikipedia.................................................................................4, xii, 17, 21, 37
**Z**
Zalgaller, Victor...............................................................................................1

www.ingramcontent.com/pod-product-compliance
Lightning Source LLC
Chambersburg PA
CBHW071353170526
45165CB00001B/20